Enzyme Handbook 12

Springer-Verlag Berlin Heidelberg GmbH

Attention all "Enzyme Handbook" users:

A file with the complete volume indexes Vols. 1 through 12 in delimited ASCII format is available for downloading at no charge from the Springer EARN mailbox. Delimited ASCII format can be imported into most databanks.

The file has been compressed using the popular shareware program "PKZIP" (Trademark of PKware inc., PKZIP is available from most BBS and shareware distributors).

This file is distributed without any expressed or implied warranty.

To receive this file send an e-mail message to:
SVSERV@DHDSPRI6.BITNET

The message must be:
GET /CHEMISTRY/ENZ_HB.ZIP

SVSERV is an automatic data distribution system. It responds to your message. The following commands are available:

HELP	returns a detailed instruction set for the use of SVSERV,
DIR *(name)*	returns a list of files available in the directory "name",
INDEX *(name)*	same as "DIR"
CD <*name*>	changes to directory "name",
SEND <*filename*>	invokes a message with the file "filename",
GET <*filename*>	same as "SEND".

D. Schomburg · D. Stephan (Eds.)
GBF – Gesellschaft für Biotechnologische Forschung

Enzyme Handbook

Class 2.3.2 – 2.4 Transferases

12

Springer

Professor Dr. Dietmar Schomburg
Dr. Dörte Stephan

GBF – Gesellschaft für Biotechnologische Forschung mbH
Mascheroder Weg 1
38124 Braunschweig
FRG

This collection of datasheets was generated from the database „BRENDA"

Library of Congress Cataloging-in-Publication Data

Enzyme handbook/ D. Schomburg, M. Salzmann (eds.). v. (1–2,4–10); 23 cm. Vols. 6–7 edited by D. Schomburg, M. Salzmann, D. Stephan. Vols. 9–10 edited by D. Schomburg, D. Stephan.
Loose-leaf.
Includes bibliographical references and indexes.
Contents: 1. Claas 4: Lyases – 2. Class 5: Isomerases. Class 6: Ligases – 4–5. Class 3: Hydrolases – 6. Class 1.2–1.4, Oxidoreductases – 7. Class 1.5–1.12, Oxidoreductases – 8. Class 1.13–1.97, Oxidoreductases – 9 Class 1.1, Oxidoreductases, EC 1.1.1.150 – EC 1.1.99.26 – v. 10. Class 1.1, Oxidoreductases, EC 1.1.1.150 – EC 1.1.99.26.

ISBN 978-3-642-47805-5   ISBN 978-3-642-61117-9 (eBook)
DOI 10.1007/978-3-642-61117-9

1. Enzymes-Handbooks, manuals, etc. I. Schomburg, D. (Dietmar)   II. Salzmann, M. (Margit)   III. Stephan, D. (Dörte)
QP601-E5158   1990
660'.634–dc20

© Springer-Verlag Berlin Heidelberg 1996
Originally published by Springer-Verlag Berlin Heidelberg New York Tokyo in 1996
Softcover reprint of the hardcover 1st edition 1996

Media conversion, printing and bookbinding: Brühlsche Universitätsdruckerei, Giessen
Production of the plasticfiles: Lux-Plastik oHG, Murnau
SPIN: 10076281   51/3020 - 5 4 3 2 1 0 - Printed on acid-free paper

Preface

Recent progress on enzyme immobilisation, enzyme production, coenzyme regeneration and enzyme engineering has opened up fascinating new fields for the potential application of enzymes in a large range of different areas. As more progress in research and application of enzymes has been made the lack of an up-to-date overview of enzyme molecular properties has become more apparent. Therefore, we started the development of an enzyme data information system as part of protein-design activities at GBF. The present book "Enzyme Handbook" represents the printed version of this data bank. In future a computer searchable version will be also available.

The enzymes in this Handbook are arranged according to the Enzyme Commission list of enzymes. Some 3000 "different" enzymes will be covered. Frequently enzymes with very different properties are included under the same EC number. Although we intend to give a representative overview on the characteristics and variability of each enzyme the Handbook is not a compendium. The reader will have to go to the primary literature for more detailed information. Naturally it is not possible to cover all the numerous literature references for each enzyme (for special enzymes up to 40000) if the data representation is to be concise as is intended.

It should be mentioned here that the literature data are extracted from literature and critically evaluated by qualified scientists. On the other hand the original authors' nomenclature for enzyme forms and subunits is retained as is their nomenclature for organisms and strains even if the organism is reclassified in the meantime. The cross references to the protein sequence data bank and to the Brookhaven protein 3D structure data bank are taken directly from their data files without further verification by the authors. In order to keep the tables concise redundant information is avoided as far as possible (e.g. if K_m values are measured in the presence of an obvious cosubstrate, only the name of the cosubstrate is given in parentheses as a commentary without reference to its specific role).

The authors are grateful to the following biologists and chemists for invaluable help in the compilation of data: Cornelia Munaretto, Dr. Astrid Beermann, Dr. Ida Schomburg and Dr. Astrid Haberz. In addition we would like to thank Mrs. C. Munaretto and Dr. I. Schomburg for the correction of the final manuscript.

Braunschweig, 1996   Dörte Stephan
                                                     Dietmar Schomburg

V

BRENDA – Compilation of Enzyme Data

To collect basic characteristics of enzymes – is that not a kind of archaic activity in the times of molecular biology and computer-aided data banks providing sequences of nucleic acids and proteins with little more delay than a few days as well as their three-dimensional structures? What should be the purpose of compiling turnover numbers, Michaelis constants, substrate specificities, sources, synonyms etc. of enzymes from sometimes remote publications? The answer sounds as simple as surprising: The aim of the compilation of data is to make use of the overwhelming abundance of structural knoweldge we owe to the new techniques of molecular biology.

Admittedly, it was not primarily enzymology which caused the explosion of knowledge in biology during the last decade. This was due to the advance of molecular biology which enabled us to isolate genes, to amplify them *ad libidum* and to elucidate their primary structure within days only. Also, the optimization and automatization of techniques for the analysis of macromolecules has provided detailed insights into a large variety of complex biomolecules nobody would have anticipated in the early seventies. Due to powerful computers it has now become feasible to propose fairly realistic models of macromolecules based solely on primary structures and homology considerations.

Nevertheless – or therefore – it appears as mandatory as rewarding to know the brave world of enzymology in which one had and often still has to come along without any detailed structural knowledge. We should not ignore that nature has not generated the multiplicity of structures, because it simply felt obliged to the principle of diversification or because it wanted to test our computing capacity to handle sequence data. It had to create new structures to cope with the steadily changing demands of a variable environment. Thus, amino acid sequences, folding of peptide chains and conformational details are only the technical tools of nature to catalyse specific biological functions. In consequence, *it is the functional profile of an enzyme which enables a biologist or physician to analyze a metabolic pathway and its disturbance; it is the substrate specificity of an enzyme which tells an analytical biochemist how to design an assay; it is the stability, specificity and efficiency of an enzyme which determines its usefulness in the biotechnical transformation of a molecule.* And the sum of all these functional data will have to be considered when the designer of artificial biocatalysts has to choose the optimum prototype to start with.

Unfortunately, it is by no means as simple to design (organize) a meaningful and systematic compilation of functional enzymological data as to enter sequences of amino acids or nucleotides into a data base. Functional data are less well defined, are never devoid of a trace of ambiguity, their selection remains inevitably subjective, and their complexity requires simplification. The present compilation of enzymological data, therefore, can and will not be a substitute for original publications but rather offer a key to the literature. But I do think that the Enzyme Handbook is indeed an excellent key to open or reopen the mysterious world of enzyme to all those who there have to find the solutions of their problems: to biologists, physicians, structural biochemists, biochemical analysts, biotechnologists and also to the molecular biologists.

Braunschweig, Spring 1993   Leopold Flohé

List of Abbreviations

A	adenosine	ER	endoplasmic reticulum
Ac	acetyl	Et	ethyl
ACP	acyl-carrier-protein	EXAFS	extended X-ray absorption
ADP	adenosine 5'-diphosphate		fine structure
Ala	alanine	FAD	flavin-adenine dinucleotide
All	allose	FMN	flavin mononucleotide (ribo-
Alt	altrose		flavin 5'-monophosphate)
AMP	adenosine 5'-monophosphate	FPLC	fast protein liquid chroma-
Ara	arabinose		tography
Arg	arginine	Fru	fructose
Asn	asparagine	Fuc	fucose
Asp	aspartic acid	G	guanosine
ATP	adenosine 5'-triphosphate	GABA	4-aminobutanoic acid
Bicine	N,N'-bis(2-hydroxyethyl)	Gal	galactose
	glycine	GDP	guanosine 5'-diphosphate
C	cytidine	Glc	glucose
cal	calorie	GlcN	glucosamine
CDP	cytidine 5'-diphosphate	GlcNAc	N-acetylglucosamine
CDTA	trans-1,2-diaminocyclo-hexa-	Gln	glutamine
	ne-N,N,N,N-tetra-aceticacid	Glu	glutamic acid
CHAPS	3-[(3-cholamidopropyl)-	Gly	glycine
	dimethylammonio]-1-	Glygly	glycylglycine
	propanesulfonate	GMP	guanosine
CHAPSO	3-[(3-cholamidopropyl)-		5'-monophosphate
	dimethylammonio]-	GSH	glutathione
	2-hydroxy-1-propane-	GSSG	oxidized glutathione
	sulfonate	GTP	guanosine 5'-triphosphate
CMP	cytidine 5'-monophosphate	Gul	gulose
CoA	coenzyme A	h	hour
CTP	cytidine 5'-triphosphate	H_4	tetrahydro
Cys	cysteine	HEPES	4-(2-hydroxyethyl)-1-piper-
d	deoxy-		azineethane sulfonic acid
D- and L-	prefixes indicating	His	histidine
	configuration	HPLC	high performance liquid
Dap	diaminopimelic acid		chromatography
DFP	diisopropylfluorophosphate	Hyl	hydroxylysine
DNA	deoxyribonucleic acid	Hyp	hydroxyproline
DPN	diphosphopyridinium	IAA	iodoacetamide
	nucleotide (now NAD)	Ig	immunoglobulin
DTNB	5,5'-dithiobis(2-nitrobenzoate)	Ile	isoleucine
DTT	dithiothreitol (i.e. Cleland's	Ido	idose
	reagent)	IDP	inosine 5'-diphosphate
e	electron	IMP	inosine 5'-monophosphate
EC	number of enzyme in Enzyme	ir	irreversible
	Commission's system	ITP	inosine 5'-triphosphate
E. coli	Escherichia coli	K_m	Michaelis constant
EDTA	ethylene diaminetetraacetate	L-	see D-
EGTA	ethylene glycol bis (β-amino-	Leu	leucine
	ethylether) tetraacetate	Lys	lysine
EPR	electron paramagnetic	Lyx	lyxose
	resonance	M	mol/l

m-	meta-	mRNA	messenger RNA
Man	mannose	rRNA	ribosomal RNA
MES	2-(N-morpholino)ethane	tRNA	transfer RNA
	sulfonate	Sar	N-methylglycine
Met	methionine		(sarcosine)
min	minute	SDS-PAGE	sodium dodecyl sulphate
MOPS	3-(N-morpholino)		polyacrylamide gel
	propane sulfonate		electrophoresis
Mur	muramic acid	Ser	serine
MW	molecular weight	SFK-525A	2-diethylaminoethyl-2,2-
NAD	nicotinamide-adenine		diphenylvalerate
	dinucleotide	sp.	species
NADH	reduced NAD	T	ribosylthymine
NADP	NAD phosphate	$t\frac{1}{2}$	time for half-completion
NADPH	reduced NADP		of reaction
NAD(P)H	indicates either NADH	Tal	talose
	or NADPH	TDP	ribosylthymine
NDP	nucleoside 5'-diphosphate		5'-diphosphate
NEM	N-ethylmaleimide	TEA	triethanolamine
Neu	neuraminic acid	TES	N-tris[hydroxymethyl]-
Nle	norleucine		methyl-2-amino-
NMN	nicotinamide		ethanesulfonic acid
	mononucleotide	THF	tetrahydrofolate
NMP	nucleoside	Thr	threonine
	5'-monophosphate	TMP	ribosylthymine
NTP	nucleoside 5'-triphosphate		5'-monophosphate
o-	ortho-	Tos-	tosyl-(p-toluenesulfonyl-)
OMP	orotidine 5-monophosphate	TPN	triphosphopyridinium
Orn	ornithine		nucleotide (now NADP)
p-	para-	Tris	tris(hydroxymethyl)-
PAPS	3'-phosphoadenylylsulfate		aminomethane
PCMB	p-chloro-mercuribenzoate	Trp	tryptophan
PEG	polyethylene glycol	TTP	ribosylthymine
PEP	phosphoenolpyruvate		5'-triphosphate
pH	$-\log_{10} [H^+]$	Tyr	tyrosine
Ph	phenyl	U	uridine
Phe	phenylalanine	U/mg	μmol/(mg·min)
PIXE	proton-induced	UDP	uridine 5'-diphosphate
	X-ray emission	UMP	uridine 5'-monophosphate
PMSF	phenylmethane-	UTP	uridine 5'-triphosphate
	sulfonylfluoride	UV	ultraviolet
Pro	proline	Val	valine
Q_{10}	factor for the change in	Xaa	symbol for an amino
	reaction rate for a 10°		acid of unknown consti-
	temperature increase		tution in peptide formula
r	reversible	XAS	X-ray absorption
Rha	rhamnose		spectroscopy
Rib	ribose	XTP	xanthosine 5'-triphosphate
RNA	ribonucleic acid	Xyl	xylose

Index
(Alphabetical order of Enzyme names)

EC-No.	Name	EC-No.	Name

2.4.1.20 Cellobiose phosphorylase

2.4.1.49 Cellodextrin phosphorylase

2.4.1.29 Cellulose synthase (GDP-forming)

2.4.1.12 Cellulose synthase (UDP-forming)

2.4.1.80 Ceramide glucosyltransferase

2.4.1.16 Chitin synthase

2.4.1.142 Chitobiosyldiphosphodolichol alpha-mannosyltransferase

2.4.1.177 Cinnamate beta-D-glucosyltransferase

2.4.1.111 Coniferyl-alcohol glucosyltransferase

2.4.1.114 2-Coumarate O-beta-glucosyltransferase

2.4.1.116 Cyanidin-3-rhamnosylglucoside 5-O-glucosyltransferase

2.4.1.85 Cyanohydrin beta-glucosyltransferase

2.4.1.19 Cyclomaltodextrin glucanotransferase

2.4.1.118 Cytokinin 7-beta-glucosyltransferase

2.4.2.23 Deoxyuridine phosphorylase

2.4.1.5 Dextransucrase

2.4.1.2 Dextrin dextranase

2.4.1.46 1,2-Diacylglycerol 3-beta-galactosyltransferase

2.4.1.157 1,2-Diacylglycerol 3-glucosyltransferase

2.4.1.104 o-Dihydroxycoumarin 7-O-glucosyltransferase

2.4.1.202 2,4-Dihydroxy-7-methoxy-2H-1,4-benzoxazin-3(4H)-one 2-D-glucosyltransferase

2.4.2.20 Dioxotetrahydropyrimidine phosphoribosyltransferase

2.4.1.26 DNA alpha-glucosyltransferase

2.4.1.27 DNA beta-glucosyltransferase

2.4.1.119 Dolichyl-diphosphooligosaccharide-protein glycotransferase

2.4.1.153 Dolichyl-phosphate alpha-N-acetylglucosaminyltransferase

2.4.1.117 Dolichyl-phosphate beta-glucosyltransferase

2.4.1.130 Dolichyl-phosphate-mannose-glycolipid alpha-mannosyltransferase

2.4.1.109 Dolichyl-phosphate-mannose-protein mannosyltransferase

2.4.1.83 Dolichyl-phosphate beta-D-mannosyltransferase

2.4.2.32 Dolichyl-phosphate D-xylosyltransferase

2.4.2.33 Dolichyl-xylosyl-phosphate-protein xylosyltransferase

2.4.2.27 dTDPdihydrostreptose-streptidine-6-phosphate dihydrostreptosyltransferase

2.4.1.185 Flavanone 7-O-beta-glucosyltransferase

2.4.2.25 Flavone apiosyltransferase

2.4.1.81 Flavone 7-O-beta-glucosyltransferase

2.4.1.159 Flavonol-3-O-glucoside L-rhamnosyltransferase

2.4.1.91 Flavonol 3-O-glucosyltransferase

2.4.2.35 Flavonol-3-O-glycoside xylosyltransferase

2.4.1.100 1,2-beta-Fructan 1^F-fructosyltransferase

2.4.1.40 Fucosylgalactose alpha-N-acetylgalactosaminyltransferase

2.4.1.37 Fucosylglycoprotein 3-alpha-galactosyltransferase

EC-No.	Name	EC-No.	Name

2.4.1.44 Lipopolysaccharide 3-alpha-galactosyltransferase

2.4.1.58 Lipopolysaccharide glucosyltransferase I

2.4.1.73 Lipopolysaccharide glucosyltransferase II

2.4.1.191 Luteolin-7-O-diglucuronide 4'-O-glucuronosyltransferase

2.4.1.190 Luteolin-7-O-glucuronide 7-O-glucuronosyltransferase

2.4.1.189 Luteolin 7-O-glucuronosyltransferase

2.3.2.3 Lysyltransferase

2.4.1.8 Maltose phosphorylase

2.4.1.139 Maltose synthase

2.4.1.101 alpha-1,3-Mannosyl-glyco-protein beta-1,2-N-acetyl-glucosaminyltransferase

2.4.1.143 alpha-1,6-Mannosyl-glyco-protein beta-1,2-N-acetyl-glucosaminyltransferase

2.4.1.145 alpha-1,3-Mannosylglyco-protein beta-1,4-N-acetyl-glucosaminyltransferase

2.4.1.144 beta-1,4-Mannosyl-glyco-protein beta-1,4-N-acetyl-glucosaminyltransferase

2.4.1.201 Mannosyl-glycoprotein beta-1,4-N-acetylglucosaminyl-transferase

2.4.1.155 alpha-1,3(6)-Mannosylgly-coprotein beta-1,6-N-ace-tylglucosaminyltransferase

2.4.1.199 beta-Mannosylphosphodecaprenol-mannooligo-saccharide 6-mannosyl-transferase

2.4.1.138 Mannotetraose 2-alpha-N-acetylglucosaminyltrans-ferase

2.4.1.171 Methyl-ONN-azoxymethanol beta-D-glucosyltransferase

2.4.2.28 5'-Methylthioadenosine phosphorylase

2.4.99.2 Monosialoganglioside sialyltransferase

2.4.1.127 Monoterpenol beta-glucosyltransferase

2.4.2.30 NAD+ ADP-ribosyltrans-ferase

2.4.2.37 NAD+-dinitrogen-reductase ADP-D-ribosyltransferase

2.4.2.36 NAD+-diphthamide ADP-ribosyltransferase

2.4.2.31 NAD(P)+-arginine ADP-ribosyltransferase

2.4.99.10 Neolactotetraosylceramide alpha-2,3-sialyltransferase

2.4.2.12 Nicotinamide phosphoribo-syltransferase

2.4.1.196 Nicotinate glucosyltrans-ferase

2.4.2.21 Nicotinate-nucleotide-dime-thylbenzimidazole phos-phoribosyltransferase

2.4.2.19 Nicotinate-nucleotide pyro-phosphorylase (carboxylating)

2.4.2.11 Nicotinate phosphoribosyl-transferase

2.4.1.192 Nuatigenin 3beta-glucosyl-transferase

2.4.2.6 Nucleoside deoxyribosyl-transferase

2.4.2.5 Nucleoside ribosyltrans-ferase

2.4.1.30 1,3-beta-Oligoglucan phosphorylase

2.4.1.161 Oligosaccharide 4-alpha-D-glucosyltransferase

2.4.2.10 Orotate phosphoribosyl-transferase

2.4.1.129 Peptidoglycan glycosyl-transferase

2.3.2.12 Peptidyltransferase

2.4.1.35 Phenol beta-glucosyl-transferase

2.4.1.198 Phosphatidylinositol N-ace-tylglucosaminyltransferase

2.4.1.57 Phosphatidyl-myo-inositol alpha-mannosyltrans-ferase

EC-No.	Name	EC-No.	Name

EC-No.	Name	EC-No.	Name

2.4.1.169 Xyloglucan 6-xylosyltransferase

2.4.1.133 Xylosylprotein 4-beta-galactosyltransferase

2.4.1.203 Zeatin O-beta-D-glucosyltransferase

2.4.1.204 Zeatin O-beta-D-xylosyltransferase

1 NOMENCLATURE

EC number
2.4.1.123

Systematic name
UDPgalactose:myo-inositol 1-alpha-D-galactosyltransferase

Recommended name
Inositol 1-alpha-galactosyltransferase

Synonyms
Galactinol synthase
Galactosyltransferase, uridine diphosphogalactose-inositol
UDP-D-galactose:inositol galactosyltransferase [1]
More (cf. EC 2.4.1.67 and EC 2.4.1.82) [1]

CAS Reg. No.
79955-89-8

---

2 REACTION AND SPECIFICITY

Catalysed reaction
UDPgalactose + myo-inositol →
→ UDP + 1-O-alpha-D-galactosyl-D-myo-inositol

Reaction type
Hexosyl group transfer

Natural substrates
UDPgalactose + myo-inositol (responsible for galactinol synthesis, involved in raffinose and stachyose biosynthesis, cf. EC 2.4.1.67 and EC 2.4.1.82) [1]

Substrate spectrum
1 UDPgalactose + myo-inositol [1]

Product spectrum
1 UDP + 1-O-alpha-D-galactosyl-D-myo-inositol [1]

Inhibitor(s)

Cofactor(s)/prosthetic group(s)/activating agents

Metal compounds/salts

Turnover number (min^{-1})

Specific activity (U/mg)
  More [1]

K$_m$-value (mM)

pH-optimum
  More (Mn^{2+}-concentration influences pH-optimum) [1]; 5.5 (at 7 mM Mn^{2+})
  [1]; 6.2 (at 2 mM Mn^{2+}) [1]; 7.0 (at 0.2 mM Mn^{2+}) [1]

pH-range
  5–7 (about half-maximal activity at pH 5 and 7, at 7 mM Mn^{2+}) [1]; 5.1–7.8
  (about half-maximal activity at pH 5.1 and 7.8, at 2.0 mM Mn^{2+}) [1]; 5.2–8.6
  (about half-maximal activity at pH 5.2 and 8.6, at 0.2 mM Mn^{2+}) [1]

Temperature optimum (°C)
  30 (assay at) [1]

Temperature range (°C)

---

3 ENZYME STRUCTURE

Molecular weight

Subunits

Glycoprotein/Lipoprotein
  –

---

4 ISOLATION/PREPARATION

Source organism
  Cucumis sativus (cucumber, cv. Chipper) [1]

Source tissue
  Leaf [1]

Localization in source

Purification
  Cucumis sativus (partial) [1]

Crystallization
  –

Cloned
  –

Renatured
  –

2

5 STABILITY

pH

Temperature (°C)

Oxidation

Organic solvent

General stability information

Storage

6 CROSSREFERENCES TO STRUCTURE DATABANKS

PIR/MIPS code

Brookhaven code

7 LITERATURE REFERENCES

[1] Pharr, D.M., Sox, H.N., Locy, R.D., Huber, S.C.: Plant Sci. Lett.,23,25–33 (1981)

1 NOMENCLATURE

EC number
2.4.1.124

Systematic name
UDPgalactose:beta-D-galactosyl-1,4-N-acetyl-D-glucosamine 3-alpha-D-galactosyltransferase

Recommended name
N-Acetyllactosamine 3-alpha-galactosyltransferase

Synonyms
Galactosyltransferase, uridine diphosphogalactose-acetyllactosamine
UDP-galactose-acetyllactosamine alpha-D-galactosyltransferase
UDP-Gal:N-acetyllactosaminide alpha-1,3-D-galactosyltransferase [2]

CAS Reg. No.
78642-28-1

2 REACTION AND SPECIFICITY

Catalysed reaction
UDPgalactose + beta-D-galactosyl-1,4-N-acetyl-D-glucosamine →
→ UDP + alpha-D-galactosyl-1,3-beta-D-galactosyl-1,4-N-acetyl-D-glucosamine

Reaction type
Hexosyl group transfer

Natural substrates
More (synthesis of Ehrlich ascites tumor cell glycoproteins [1], functions in biosynthesis of calf thymocyte cell-surface glycoconjugates including glycoproteins [2]) [1, 2]

Substrate spectrum
1 UDPgalactose + beta-D-galactosyl-1,4-N-acetyl-D-glucosamine (i.e. N-acetyllactosamine) [1–3]
2 Glycoprotein + UDPgalactose (the nonreducing terminal N-acetyllactosamine residue of glycoproteins can act as acceptor [1, 2], e.g. asialo-alpha$_1$-acid glycoprotein [2]) [1–3]
3 More (specificity, most active acceptors have the structure beta-D-Gal-(1→4)-beta-D-GlcNAc(1→) at their nonreducing termini) [1]

Product spectrum
  1 UDP + alpha-D-galactosyl-1,3-beta-D-galactosyl-1,4-N-acetyl-D-glucosa-
     mine [1–3]
  2 ?
  3 ?

Inhibitor(s)
  EDTA [2]

Cofactor(s)/prosthetic group(s)/activating agents
  Triton X-100 (activates, 0.8% v/v stimulates 4fold) [2]

Metal compounds/salts
  Mn^{2+} (pronounced requirement, K_m: 6.1 mM) [2]; Mg^{2+} (some activation) [2]

Turnover number (min^{-1})

Specific activity (U/mg)

K_m-value (mM)
  0.2 (UDPgalactose) [2]; 2.7 (N-acetyllactosamine) [2]; 3.7 (asialo-alpha$_1$-
  acid glycoprotein) [2]

pH-optimum
  5.5–7.0 [2]

pH-range

Temperature optimum (°C)
  37 (assay at) [2, 3]

Temperature range (°C)

---

3 ENZYME STRUCTURE

Molecular weight

Subunits

Glycoprotein/Lipoprotein
  —

---

4 ISOLATION/PREPARATION

Source organism
  Mouse [1]; Bovine (calf) [2, 3]

Source tissue
  Ehrlich ascites tumor cells [1]; Thymus [2, 3]

Localization in source
  Membrane (bound [2]) [1, 2]

Purification

Crystallization
–

Cloned
–

Renatured
–

---

5 STABILITY

pH

Temperature (°C)

Oxidation

Organic solvent

General stability information

Storage

---

6 CROSSREFERENCES TO STRUCTURE DATABANKS

PIR/MIPS code
  PIR2:A44785 (bovine)

Brookhaven code

---

7 LITERATURE REFERENCES

[1] Blake, D.A., Goldstein, I.J.: J. Biol. Chem.,256,5387–5393 (1981)
[2] van den Eijnden, D.H., Blanken, W.M., Winterwerp, H., Schiphorst, W.E.C.M.:
    Eur. J. Biochem.,134,523–530 (1983)
[3] van Halbeek, H., Vliegenthart, J.F.G.: Biochem. Biophys. Res. Commun.,110,
    124–131 (1983)

1 NOMENCLATURE

EC number
2.4.1.125

Systematic name
Sucrose:1,6-alpha-D-glucan 3(6)-alpha-D-glucosyltransferase

Recommended name
Sucrose-1,6-alpha-glucan 3(6)-alpha-glucosyltransferase

Synonyms
GTF-S [4]
Water-soluble-glucan synthase
Glucosyltransferase, sucrose-1,6-alpha-glucan 3(6)-alpha-
Sucrose:1,6-alpha-D-glucan 3-alpha- and 6-alpha-glucosyltransferase [1, 2, 5]
Sucrose:1,6-, 1,3-alpha-D-glucan 3-alpha- and 6-alpha-D-glucosyltransferase [3]

CAS Reg. No.
81725-87-3

---

2 REACTION AND SPECIFICITY

Catalysed reaction
Sucrose + (1,6-alpha-D-glucosyl)$_n$ →
→ D-fructose + (1,6-alpha-D-glucosyl)$_{n+1}$

Reaction type
Hexosyl group transfer

Natural substrates

Substrate spectrum
1 Sucrose + (1,6-alpha-D-glucosyl)$_n$ (also transfers glucosyl residues to the
3-position on glucose residues in glucans, producing a highly-branched
1,6-alpha-D-glucan [1–3, 5]) [1–6]

Product spectrum
1 D-Fructose + (1,6-alpha-D-glucosyl)$_{n+1}$ (1,6-alpha-D-glucan with highly
(35%) branched structure of 1,3,6-linked glucose residues [1], 1,6-alpha-
D-glucan with 17.7% of 1,3,6-branching structure [2], glucan consists of
49.1 mol% 1,6-alpha-linked glucose and 33.9 mol% 1,3-alpha-linked glucose with 13.6 mol% terminal glucose and 3.3 mol% 1,3,6-alpha-bran-
ched glucose [3], 1,6-alpha-D-glucan with 20 and 24.5 mol% 1,3,6-branch
points [5], enzyme exhibits 87% 1,6-alpha-bond-, 6% 1,3-alpha-bond- and
7% 1,3,6-branch-forming activities [6]) [1–6]

Inhibitor(s)
  6-Deoxysucrose (competitive) [4]; 6-Thiosucrose [4]

Cofactor(s)/prosthetic group(s)/activating agents
  Dextran T10 (activation [2], no stimulation in the range 0.01–2.0 mg/ml [3])
  [2]

Metal compounds/salts

Turnover number (min^{-1})

Specific activity (U/mg)
  5.86–6.8 [5]; 9.8 [2]; 34.9 [1]; 89.7 [3]

K_m-value (mM)
  0.0071 (dextran) [2]; 1.3 (sucrose) [5]; 2.4 (sucrose) [1]; 4.3 (sucrose) [2];
  4.9 (sucrose) [3]

pH-optimum
  5.5 [1, 5]; 6.0 [3]; 6.5 [2]

pH-range

Temperature optimum (°C)
  37 (assay at) [1, 3]

Temperature range (°C)

3 ENZYME STRUCTURE

Molecular weight
  149000 (Streptococcus mutans, sedimentation equilibrium studies) [1]

Subunits
  ? (x × 151000, Streptococcus mutans, SDS-PAGE [2], x × 159000, Streptococcus mutans, SDS-PAGE [3], x × 161000, Streptococcus mutans, enzyme
  I, SDS-PAGE [5], x × 174000, Streptococcus mutans, enzyme II, SDS-PAGE
  [5]) [2, 3, 5]

Glycoprotein/Lipoprotein
  More (carbohydrate content: less than 1% w/w [1], 1.5% [3]) [1, 3]

4 ISOLATION/PREPARATION

Source organism
  Streptococcus mutans (6715, serotype g [1, 4, 6], Ingbritt, serotype c [2],
  HS6, serotype a [3, 5]) [1–6]

Source tissue
Culture supernatant [1, 3, 5]

Localization in source
Extracellular [1, 3, 5, 6]

Purification
Streptococcus mutans (6715 [1], Ingbritt, serotype c, 3 isoenzymes [2],
HS6, serotype a [3, 5], 2 isoenzymes: I and II [5]) [1–3, 5]

Crystallization
–

Cloned
–

Renatured
–

---

5 STABILITY

pH

Temperature (°C)

Oxidation

Organic solvent

General stability information

Storage

---

6 CROSSREFERENCES TO STRUCTURE DATABANKS

PIR/MIPS code

Brookhaven code

---

7 LITERATURE REFERENCES

[1] Shimamura, A., Tsumori, H., Mukasa, H.: Biochim. Biophys. Acta,702,72–80 (1982)
[2] Mukasa, H., Shimamura, A., Tsumori, H.: Biochim. Biophys. Acta,719,81–89 (1982)
[3] Tsumori, H., Shimamura, A., Mukasa, H.: J. Gen. Microbiol.,131,3347–3353 (1985)
[4] Binder, T.P., Robyt, J.F.: Carbohydr. Res.,140,9–20 (1985)
[5] Tsumori, H., Shimamura, A., Mukasa, H.: J. Gen. Microbiol.,129,3251–3259 (1983)
[6] Shimamura, A., Tsumori, H., Mukasa, H.: FEBS Lett.,157,79–84 (1983)

1 NOMENCLATURE

EC number
2.4.1.126

Systematic name
UDPglucose:trans-4-hydroxycinnamate 4-O-beta-D-glucosyltransferase

Recommended name
Hydroxycinnamate 4-beta-glucosyltransferase

Synonyms
Glucosyltransferase, uridine diphosphoglucose-hydroxycinnamate
UDP-glucose-hydroxycinnamate glucosyltransferase
Hydroxycinnamoyl glucosyltransferase [1]
More (cf. EC 2.4.1.120)

CAS Reg. No.
77848-85-2

---

2 REACTION AND SPECIFICITY

Catalysed reaction
UDPglucose + trans-4-hydroxycinnamate →
→ UDP + 4-O-beta-D-glucosyl-4-hydroxycinnamate

Reaction type
Hexosyl group transfer

Natural substrates

Substrate spectrum
1 UDPglucose + ferulic acid [1]
2 UDPglucose + p-coumaric acid [1]
3 UDPglucose + caffeic acid [1]
4 UDPglucose + sinapic acid [1]
5 More (acts also on other phenolic substrates, such as kaempferol,
   quercetin, aesculin, daphnetin, scopoletin) [1]

Product spectrum

1 UDP + feruloylglucose + 4-O-glucosyl-ferulic acid (a mixture of the glucoside and the glucose ester is formed) [1]
2 UDP + p-coumaroylglucose + 4-O-glucosyl-p-coumaric acid (a mixture of the glucoside and the glucose ester is formed) [1]
3 UDP + 4-O-glucosyl-caffeic acid (no glucose ester is formed) [1]
4 UDP + sinapoylglucose + 4-O-glucosyl-sinapic acid (a mixture of the glucoside and the glucose ester is formed) [1]
5 ?

Inhibitor(s)

EDTA (0.001 mM: 30% inhibition, 0.01 mM: complete inactivation) [1]; Iodoacetate (0.001–0.01 mM: 50–70% inhibition) [1]; p-Chloromercuribenzoate (0.001–0.01 mM: 50–70% inhibition) [1]; Ca^{2+} (0.1–10 mM: 20–50% inhibition) [1]; Mg^{2+} (0.1–10 mM: 20–50% inhibition) [1]; Mn^{2+} (0.1–10 mM: 20–80% inhibition) [1]; UDP (strong) [1]

Cofactor(s)/prosthetic group(s)/activating agents

Metal compounds/salts

$MnCl_2$ (0.006–0.01 mM: stimulation of about 10%) [1]

Turnover number (min^{-1})

Specific activity (U/mg)

0.00112 (ferulate) [1]

K_m-value (mM)

0.0008 (p-coumaric acid) [1]; 0.0014 (ferulic acid) [1]; 0.0015 (caffeic acid) [1]; 0.0025 (sinapic acid) [1]; 0.01 (UDPglucose (+ ferulic acid)) [1]

pH-optimum

7.0 (formation of glucose ester of p-coumaric acid) [1]; 7.5 (assay at, pH favorizes glucosylation over glucose ester formation) [1]; 8.0 (formation of glucoside of p-coumaric acid) [1]

pH-range

Temperature optimum (°C)

30 (assay at) [1]

Temperature range (°C)

3 ENZYME STRUCTURE

Molecular weight

Subunits

Glycoprotein/Lipoprotein
  –

4 ISOLATION/PREPARATION

Source organism
  Lycopersicon esculentum var. cerasiforme (tomato) [1]

Source tissue
  Fruit [1]

Localization in source

Purification
  Lycopersicon esculentum var. cerasiforme (partial) [1]

Crystallization
  –

Cloned
  –

Renatured
  –

5 STABILITY

pH

Temperature (°C)

Oxidation

Organic solvent

General stability information
  beta-Mercaptoethanol, 10 mM, stabilizes, 80% restorage of activity by its
  addition after treatment with p-chloromercuribenzoate [1]

Storage
  –20°C, purified enzyme, 12–24 h: complete loss of activity [1]; 0–4°C, purified enzyme, 24 h: 40% loss of activity [1]; 0–4°C, purified enzyme, 48 h:
  70% loss of activity [1]

6 CROSSREFERENCES TO STRUCTURE DATABANKS

PIR/MIPS code

Brookhaven code

7 LITERATURE REFERENCES

[1] Fleuriet, A., Macheix, J.J.: Z. Naturforsch.,35c,967–972 (1980)

1 NOMENCLATURE

EC number
2.4.1.127

Systematic name
UDPglucose:(-)-menthol O-beta-D-glucosyltransferase

Recommended name
Monoterpenol beta-glucosyltransferase

Synonyms
Glucosyltransferase, uridine diphosphoglucose-monoterpenol
UDPglucose:monoterpenol glucosyltransferase [1]

CAS Reg. No.
78990-64-4

2 REACTION AND SPECIFICITY

Catalysed reaction
UDPglucose + (-)-menthol →
→ UDP + (-)-menthyl O-beta-D-glucoside

Reaction type
Hexosyl group transfer

Natural substrates

Substrate spectrum
1 UDPglucose + l-menthol [1]
2 UDPglucose + d-neomenthol [1]

Product spectrum
1 UDP + l-menthyl-beta-D-glucoside [1]
2 UDP + d-neomenthyl-beta-D-glucoside [1]

Inhibitor(s)
Co^{2+} (5 mM, strong) [1]; Ni^{2+} (5 mM: strong) [1]; Ca^{2+} (5 mM: strong) [1];
Zn^{2+} (5 mM: strong) [1]; p-Hydroxymercuribenzoate (strong) [1]; $CdCl_2$
(strong) [1]; $HgCl_2$ (strong) [1]

Cofactor(s)/prosthetic group(s)/activating agents
Mercaptoethanol (without addition 73% decrease of activity, suggesting a
thiol function necessary for activity) [1]

Metal compounds/salts
$MgCl_2$ (about 30% stimulation) [1]; $MnCl_2$ (slight stimulation) [1]

Turnover number (min^{-1})

Specific activity (U/mg)

K_m-value (mM)
0.048 (d-neomenthol) [1]; 0.059 (l-menthol) [1]; 0.2 (UDPglucose (+ d-neo-menthol)) [1]; 0.25 (UDPglucose (+ l-menthol)) [1]

pH-optimum
7.3–7.7 (depending on buffer) [1]

pH-range

Temperature optimum (°C)
30 (assay at) [1]

Temperature range (°C)

---

3 ENZYME STRUCTURE

Molecular weight
46000 (Mentha piperita, gel permeation chromatography) [1]

Subunits

Glycoprotein/Lipoprotein
–

---

4 ISOLATION/PREPARATION

Source organism
Mentha piperita L. cv. Black Mitcham (peppermint) [1]

Source tissue
Leaf [1]

Localization in source

Purification
Mentha piperita L. cv. Black Mitcham (partial) [1]

Crystallization
–

Cloned

–

Renatured

–

5 STABILITY

pH

Temperature (°C)

Oxidation

Organic solvent

General stability information
    Mercaptoethanol stabilizes [1]

Storage

6 CROSSREFERENCES TO STRUCTURE DATABANKS

PIR/MIPS code

Brookhaven code

7 LITERATURE REFERENCES

[1] Martinkus, C., Croteau, R.: Plant Physiol.,68,99–106 (1981)

Enzyme Handbook © Springer-Verlag Berlin Heidelberg 1996
Duplication, reproduction and storage in data banks are only
allowed with the prior permission of the publishers   3

1 NOMENCLATURE

EC number
2.4.1.128

Systematic name
Scopoletin glucosyltransferase

Recommended name
UDPglucose:scopoletin O-beta-D-glucosyltransferase

Synonyms
Glucosyltransferase, uridine diphosphoglucose-scopoletin
UDPglucose:scopoletin glucosyltransferase [1]
SGTase [1]

CAS Reg. No.
81210-69-7

2 REACTION AND SPECIFICITY

Catalysed reaction
UDPglucose + scopoletin →
→ UDP + scopolin

Reaction type
Hexosyl group transfer

Natural substrates

Substrate spectrum
1 UDPglucose + scopoletin [1]

Product spectrum
1 UDP + scopolin [1]

Inhibitor(s)

Cofactor(s)/prosthetic group(s)/activating agents
2,4-Dichlorophenoxyacetic acid (stimulation) [1]

Metal compounds/salts

Turnover number (min^{-1})

Specific activity (U/mg)
0.0116 [1]

K_m-value (mM)

pH-optimum
    7.5 (assay at) [1]

pH-range

Temperature optimum (°C)
    30 (assay at) [1]

Temperature range (°C)

3 ENZYME STRUCTURE

Molecular weight
    45000 (Nicotiana tabacum, PAGE) [1]

Subunits

Glycoprotein/Lipoprotein
    –

4 ISOLATION/PREPARATION

Source organism
    Nicotiana tabacum L. ("Bright Yellow") [1]

Source tissue
    Callus (cell culture) [1]

Localization in source

Purification
    Nicotiana tabacum (partial) [1]

Crystallization
    –

Cloned
    –

Renatured
    –

5 STABILITY

pH

Temperature (°C)

Oxidation

Organic solvent

General stability information

Storage

6 CROSSREFERENCES TO STRUCTURE DATABANKS

PIR/MIPS code

Brookhaven code

7 LITERATURE REFERENCES

[1] Hino, F., Okazaki, M., Miura, Y.: Plant Physiol.,69,810–813 (1982)

1 NOMENCLATURE

EC number
2.4.1.129

Systematic name
Bactoprenyldiphospho-N-acetylmuramoyl-(N-acetyl-D-glucosaminyl)-penta-peptide:peptidoglycan N-acetylmuramoyl-N-acetyl-D-glucosaminyltrans-ferase

Recommended name
Peptidoglycan glycosyltransferase

Synonyms
Glycosyltransferase, peptidoglycan
Peptidoglycan transglycosylase
PG-II [1]
Penicillin binding protein (3 [2] or 1B [4]) [1, 2, 4]

CAS Reg. No.
79079-04-2

2 REACTION AND SPECIFICITY

Catalysed reaction
n Bactoprenyldiphospho-N-acetylmuramyl-(N-acetyl-D-glucosaminyl)-L-alanyl-D-glutamyl-meso-diaminopimelyl-D-alanyl-D-alanine →
→ a polymer with n linked disaccharide subunits

Reaction type
Hexosyl group transfer

Natural substrates
n N-Acetylglucosaminyl-N-acetylmuramyl(-pentapeptide)-diphosphoryl un-decaprenyl (i.e. bactoprenyldiphosphoryl-N-acetyl-muramic acid-N-ace-tyl-glucosamine-pentapeptide or lipid linked precursor, involved in bacterial cell wall biosynthesis) [1–5]

Substrate spectrum
1 n N-Acetylglucosaminyl-N-acetylmuramyl(-pentapeptide)-diphosphoryl un-decaprenyl (i.e. bactoprenyldiphosphoryl-N-acetyl-muramic acid-N-ace-tyl-glucosamine-pentapeptide or lipid linked precursor, the enzyme has penicillin binding properties (not [3, 5]) [1, 2, 4], bifunctional protein with transglycosylase and transpeptidase activities [2, 4]) [1–5]

Product spectrum
1 Peptidoglycan (lysozyme sensitive) [1–5]

Inhibitor(s)
Enramycin [4]; Macarbomycin (Staphylococcus aureus: strong, Micro-coccus luteus: weak [3]) [2–4]; Moenomycin (not Micrococcus luteus [3]) [2–5]; EDTA (in the absence of detergents, stimulates in the presence of high concentrations of methanol and detergents) [4]; Sodium 1,2-cyclo-hexanediamine-N,N,N',N'-tetraacetic acid (in the absence of detergents, stimulates in the presence of high concentrations of methanol and deter-gents) [4]; Dimethylsulfoxide (in the presence of 0.05% sarkosyl) [4]; Ca^{2+} [4]; Co^{2+} [4]; Fe^{2+} [4]; Mn^{2+} [4]; Ni^{2+} [4]; Zn^{2+} [4]; Vancomycin [1, 4, 5]; Triton X-100 (activation, 0.05% [4], up to 0.6% [5], inhibits at 0.1% [4]) [4]; Sodium deoxycholate (activation, 0.05–0.1%, in the presence of methanol, inhibits at 0.5%) [4]; More (no inhibitors: bacitracin, tunicamycin, sulfhydryl inhibitors [1], penicillin G [1, 2], apalcillin, cephalexin, N-formimidoyl thienamycin, no-cardicin A [2], PCMB, DTNB [4]) [1, 2, 4]

Cofactor(s)/prosthetic group(s)/activating agents
Benzylpenicillin (stimulation, in the presence of 15% methanol, not at higher methanol concentrations or in the presence of deoxycholate) [4]; Imipenem (stimulation, in the presence of 15% methanol, not at higher methanol con-centrations or in the presence of deoxycholate) [4]; Dimethylsulfoxide (acti-vation, in the absence of methanol, inhibits in the presence of 0.05% sarko-syl) [4]; Sarkosyl (activation) [4]; Triton X-100 (activation, 0.05% [4], up to 0.6% [5], inhibits at 0.1% [4]) [4, 5]; Sodium deoxycholate (activation, 0.05–0.1%, in the presence of methanol, inhibits at 0.5%) [4]; EDTA (stimula-tion in the precence of high concentrations of methanol and detergents) [4]; Sodium 1,2-cyclohexanediamine-N,N,N',N'-tetraacetic acid (stimulation in the presence of high concentrations of methanol and detergents) [4]; More (no activation by cephalexin, nocardicin A or mecillinam) [4]

Metal compounds/salts
Mg^{2+} (slight stimulation [4], not [1, 3]) [4]; More (no divalent cation require-ment) [4]

Turnover number (min^{-1})

Specific activity (U/mg)
More [1]; 0.00000316 (Micrococcus luteus) [3]; 0.00000392 (Staphylococ-cus aureus) [3]; 0.000591 [4]

K_m-value (mM)

pH-optimum
7–8.2 (Micrococcus luteus) [3]; 7–9 (broad, in the presence of 0.1% deoxycholate) [4]; 7.2–7.5 (Staphylococcus aureus) [3]; 8.5 (Tris-HCl buffer without deoxycholate) [4]

pH-range

Temperature optimum (°C)
37 [4]

Temperature range (°C)
20–45 (about 60% of maximal activity at 20°C and about half-maximal activity at 45°C) [4]

---

3 ENZYME STRUCTURE

Molecular weight
61000 (Bacillus megaterium, gel filtration) [1]

Subunits
? ($x \times 60000$, E. coli, SDS-PAGE [2], $x \times 90000$, E. coli penicillin binding protein 1B alpha, beta or gamma [4]) [2, 4]

Glycoprotein/Lipoprotein
–

---

4 ISOLATION/PREPARATION

Source organism
Bacillus megaterium (strain 899) [1]; E. coli K12 (strains JST975srev61/pLC26–6 (F⁻ mrcB mreA recA) [2] and JA200/pLC19–19 [4]) [2, 4]; Micrococcus luteus (strain SM1, synonym M. lysodeikticus) [3]; Staphylococcus aureus (strain SAK 101) [3]; Streptococcus pneumoniae (strain R6cwl) [5]

Source tissue
Cell [1–5]

Localization in source
Protoplast membrane [1–5]

Purification
Bacillus megaterium [1]; E. coli (3 isozymes: penicillin-binding protein 1B alpha, beta and gamma, solubilized with sarkosyl, affinity chromatography with ampicillin-Sepharose [4], penicillin-binding protein 3 [2]) [2, 4]; Micrococcus luteus (partial) [3]; Staphylococcus aureus (partial) [3]; Streptococcus pneumoniae (partial) [5]

Crystallization
–

Cloned
   (E. coli structural gene mrcB, recloned from plasmid pLC19–19 to high copy
   number plasmid pBr322, yielding plasmid pTM13) [4]

Renatured
   (E. coli penicillin binding protein 1B alpha, beta and gamma) [4]

---

5 STABILITY

pH

Temperature (°C)
   60 (crude: 10 min stable, purified: 41–87% loss of activity within 10 min) [4]

Oxidation

Organic solvent

General stability information

Storage
   –80°C, in concentrated PEG 6000 solution, several months [4]

---

6 CROSSREFERENCES TO STRUCTURE DATABANKS

PIR/MIPS code

Brookhaven code

---

7 LITERATURE REFERENCES

[1] Taku, A., Stuckey, M., Fan, D.P.: J. Biol. Chem.,257,5018–5022 (1982)
[2] Ishino, F., Matsuhashi, M.: Biochem. Biophys. Res. Commun.,101,905–911 ((1981)
[3] Park, W., Matsuhashi, M.: J. Bacteriol.,157,538–544 (1984)
[4] Nakagawa, J., Tamaki, S., Tomioka, S., Matsuhashi, M.: J. Biol. Chem.,259,
    13937–13946 (1984)
[5] Park, W., Seto, H., Hakenbeck, R., Matsuhashi, M.: FEMS Microbiol. Lett.,27,45–48
    (1985)

1 NOMENCLATURE

EC number
2.4.1.130

Systematic name
Dolichyl-phosphate-D-mannose:glycolipid alpha-D-mannosyltransferase

Recommended name
Dolichyl-phosphate-mannose-glycolipid alpha-mannosyltransferase

Synonyms
Mannosyltransferase, dolichol phosphomannose-oligosaccharide-lipid
Oligomannosylsynthase

CAS Reg. No.
77967-76-1

2 REACTION AND SPECIFICITY

Catalysed reaction
Transfers an alpha-D-mannosyl residue from dolichyl-phosphate D-mannose
into membrane lipid-linked oligosaccharide

Reaction type
Hexosyl group transfer

Natural substrates
Dolichol-phosphomannose + lipid-linked oligosaccharide (pathway leading
to biosynthesis of asparagine-linked oligosaccharides of mammalian glycolipids) [1]

Substrate spectrum
1 Dolichol-phosphomannose + lipid-linked oligosaccharide (strict donor
  specificity, less specific for oligosaccharide acceptor, transfers 4 man-
  nosyl-residues to lipid-linked oligosaccharide) [1]

Product spectrum
1 Dolichol phosphate + lipid-linked mannosyl-oligosaccharide [1]

Inhibitor(s)

Cofactor(s)/prosthetic group(s)/activating agents

Metal compounds/salts

Turnover number (min^{-1})

Specific activity (U/mg)

K$_m$-value (mM)

pH-optimum
    7.4 (assay at) [1]

pH-range

Temperature optimum (°C)
    37 (assay at) [1]

Temperature range (°C)

---

3 ENZYME STRUCTURE

Molecular weight

Subunits

Glycoprotein/Lipoprotein
    –

---

4 ISOLATION/PREPARATION

Source organism
    Mouse (Thy-1–mutant) [1]

Source tissue
    Lymphoma cells (cell line BW5147.3.(Thy-1 E)10) [1]

Localization in source
    Membrane-bound [1]

Purification

Crystallization
    –

Cloned
    –

Renatured
    –

5 STABILITY

pH

Temperature (°C)

Oxidation

Organic solvent

General stability information

Storage

6 CROSSREFERENCES TO STRUCTURE DATABANKS

PIR/MIPS code

Brookhaven code

7 LITERATURE REFERENCES

[1] Rearick, J.I., Fujimoto, K., Kornfeld, S.: J. Biol. Chem.,256,3762–3769 (1981)

1 NOMENCLATURE

EC number
2.4.1.131

Systematic name
GDPmannose:glycolipid 1,2-alpha-D-mannosyltransferase

Recommended name
Glycolipid 2-alpha-mannosyltransferase

Synonyms
Mannosyltransferase, guanosine diphosphomannose-oligosaccharide-lipid
GDP-mannose-oligosaccharide-lipid mannosyltransferase
Oligosaccharide-lipid mannosyltransferase

CAS Reg. No.
74506-43-7

2 REACTION AND SPECIFICITY

Catalysed reaction
Transfers an alpha-D-mannosyl residue from GDPmannose into lipid-linked
oligosaccharide, forming an alpha-1,2-D-mannosyl-D-mannose linkage

Reaction type
Hexosyl group transfer

Natural substrates
GDPmannose + oligosaccharide-lipid acceptor (involved in biosynthesis of
lipid-linked oligosaccharides up to $(Man)_5$-$(GlcNAc)_2$ [4], involved in the production of two 1,2-linked mannosyl residues in mammalian lipid-linked
tetradecasaccharide of the structure $Glc_3Man_9GlcNAc_2$ (major precursor for
eukaryotic en-bloc-glycosylation of asparagin-linked glycoproteins) [4]) [1, 4]

Substrate spectrum
1 GDPmannose + oligosaccharide-lipid acceptor (presumably multiple
  mannosyltransferases involved [4], elongation reaction via penta- and
  hexasaccharide intermediates [4], acceptor isolated from pig liver [1, 2]
  or baby hamster kidney (i.e. BHK) cells [4], acceptors can be mannose
  or Man-$(GlcNAc)_2$ to $(Man)_5$-$(GlcNAc)_2$ [4], no substrates: dolichol-
  phosphomannose [1, 2] or mannosylphosphorylretinol [4]) [1–4]

Product spectrum
1 GDP + alpha-mannosyl-1,2-oligosaccharide-lipid (presumably a
  heptasaccharide with a 1,6-linked branched glycosyl unit [1]) [1–4]

Inhibitor(s)
    Deoxycholate [1]; Taurocholate [1]; GDP (strong, 0.02 mM) [1]; GTP (about
    50% as effective as GDP, 0.02 mM) [1]; GMP (about 30% as effective as
    GDP, 0.02 mM) [1]; ADP (weak, 2 mM) [1]; ATP (weak, 2 mM) [1]; UDP
    (weak, 2 mM) [1]; UDPglucose (weak, 2 mM) [1]; UDPgalactose (weak,
    2 mM) [1]; UDP-N-acetylglucosamine (weak, 2 mM) [1]; Cu^{2+} (strong) [4];
    Ca^{2+} (weak) [4]; Mn^{2+} [4]; More (no inhibition by amphomycin [2, 4] or
    EDTA) [1, 2, 4]

Cofactor(s)/prosthetic group(s)/activating agents
    Endogen lipids (soluble in $CHCl_3/CH_3OH$ (2:1), activation) [4]; EDTA (slight
    stimulation) [4]

Metal compounds/salts
    More (no divalent cation requirement) [1, 2, 4]

Turnover number (min^{-1})

Specific activity (U/mg)
    More [1, 3]

K_m-value (mM)
    0.00064 (oligosaccharide-lipid acceptor) [1]; 0.0047 (GDPmannose) [1]

pH-optimum
    6.5–7.5 [1]; 6.8–7.6 (broad) [4]

pH-range

Temperature optimum (°C)
    37 (assay at) [1–3]

Temperature range (°C)

3 ENZYME STRUCTURE

Molecular weight

Subunits

Glycoprotein/Lipoprotein
    –

4 ISOLATION/PREPARATION

Source organism
    Rabbit (New Zealand White [1]) [1, 3]; Pig [2]; Bovine (cow) [4]

Source tissue
Liver [1, 3]; Aorta tissue (intimal layer) [2]; Mammary tissue (lactating) [4]

Localization in source
Microsomes [1, 3, 4]; Membrane-bound [1–4]

Purification
Rabbit (solubilized with 0.15% v/v Nonidet P-40, partially purified) [1]

Crystallization
–

Cloned
–

Renatured
–

---

5 STABILITY

pH

Temperature (°C)

Oxidation

Organic solvent

General stability information

Storage
0°C, partially purified, 5 days [1]; 0–4°C, solubilized enzyme, at least
2 weeks [1]; 4°C, solubilized enzyme preparation, about 40% loss of activity
within 2 weeks and about 20% loss of activity within 7 days [4]

---

6 CROSSREFERENCES TO STRUCTURE DATABANKS

PIR/MIPS code

Brookhaven code

---

7 LITERATURE REFERENCES

[1] Schutzbach, J.S., Springfield, J.D., Jensen, J.W.: J. Biol. Chem.,255,4170–4175
    (1980)
[2] Spencer, J.P., Elbein, A.D.: Proc. Natl. Acad. Sci. USA,77,2524–2527 (1980)
[3] Jensen, J.W., Springfield, J.D., Schutzbach, J.S.: J. Biol. Chem.,255,11268–11272
    (1980)
[4] Prakash, C., Katial, A., Kang, M.S., Vijay, I.K.: Eur. J. Biochem.,139,87–93 (1984)

---

1 NOMENCLATURE

EC number
2.4.1.132

Systematic name
GDPmannose:glycolipid 1,3-alpha-D-mannosyltransferase

Recommended name
Glycolipid 3-alpha-mannosyltransferase

Synonyms
Mannosyltransferase, guanosine diphosphomannose-oligosaccharide-lipid II
GDP-mannose-oligosaccharide-lipid mannosyltransferase II
Mannosyltransferase II

CAS Reg. No.
81181-76-2

---

2 REACTION AND SPECIFICITY

Catalysed reaction
Transfers an alpha-D-mannosyl residue from GDPmannose into lipid-linked oligosaccharide, forming an alpha-1,3-D-mannosyl-D-mannose linkage

Reaction type
Hexosyl group transfer

Natural substrates
GDPmannose + tetrasaccharide-diphosphoryl-lipid (involved in the bio-synthesis of asparagine-linked saccharide chains of mammalian glycopro-teins [1, 2], the 1,3-linked mannosyl residue in mammalian lipid-linked oligosaccharide of the structure $Glc_3Man_9GlcNAc_2$ is produced by this enzyme) [1, 2]

Substrate spectrum
1 GDPmannose + tetrasaccharide-diphosphoryl-lipid (no acceptor is dolichol-phosphate [1]) [1–3]

Product spectrum
1 GDP + mannosyl-alpha-1,3-tetrasaccharide-diphosphoryl-lipid [1–3]

Inhibitor(s)
EDTA [1]; ZnCl$_2$ [1]; NiCl$_2$ [1]; Glycerol (1–2% v/v) [1]; GDP [1]; GTP [1]; GMP [1]; GDPglucose [1]; Phospholipids (stable bilayer forming, e.g. phosphatidylcholine, restorable by addition of cholesterol, dolichol and dolichol-derivatives) [2]; ADP (weak) [1]; ATP (weak) [1]; UDP (weak) [1]; UTP (weak) [1]; UDPglucose (weak) [1]; UDP-N-acetylglucosamine (weak) [1]; More (no inhibition by amphomycin) [1]

Cofactor(s)/prosthetic group(s)/activating agents
Nonidet P-40 (activation, up to 0.0225% v/v) [1]; Phosphatidylethanolamine (activation [2, 3], phosphoethanolamine (containing unsaturated acylchains [3], from various eu- and prokaryotic sources [3]) and mixtures with other phospholipids forming non-bilayer phospholipid phases (e.g. hexagonal phases [3]) [2, 3], in the absence of detergent [3], enzyme-phospholipid-complex formation [3]. Not phosphatidylcholine, phosphatidylinositol, phosphatidylglycerol, sphingomyelin, lysophosphocholine or lysophos-phoethanolamine [3]) [2, 3]; Phosphatidylserine (activation, 16–43% as effective as phosphatidylethanolamine) [3]; Cardiolipin (activation, 35% as effective as phosphatidylethanolamine) [3]

Metal compounds/salts
Mg^{2+} (requirement [1, 3], 10 mM [1]) [1, 3]; Ca^{2+} (activation, about 60% as effective as Mg^{2+}) [1]; Mn^{2+} (activation, about 40% as effective as Mg^{2+}) [1]; Co^{2+} (activation, about 4% as effective as Mg^{2+}) [1]

Turnover number (min^{-1})

Specific activity (U/mg)
More [1]

K$_m$-value (mM)
0.00021 (tetrasaccharide-diphosphoryl-lipid, detergent-stimulated) [1]; 0.00065 (GDPmannose, detergent-stimulated) [1]; 0.00096 (tetrasaccha-ride-diphosphoryl-lipid) [3]; 0.00142 (GDPmannose) [3]

pH-optimum
6.8–7.3 [1]

pH-range

Temperature optimum (°C)
37 (assay at) [1–3]

Temperature range (°C)

3 ENZYME STRUCTURE

Molecular weight

Subunits

Glycoprotein/Lipoprotein
–

4 ISOLATION/PREPARATION

Source organism
Rabbit [1–3]

Source tissue
Liver [1–3]

Localization in source
Microsomes [1–3]

Purification
Rabbit (partial) [1]

Crystallization
–

Cloned
–

Renatured
–

5 STABILITY

pH

Temperature (°C)
37 (5 min, in 0.0225% Nonidet P-40 without substrate, about 80% loss of activity. In phosphatidylethanolamine without substrate: at least 5 min stable, after 25 min about 15% loss of activity) [3]; More (phosphatidylethanolamine increases heat-stability) [3]

Oxidation

Organic solvent

General stability information
Glycerol, 0.5% v/v, stabilizes during storage and purification [1]

Storage

–20°C, in 0.1% v/v Nonidet P-40 and 10% v/v glycerol, at least 3 months [3];
0–4°C, in 0.1% v/v Nonidet P-40 and 10% v/v glycerol, 24 h [3]; Glycerol,
10% v/v, stabilizes during storage [2, 3]

6 CROSSREFERENCES TO STRUCTURE DATABANKS

PIR/MIPS code

Brookhaven code

7 LITERATURE REFERENCES

[1] Jensen, J.W., Schutzbach, J.S.: J. Biol. Chem.,256,12899–12904 (1981)
[2] Jensen, J.W., Schutzbach, J.S.: Biochemistry,23,1115–1119 (1984)
[3] Jensen, J.W., Schutzbach, J.S.: J. Biol. Chem.,257,9025–9029 (1982)

1 NOMENCLATURE

EC number
2.4.1.133

Systematic name
UDPgalactose:O-beta-D-xylosylprotein 4-beta-D-galactosyltransferase

Recommended name
Xylosylprotein 4-beta-galactosyltransferase

Synonyms
Galactosyltranferase I
Galactosyltransferase, uridine diphosphogalactose-xylose
UDP-D-galactose:D-xylose galactosyltransferase
UDP-D-galactose:xylose galactosyltransferase
More (cf. EC 2.4.1.134 and EC 2.4.1.135)

CAS Reg. No.
52227-72-2

---

2 REACTION AND SPECIFICITY

Catalysed reaction
UDPgalactose + O-beta-D-xylosylprotein →
→ UDP + 4-beta-D-galactosyl-O-beta-D-xylosylprotein

Reaction type
Hexosyl group transfer

Natural substrates
UDPgalactose + O-beta-D-xylosylprotein (involved in chondroitin sulfate biosynthesis) [1–3]

Substrate spectrum
1 UDPgalactose + xylose [1–3]
2 UDPgalactose + xylosylserine [2]

Product spectrum
1 UDP + 4-O-beta-D-galactosyl-O-beta-D-xylose [1–3]
2 ?

Inhibitor(s)

Cofactor(s)/prosthetic group(s)/activating agents
    Mn^{2+} (requirement, 15 mM) [1, 2]

Metal compounds/salts

Turnover number (min^{-1})

Specific activity (U/mg)
    More [1]

K_m-value (mM)
    0.05 (UDPgalactose) [1]

pH-optimum
    6.5 (assay at) [3]

pH-range

Temperature optimum (°C)
    37 (assay at) [1–3]

Temperature range (°C)

---

3 ENZYME STRUCTURE

Molecular weight

Subunits

Glycoprotein/Lipoprotein
    –

---

4 ISOLATION/PREPARATION

Source organism
    Chicken [1–3]

Source tissue
    Cartilage (epiphyses from embryonic femurs and tibiae [1, 2]) [1–3]

Localization in source
    Membrane-bound [1–3]

Purification
    Chicken (solubilized and partially purified, affinity chromatography) [1–3]

Crystallization
    –

Cloned

–

Renatured

–

5 STABILITY

pH

Temperature (°C)

Oxidation

Organic solvent

General stability information
Nonidet P-40 stabilizes solubilized enzyme [2]; Dialysis against detergent-free buffer, +/- 0.25 M KCl inactivates, dialysis against 1% Nonidet P-40 and 0.25 M KCl restores activity, even after a period of 18 h in the absence of detergent [2]

Storage
–17°C, concentrated enzyme preparation of 1 mg/ml, at least 6 months [1]

6 CROSSREFERENCES TO STRUCTURE DATABANKS

PIR/MIPS code

Brookhaven code

7 LITERATURE REFERENCES

[1] Schwartz, N.B., Rodén, L.: J. Biol. Chem.,250,5200–5207 (1975)
[2] Schawrtz, N.: J. Biol. Chem.,251,285–291 (1976)
[3] Schwartz, N.B., Rodén, L., Dorfman, A.: Biochem. Biophys. Res. Commun.,56, 717–724 (1974)

1 NOMENCLATURE

EC number
2.4.1.134

Systematic name
UDPgalactose:4-beta-D-galactosyl-O-beta-D-xylosylprotein 3-beta-D-galac-
tosyltransferase

Recommended name
Galactosylxylosylprotein 3-beta-galactosyltransferase

Synonyms
Galactosyltransferase II
Galactosyltransferase, uridine diphosphogalactose-galactosylxylose
More (cf. EC 2.4.1.133 and EC 2.4.1.135)

CAS Reg. No.
56626-19-8; 56626-21-2

---

2 REACTION AND SPECIFICITY

Catalysed reaction
UDPgalactose + 4-beta-D-galactosyl-O-beta-D-xylosylprotein →
→ UDP + 3-beta-D-galactosyl-4-beta-D-galactosyl-O-beta-D-xylosylprotein

Reaction type
Hexosyl group transfer

Natural substrates
UDPgalactose + 4-beta-D-galactosyl-O-beta-D-xylosylprotein (involved in
biosynthesis of chondroitin sulfate) [1]

Substrate spectrum
1 UDPgalactose + 4-beta-D-galactosyl-O-beta-D-xylose [1, 2]
2 UDPgalactose + galactosyl-xylosylserine [2]
3 UDPgalactose + beta-galactosides (e.g. phenyl or pyridine 3-O-beta-ga-
lactosides, poor substrate: pyridine 2-S-beta-thiogalactoside, no substrate: D-galactose) [2]

Product spectrum
1 UDP + 3-beta-D-galactosyl-4-beta-D-galactosyl-O-beta-D-xylose [1]
2 UDP + 3-beta-D-galactosyl-4-beta-D-galactosyl-O-beta-D-xylosylserine [2]
3 ?

Inhibitor(s)

Cofactor(s)/prosthetic group(s)/activating agents

Metal compounds/salts
Mn^{2+} (requirement) [1]

Turnover number (min^{-1})

Specific activity (U/mg)

K$_m$-value (mM)
1.05 (UDPgalactose (+ pyridine 3-O-beta-galactoside)) [2]; 100 (above, pyridine 3-O-beta-galactoside) [2]

pH-optimum
6 (broad, beta-galactosides) [2]

pH-range

Temperature optimum (°C)
37 (assay at) [1, 2]

Temperature range (°C)

---

3 ENZYME STRUCTURE

Molecular weight

Subunits

Glycoprotein/Lipoprotein
–

---

4 ISOLATION/PREPARATION

Source organism
Chicken [1, 2]

Source tissue
Cartilage (epiphyses from embryonic femurs and tibiae) [1, 2]

Localization in source
Membrane-bound [1]

Purification
Chicken (partial) [1]

Crystallization
–

Cloned

–

Renatured

–

5 STABILITY

pH

Temperature (°C)
50 (60 min, about 40% loss of activity, beta-galactosides) [2]; 60 ($t_{1/2}$: 7 min, beta-galactosides) [2]

Oxidation

Organic solvent

General stability information

Storage
–20°C, microsomal enzyme preparation, at least 12 months [2]

6 CROSSREFERENCES TO STRUCTURE DATABANKS

PIR/MIPS code

Brookhaven code

7 LITERATURE REFERENCES

[1] Schwartz, N.B., Rodén, L.: J. Biol. Chem.,250,5200–5207 (1975)
[2] Robinson, J.A., Robinson, H.C.: Biochem. J.,227,805–814 (1985)

Enzyme Handbook © Springer-Verlag Berlin Heidelberg 1996
Duplication, reproduction and storage in data banks are only
allowed with the prior permission of the publishers

1 NOMENCLATURE

EC number
2.4.1.135

Systematic name
UDPglucuronate:3-beta-D-galactosyl-4-beta-D-galactosyl-O-beta-D-xylosyl-
protein D-glucuronosyltransferase

Recommended name
Galactosylgalactosylxylosylprotein 3-beta-glucuronosyltransferase

Synonyms
Glucuronosyltransferase I
More (cf. EC 2.4.1.133 and EC 2.4.1.134)

CAS Reg. No.

2 REACTION AND SPECIFICITY

Catalysed reaction
UDPglucuronate + 3-beta-D-galactosyl-4-beta-D-galactosyl-O-beta-D-xylosyl-
protein →
→ UDP + 3-beta-D-glucuronosyl-3-beta-D-galactosyl-4-beta-D-galactosyl-
O-beta-D-xylosylprotein

Reaction type
Hexosyl group transfer

Natural substrates
UDPglucuronate + 3-beta-D-galactosyl-4-beta-D-galactosyl-O-beta-D-xylosyl-
protein (involved in the biosynthesis of the heparin-polypeptide linkage region [2] and the polysaccharide-protein-linkage region of proteochondroitin
sulfate [1]) [1, 2]

Substrate spectrum
1 UDPglucuronate + 3-beta-D-galactosyl-D-galactose (transfer to non-reducing terminal [1], acceptors are similar, larger fragments from poly-
   saccharide/protein linkage region of chondroitin sulfate [1], high acceptor
   specificity [2], less effective acceptor substrates are galactosyl-beta-
   (1→4) galactose [1], galactosyl-beta-(1→6)galactose [1, 2], galactosyl-
   beta-(1→4)xylose, galactosyl-beta-(1→4)glucose [2]. No acceptors are
   raffinose, xylose, fucose, N-acetylglucosamine [1], glucose [1, 2], galactose or N-acetylglucosaminyl residues [2]) [1–3]
2 UDPglucuronate + 3-beta-D-galactosyl-D-galactosyl-4-xylose [1]
3 UDPglucuronate + 3-beta-D-galactosyl-D-galactosyl-4-xylosylserine [1]

Product spectrum
1 UDP + 3-beta-D-glucuronosyl-3-beta-D-galactosyl-4-beta-D-galactose [1, 2]
2 UDP + 3-beta-D-glucuronosyl-3-beta-D-galactosyl-D-galactosyl-4-xylose [1]
3 UDP + 3-beta-D-glucuronosyl-3-beta-D-galactosyl-D-galac-
tosyl-4-xylosylserine [1]

Inhibitor(s)
3-O-beta-D-Galactosyl-galactose (with 3-galactosyl-4-galactosyl-xylose
as substrate) [1]; More (no inhibition by chondroitin 6-sulfate tri- or pen-
tasaccharides containing N-acetylglucosamine 6-sulfate at their non-
reducing termini) [1]

Cofactor(s)/prosthetic group(s)/activating agents

Metal compounds/salts
Mn^{2+} (requirement [1–3], 15 mM [1], 20 mM [2]) [1–3]

Turnover number (min^{-1})

Specific activity (U/mg)
More [2]

K_m-value (mM)
0.00025 (UDPglucuronate); 0.0091 (3-beta-galactosyl-galactose) [2]

pH-optimum
5–7.5 (broad) [1]; 5.4 (assay at, due to UDPglucuronate instability at higher
pH values) [1]; 7.5 [2]

pH-range
6.5–8.5 (about half-maximal activity at pH 6.5 and 8.5) [2]

Temperature optimum (°C)
37 (assay at) [1, 3]

Temperature range (°C)

3 ENZYME STRUCTURE

Molecular weight

Subunits

Glycoprotein/Lipoprotein
–

4 ISOLATION/PREPARATION

Source organism
Chicken [1, 3]; Mouse [2]

Source tissue
Cartilage (embryonic, epiphyses from femurs and tibiae [3]) [1, 3]; Mastocytoma tissue (heparin-producing) [2]

Localization in source
Membrane-bound [1–3]; Microsomes [2]

Purification
Mouse (partial) [2]

Crystallization
–

Cloned
–

Renatured
–

5 STABILITY

pH

Temperature (°C)

Oxidation

Organic solvent

General stability information

Storage

6 CROSSREFERENCES TO STRUCTURE DATABANKS

PIR/MIPS code

Brookhaven code

7 LITERATURE REFERENCES

[1] Helting, T., Rodén, L.: J. Biol. Chem.,244,2799–2805 (1969)
[2] Helting, T.: J. Biol. Chem.,247,4327–4332 (1972)
[3] Schwartz, N.B., Rodén, L.: J. Biol. Chem.,250,5200–5207 (1975)

1 NOMENCLATURE

EC number
2.4.1.136

Systematic name
UDPglucose:gallate beta-D-glucosyltransferase

Recommended name
Gallate 1-beta-glucosyltransferase

Synonyms
Glucosyltransferase, uridine diphosphoglucose-vanillate 1-
UDPglucose-vanillate 1-glucosyltransferase
UDPglucose:vanillate 1-O-glucosyltransferase [1]
UDPglucose:gallate glucosyltransferase [2]

CAS Reg. No.
89700-30-1

---

2 REACTION AND SPECIFICITY

Catalysed reaction
UDPglucose + gallate →
→ UDP + 1-galloyl-beta-D-glucose

Reaction type
Hexosyl group transfer

Natural substrates
UDPglucose + gallate [1]
More (physiological role probably is the formation of beta-glucogallin, the
putative first intermediate in the biosynthesis of gallotannins) [1, 2]

Substrate spectrum
1 UDPglucose + gallate (r [1, 2], 52% of activity compared to vanillate [1])
  [1, 2]
2 UDPglucose + vanillate (best substrate) [1]
3 UDPglucose + veratrate (71% of activity compared to vanillate) [1]
4 UDPglucose + benzoate (16% of activity compared to vanillate) [1]
5 UDPglucose + protocatechuate (22% of activity compared to vanillate)
  [1]
6 UDPglucose + anisate (38% of activity compared to vanillate) [1]
7 UDPglucose + 3,4,5-trimethoxybenzoate (7% of activity compared to
  vanillate) [1]

8 UDPglucose + p-hydroxybenzoate (48% of activity compared to vanillate) [1]
9 UDPglucose + cinnamate (9% of activity compared to vanillate) [1]
10 UDPglucose + m-coumarate (14% of activity compared to vanillate) [1]
11 UDPglucose + p-coumarate (32% of activity compared to vanillate) [1]
12 UDPglucose + ferulate (35% of activity compared to vanillate) [1]
13 UDPglucose + sinapate (13% of activity compared to vanillate) [1]

Product spectrum
1 UDP + 1-O-galloyl-beta-D-glucose (i.e. beta-glucogallin) [1, 2]
2 ?
3 ?
4 ?
5 ?
6 ?
7 ?
8 ?
9 ?
10 ?
11 ?
12 ?
13 ?

Inhibitor(s)
UDP (strong) [1]; Dithiothreitol (4 mM, slight) [1]

Cofactor(s)/prosthetic group(s)/activating agents

Metal compounds/salts

Turnover number (min^{-1})

Specific activity (U/mg)
0.0117 (gallate) [1]

K_m-value (mM)
0.57 (vanillate) [1]; 0.72 (veratrate) [1]; 1.1 (gallate) [2]; 1.11 (gallate) [1];
2.3 (UDPglucose (+ gallate)) [2]; 3.45 (p-hydroxybenzoate) [1]

pH-optimum
6.5–7.0 [1]; 7.0 [2]

pH-range
4.7–9.0 (50% of maximal activity at pH 4.7 and 9.0) [1]; 5.5–8.5 (50% of
maximal activity at pH 5.5 and 8.5) [2]

Temperature optimum (°C)
30 (assay at) [1, 2]; 42 [1]

Temperature range (°C)

---

3 ENZYME STRUCTURE

Molecular weight
68000 (Quercus rubra, gel filtration) [1]

Subunits

Glycoprotein/Lipoprotein
–

---

4 ISOLATION/PREPARATION

Source organism
Quercus rubra (oak) [1, 2]

Source tissue
Leaf [1, 2]

Localization in source

Purification
Quercus rubra [1, 2]

Crystallization
–

Cloned
–

Renatured
–

---

5 STABILITY

pH

Temperature (°C)

Oxidation
O_2-sensitive, mercaptoethanol stabilizes [1]

Organic solvent

General stability information
Mercaptoethanol in extraction buffers stabilizes [1]; Diluted enzyme is unstable [1]

Storage
0–4°C, 1–2 weeks [1]

6 CROSSREFERENCES TO STRUCTURE DATABANKS

PIR/MIPS code

Brookhaven code

7 LITERATURE REFERENCES

[1] Gross, G.: Phytochemistry,22,2179–2182 (1983)
[2] Gross, G.: FEBS Lett.,148,67–70 (1982)

1 NOMENCLATURE

EC number

2.4.1.137

Systematic name

UDPgalactose:sn-glycerol-3-phosphate 2-alpha-D-galactosyltransferase

Recommended name

sn-Glycerol-3-phosphate 2-alpha-galactosyltransferase

Synonyms

Floridoside-phosphate synthase

UDP-galactose:sn-glycerol-3-phosphate-2-D-galactosyl transferase [1]

FPS [2]

UDP-galactose, sn-3-glycerol phosphate:1→2' galactosyltransferase [2]

Synthetase, floridoside phosphate

Floridoside phosphate synthase

More (cf. EC 2.4.1.96)

CAS Reg. No.

80747-34-8

---

2 REACTION AND SPECIFICITY

Catalysed reaction

UDPgalactose + sn-glycerol 3-phosphate →

→ UDP + 2-(alpha-D-galactosyl)-sn-glycerol 3-phosphate

Reaction type

Hexosyl group transfer

Natural substrates

UDPgalactose + sn-glycerol 3-phosphate (involved in biosynthesis of floridoside, a typical low-molecular weight red algal carbohydrate [1], key enzyme in floridoside biosynthesis [2]) [1, 2]

Substrate spectrum

1 UDPgalactose + sn-glycerol 3-phosphate (high substrate specificity [2]) [1, 2]

Product spectrum

1 UDP + 2-(alpha-D-galactosyl)-sn-glycerol 3-phosphate (the product is hydrolyzed by a phosphatase to floridoside) [1]

Inhibitor(s)
UDP [2]

Cofactor(s)/prosthetic group(s)/activating agents
No cofactor or activator found [2]

Metal compounds/salts

Turnover number (min^{-1})

Specific activity (U/mg)
More [2]

K_m-value (mM)
2.7 (UDPgalactose) [2]; 8 (sn-glycerol 3-phosphate) [2]

pH-optimum

pH-range

Temperature optimum (°C)

Temperature range (°C)

---

3 ENZYME STRUCTURE

Molecular weight
120000 (Porphyra perforata, HPLC gel filtration) [2]

Subunits

Glycoprotein/Lipoprotein
–

---

4 ISOLATION/PREPARATION

Source organism
Porphyra perforata [2]; Dumontia incrassata (O.F. Müller, Lamour.) [1];
Chondrus crispus (Stackh.) [1]; Cystoclonium purpureum (Lightf., Batt.)
[1]; Corallina officinalis (L.) [1]; Porphyra umbilicalis (L.) [1]; Lomentaria
umbellata (H. u. H., Gendo) [1]; Catenella nipea (Zan.) [1]

Source tissue
Thallus [1]

Localization in source

Purification
Porphyra perforata [2]

Crystallization

–

Cloned

–

Renatured

–

---

5 STABILITY

pH

Temperature (°C)

Oxidation

Organic solvent

General stability information

Storage

---

6 CROSSREFERENCES TO STRUCTURE DATABANKS

PIR/MIPS code

Brookhaven code

---

7 LITERATURE REFERENCES

[1] Kremer, B.P., Kirst, G.O.: Plant Sci. Lett.,23,349–357 (1981)
[2] Meng, J., Srivastava, L.M.: Phytochemistry,30,1763–1766 (1991)

1 NOMENCLATURE

EC number
2.4.1.138

Systematic name
UDP-N-acetyl-D-glucosamine:mannotetraose alpha-N-acetyl-D-glucosaminyltransferase

Recommended name
Mannotetraose 2-alpha-N-acetylglucosaminyltransferase

Synonyms
alpha-N-Acetylglucosaminyltransferase
Acetylglucosaminyltransferase, uridine diphosphoacetylglucosamine
mannoside alpha1→2-

CAS Reg. No.
81032-47-5

2 REACTION AND SPECIFICITY

Catalysed reaction
UDP-N-acetyl-D-glucosamine + 1,3-alpha-D-mannosyl-1,2-alpha-D-mannosyl-1,2-alpha-D-mannosyl-D-mannose →
→ UDP + 1,3-alpha-D-mannosyl-1,2-(N-acetyl-alpha-D-glucosaminyl-alpha-D-mannosyl)-1,2-alpha-D-mannosyl-D-mannose

Reaction type
Hexosyl group transfer

Natural substrates
UDP-N-acetyl-D-glucosamine + 1,3-alpha-D-mannosyl-1,2-alpha-D-mannosyl-1,2-alpha-D-mannosyl-D-mannose (involved in biosynthesis of mannoprotein side-chain units) [1]

Substrate spectrum
1 UDP-N-acetyl-D-glucosamine + 1,3-alpha-D-mannosyl-1,2-alpha-D-mannosyl-1,2-alpha-D-mannosyl-D-mannose (i.e. mannotetraose acceptor, acceptor substrates are mannoprotein, saccharides with more than 25 mannose-units, with 15 to 25 units, with 8 to 15 units, with 5 to 8 units and below 5 units, the latter being the best substrates) [1]

Product spectrum
1 UDP + 1,3-alpha-D-mannosyl-1,2-(N-acetyl-alpha-D-glucosaminyl-alpha-D-mannosyl)-1,2-alpha-D-mannosyl-D-mannose [1]

Inhibitor(s)

UDP (wild-type) [1]; UDPhexylamine (wild-type) [1]; NaCl (weak) [1]; NaN_3 [1]; Imidazole (wild-type) [1]; Tris (wild-type) [1]; Tween 80 [1]; Cu^{2+} (wild-type) [1]; Phosphate (wild-type) [1]; Mn^{2+} (above 10 mM, wild-type) [1]; Co^{2+} (strong, 1 mM, even in the presence of Mn^{2+}, wild-type) [1]

Cofactor(s)/prosthetic group(s)/activating agents

Metal compounds/salts

Mn^{2+} (requirement, 10 mM, inhibits at higher concentrations, Mg^{2+} or Ca^{2+} cannot replace Mn^{2+}) [1]

Turnover number (min^{-1})

Specific activity (U/mg)

0.0056 [1]

K_m-value (mM)

0.0364 (UDP-N-acetylglucosamine, mutant 2–2) [1]; 0.0447 (UDP-N-acetylglucosamine, wild-type) [1]; 11 (mannotetraose acceptor, solubilized enzyme) [1]; 13 (mannotetraose acceptor, membrane-bound enzyme) [1]

pH-optimum

More (pI: 4.9, in the presence of Triton X-100 and Mn^{2+}) [1]; 6.5 [1]

pH-range

Temperature optimum (°C)

Temperature range (°C)

3 ENZYME STRUCTURE

Molecular weight

300000 (Kluyveromyces lactis, gel filtration in the presence of Triton X-100, native MW may be lower) [1]

Subunits

Glycoprotein/Lipoprotein

Glycoprotein [1]

4 ISOLATION/PREPARATION

Source organism

Kluyveromyces lactis (wild-type strains Y-58a his 4c or his 3 and mutant strains mnn1, mnn2–1 or mnn2–2) [1]

Source tissue
Cell [1]

Localization in source
Endoplasmic reticulum (integral membrane protein) [1]

Purification
Kluyveromyces lactis (partial) [1]

Crystallization
–

Cloned
–

Renatured
–

---

5 STABILITY

pH

Temperature (°C)
35 (2 h, inactivation) [1]; 45 ($t_{1/2}$: 4 min) [1]

Oxidation

Organic solvent

General stability information
Organic solvent extraction, unstable to [1]; Triton X-100 stabilizes solubilized enzyme [1]

Storage
4°C, in imidazole-glycerol buffer, 2% Triton X-100, $t_{1/2}$: 80 days [1]

---

6 CROSSREFERENCES TO STRUCTURE DATABANKS

PIR/MIPS code

Brookhaven code

---

7 LITERATURE REFERENCES

[1] Douglas, R.H., Ballou, C.E.: Biochemistry,21,1561–1570 (1982)

1 NOMENCLATURE

EC number
2.4.1.139

Systematic name
alpha-D-Glucose-1-phosphate:alpha-D-glucose-1-phosphate
4-alpha-D-glucosyltransferase (dephosphorylating)

Recommended name
Maltose synthase

Synonyms
Synthase, maltose

CAS Reg. No.
81669-74-1

---

2 REACTION AND SPECIFICITY

Catalysed reaction
alpha-D-Glucose 1-phosphate + alpha-D-glucose 1-phosphate →
→ maltose + 2 phosphate

Reaction type
Hexosyl group transfer

Natural substrates

Substrate spectrum
1 alpha-D-Glucose 1-phosphate + alpha-D-glucose 1-phosphate [1]
2 More (exchange of phosphate group of alpha-D-glucose 1-phosphate
  with phosphate is possible, neither maltose 1-phosphate nor free phosphate can be detected as an intermediate) [1]

Product spectrum
1 Maltose + 2 phosphate [1]
2 ?

Inhibitor(s)
beta-D-Glucose 1-phosphate [1]; Glucono-1,5-lactone (competitive) [1]; ATP
[1]; Phosphate [1]; ADPglucose (strong) [1]; UDPglucose (strong) [1]; GDP-
glucose (strong) [1]

Cofactor(s)/prosthetic group(s)/activating agents

Metal compounds/salts
More (EDTA, Mg^{2+}, Mn^{2+} have no influence) [1]

Turnover number (min^{-1})

Specific activity (U/mg)

K_m-value (mM)
1.5 (alpha-D-glucose 1-phosphate) [1]

pH-optimum
6.8 [1]

pH-range
6.2–7.5 (30% of maximal activity at pH 6.2, 45% of maximal activity at pH 7.5) [1]

Temperature optimum (°C)
32 (assay at) [1]

Temperature range (°C)

---

3 ENZYME STRUCTURE

Molecular weight
95000 (Spinacia oleracea, gel filtration) [1]

Subunits

Glycoprotein/Lipoprotein
–

---

4 ISOLATION/PREPARATION

Source organism
Spinacia oleracea (spinach) [1]

Source tissue
Seeds (cotyledon) [1]

Localization in source

Purification
Spinacia oleracea (partial) [1]

Crystallization
–

Cloned

–

Renatured

–

5 STABILITY

pH

Temperature (°C)

Oxidation

Organic solvent

General stability information
During purification addition of 15 mM alpha-D-glucose 1-phosphate to the isolation buffer is necessary for stabilization [1]; Purification has to be carried out at 4°C [1]

Storage

6 CROSSREFERENCES TO STRUCTURE DATABANKS

PIR/MIPS code

Brookhaven code

7 LITERATURE REFERENCES

[1] Schilling, N.: Planta,154,87–93 (1982)

1 NOMENCLATURE

EC number
2.4.1.140

Systematic name
Sucrose:1,6(1,3)-alpha-D-glucan 6(3)-alpha-D-glucosyltransferase

Recommended name
Alternansucrase

Synonyms
Sucrose:1,6-, 1,3-alpha-D-glucan 3-alpha- and 6-alpha-D-glucosyltransferase
[4]
Glucosyltransferase, sucrose-1,6(3)-alpha-glucan 6(3)-alpha-

CAS Reg. No.
100630-46-4

2 REACTION AND SPECIFICITY

Catalysed reaction
Transfers an alpha-D-glucosyl residue alternately to the 6-position and the
3-position of the non-reducing terminal residue of an alpha-D-glucan, thus
producing a glucan having alternating alpha-1,6- and alpha-1,3-linkages

Reaction type
Hexosyl group transfer

Natural substrates

Substrate spectrum
1 Sucrose + alpha-D-glucan [1–4]
2 Sucrose + acceptors (maltose, isomaltose,isomaltotriose, methyl-
  alpha-D-glucoside) [2]

Product spectrum
1 Alternating-1,6–1,3-alpha-D-glucan (the product has quite different properties from other dextrans, has been called alternan, insoluble D-glucan
  consists of 76 mol% 1,3-alpha-linked glucose and 24 mol% 1,6-alpha-lin-
  ked glucose [3], glucan consists of 49.1 mol% 1,6-alpha-linked glucose
  and 33.9 mol% 1,3-alpha-linked glucose with 13.6 mol% terminal glucose
  and 3.3 mol% 1,3,6-alpha-branched glucose [4]) [1–4]
2 Oligoalternan (low-molecular-weight product) [2]

Inhibitor(s)
   3-Deoxy-3-fluoro-alpha-D-glucopyranosyl fluoride [1]; EDTA (disodium salt,
   slight) [1]; 2-Aminoethanol (slight) [1]; SDS [1]; Tris [1]; Octyl beta-D-glu-
   copyranoside [1]

Cofactor(s)/prosthetic group(s)/activating agents
   Dextran T10 (stimulates [3], no stimulation [4]) [3]

Metal compounds/salts

Turnover number (min^{-1})

Specific activity (U/mg)
   7.3 [3]; 89.7 [4]

K_m-value (mM)
   4.9 (sucrose) [4]; 16.3 (sucrose) [3]

pH-optimum
   5.2 (assay at) [2]; 5.5 [1]; 5.6 [2]; 6.0 [4]; 6.5 [3]

pH-range
   3.2–7.4 (about 50% of activity maximum at pH 3.2 and 7.4) [1]; 4.6–6.5
   (about 50% of activity maximum at pH 4.6 and 6.5) [2]

Temperature optimum (°C)
   30 (assay at) [2]; 40 [2]

Temperature range (°C)
   30–50 (30°C: about 65% of activity maximum, 50°C: about 55% of activity
   maximum) [2]

3 ENZYME STRUCTURE

Molecular weight

Subunits
   ? (x × 158000, Streptococcus mutans, SDS-PAGE with and without 2-mer-
   captoethanol [3], x × 159000, Streptococcus mutans, SDS-PAGE [4]) [3, 4]

Glycoprotein/Lipoprotein
   Glycoprotein (1.5% carbohydrate) [4]

4 ISOLATION/PREPARATION

Source organism
   Leuconostoc mesenteroides (NRRL B-1355) [1, 2]; Streptococcus mutans
   (Ingbritt, serotype c [3], HS6 serotype a [4]) [3, 4]

Source tissue
Culture fluid [1, 2, 4]; Cell [3]

Localization in source
Extracellular [1, 4]; Cell-associated [3]

Purification
Leuconostoc mesenteroides (NRLL B-1355) [2]; Streptococcus mutans [3, 4]

Crystallization
–

Cloned
–

Renatured
–

---

5 STABILITY

pH

Temperature (°C)
40 (pH 5.4, half-life: 2 days) [2]

Oxidation

Organic solvent

General stability information

Storage

---

6 CROSSREFERENCES TO STRUCTURE DATABANKS

PIR/MIPS code

Brookhaven code

---

7 LITERATURE REFERENCES

[1] Cote, G.L., Robyt, J.F.: Carbohydr. Res.,101,57–74 (1982)
[2] Lopez-Munguia, A., Pelenc, V., Remaud, M., Biton, J., Michel, J.M., Lang, C., Paul, F., Monsan, P.: Enzyme Microb. Technol.,15,77–85 (1993)
[3] Mukasa, H., Shimamura, A., Tsumori, H.: J. Gen. Microbiol.,135,2055–2063 (1989)
[4] Tsumori, H., Shimamura, A., Mukasa, H.: J. Gen. Microbiol.,131,3347–3353 (1985)

---

1 NOMENCLATURE

EC number
2.4.1.141

Systematic name
UDP-N-acetyl-D-glucosamine:N-acetyl-D-glucosaminyl-diphosphodolichol
N-acetyl-D-glucosaminyltransferase

Recommended name
N-Acetylglucosaminyldiphosphodolichol N-acetylglucosaminyltransferase

Synonyms
UDP-GlcNAc:dolichyl-pyrophosphoryl-GlcNAc GlcNAc transferase [2]
Acetylglucosaminyltransferase, uridine diphosphoacetylglucosamine-
dolichylacetylglucosamine pyrophosphate
N,N'-Diacetylchitobiosylpyrophosphoryldolichol synthase

CAS Reg. No.
75536-54-8

2 REACTION AND SPECIFICITY

Catalysed reaction
UDP-N-acetyl-D-glucosamine + N-acetyl-D-glucosaminyl-diphosphodolichol →
→ UDP + N,N'-diacetylchitobiosyl-diphosphodolichol

Reaction type
Hexosyl group transfer

Natural substrates
UDP-N-acetyl-D-glucosamine + N-acetyl-D-glucosaminyl-diphosphodolichol
(enzyme is involved in biosynthesis of the oligosaccharide portion of the
N-linked glycoproteins [2], 2nd enzyme of dolichol pathway [3]) [2, 3]

Substrate spectrum
1 UDP-N-acetyl-D-glucosamine + N-acetyl-D-glucosaminyl-diphospho-
   dolichol [2, 3]
2 More (rat lung microsomal preparation catalyzes following reaction
   sequence: UDP-GlcNAc + dolichol-phosphate→ GlcNAc-P-P-dolichol +
   UMP, GlcNAc-P-P-dolichol + UDPglucosyl-glucuronic acid→ gluco-
   syl-glucuronic acid-glucosyl-N-acetyl-diphosphodolichol + UDP) [1]

Product spectrum
1 UDP + N,N'-diacetylchitobiosyl-diphosphodolichol [2, 3]
2 ?

Inhibitor(s)
UDP [2, 3]; UDPglucose [2]; Diumycin [2]; Tunicamycin (0.002 mg, weak [3], not [2]) [3]

Cofactor(s)/prosthetic group(s)/activating agents
Triton X-100 (strong requirement) [2]

Metal compounds/salts
Mg^{2+} (stimulates [2], required, optimum concentration 10 mM [3]) [2, 3]; Mn^{2+} (stimulates, not as effective as Mg^{2+}) [2, 3]; Ca^{2+} (stimulates) [3]

Turnover number (min^{-1})

Specific activity (U/mg)
More [2]

K_m-value (mM)
0.00025 (UDP-GlcNAc) [2]; 0.0022 (N-acetyl-D-glucosaminyl-diphospho-dolichol) [2]; 0.01 (UDP-GlcNAc) [3]; 0.040 (N-acetyl-D-glucosaminyl-di-phosphodolichol) [3]

pH-optimum
6.0–8.0 [3]; 7.4–7.6 [2]

pH-range

Temperature optimum (°C)
22 (assay at room temperature) [3]; 37 (assay at) [2]

Temperature range (°C)

3 ENZYME STRUCTURE

Molecular weight

Subunits

Glycoprotein/Lipoprotein
–

4 ISOLATION/PREPARATION

Source organism
Human [1]; Rat [1]; Mung bean [2]; Saccharomyces cerevisiae [3]

Source tissue
Lung [1]; Seedlings [2]; SV40-transformed lung fibroblasts [1]

Localization in source
Microsomes [1, 2]; Membrane [3]

Purification
Mung bean [2]; Saccharomyces cerevisiae [3]

Crystallization
–

Cloned
–

Renatured
–

---

5 STABILITY

pH

Temperature (°C)

Oxidation

Organic solvent

General stability information

Storage
0°C, pH 7.2, 20% glycerol, 0.5 mM DTT, 20% loss of activity after 6 days [2]; –20°C, pH 7.2, 20% glycerol, 0.5 mM DTT, 10% loss of activity after 1 month [2]; Glycerol, 20%, stabilizes during storage [2]; DTT, 0.5 mM stabilizes during storage [2]

---

6 CROSSREFERENCES TO STRUCTURE DATABANKS

PIR/MIPS code

Brookhaven code

---

7 LITERATURE REFERENCES

[1] Turco, S.J., Heath, E.C.: J. Biol. Chem.,252,2918–2928 (1977)
[2] Kaushal, G.P., Elbein, A.D.: Plant Physiol.,81,1086–1091 (1986)
[3] Sharma, C.B., Lehle, L., Tanner, W.: Eur. J. Biochem.,126,319–325 (1982)

1 NOMENCLATURE

EC number
2.4.1.142

Systematic name
GDPmannose:chitobiosyldiphosphodolichol alpha-D-mannosyltransferase

Recommended name
Chitobiosyldiphosphodolichol alpha-mannosyltransferase

Synonyms
Mannosyltransferase, guanosine diphosphomannose-dolichol diphospho-chitobiose
GDP-mannose-dolichol diphosphochitobiose mannosyltransferase

CAS Reg. No.
83380-85-2

---

2 REACTION AND SPECIFICITY

Catalysed reaction
GDPmannose + chitobiosyldiphosphodolichol →
→ GDP + alpha-D-mannosylchitobiosyldiphosphodolichol

Reaction type
Hexosyl group transfer

Natural substrates
GDPmannose + dolichyl diphosphate-$(GlcNAc)_2$ (enzyme of dolichol pathway) [1, 2]

Substrate spectrum
1 GDPmannose + chitobiosyldiphosphodolichol (i.e. dolichyl diphosphate-$(GlcNAc)_2$) [1–3]

Product spectrum
1 GDP + alpha-D-mannosylchitobiosyldiphosphodolichol (i.e. dolichyldi-phosphate-$(GlcNAc)_2$-Man) [1, 2]

Inhibitor(s)
Guanosine (weak [3]) [2, 3]; Periodate-oxidized guanosine [2]; p-Chloromercuribenzenesulfonic acid (partial prevention by DTT) [2]; Zn^{2+} [3]; Hg^{2+} [3]; GDP [1–3]; GMP [1–3]; GDPglucose [2, 3]; GTP [2, 3]; More (sensitive to high ionic strength, 200 mM NaCl, KCl or Tris diminish activity by 50%) [3]

1

Cofactor(s)/prosthetic group(s)/activating agents

Phospholipid (enzyme requires either detergent or phospholipid for maximal activity, effects of these two are not additional) [3]; Detergent (enzyme requires either detergent (Triton X-100 or NP-40) or phospholipid for maximal activity, effects of these two are not additional [3], NP-40 stimulates, optimum activity at 0.1% [2]) [2, 3]

Metal compounds/salts

Mg^{2+} (divalent cation required, Mg^{2+} is the best metal ion, optimum activity at 6 mM [3], absolute requirement, optimum at 5 mM [2], stimulates [1]) [1–3]; Mn^{2+} (stimulates [1, 3], slight stimulation [2]) [1–3]; Ca^{2+} (slight stimulation) [2, 3]

Turnover number (min^{-1})

Specific activity (U/mg)

More [1–3]

K_m-value (mM)

0.0005 (GDPmannose) [2]; 0.001 (dolichyl diphosphate-(GlcNAc)$_2$) [2]; 0.0017 (GDPmannose) [3]; 0.007 (GDPmannose) [1]; 0.009 (dolichyl diphosphate-(GlcNAc)$_2$) [3]; 0.017 (dolichyl diphosphate-(GlcNAc)$_2$) [1]

pH-optimum

6.9–7.0 [3]; 7.0 [2]; 7.5 [1]

pH-range

5.5–8.5 (5.5: about 50% of activity maximum, 8.5: about 80% of activity maximum) [1]

Temperature optimum (°C)

37 (assay at) [2]

Temperature range (°C)

---

3 ENZYME STRUCTURE

Molecular weight

Subunits

Glycoprotein/Lipoprotein

–

---

4 ISOLATION/PREPARATION

Source organism

Saccharomyces cerevisiae [1]; Pig [2]; Glycine max [3]

Source tissue
Aorta [2]; Suspension cultured cells [3]

Localization in source
Membrane [1]; Microsomes [2, 3]

Purification
Saccharomyces cerevisiae [1]; Pig (partial) [2]; Glycine max [3]

Crystallization
–

Cloned
–

Renatured
–

---

5 STABILITY

pH

Temperature (°C)

Oxidation

Organic solvent

General stability information
Glycerol, 20%, + DTT, 0.5 mM, or + dolichyl-pyrophosphoryl-N,N'-diacetyl-chitobiose, stabilizes [2]

Storage
–18°C, 20% glycerol, 0.5 mM DTT, stable for 2 weeks with only slight loss of activity [2]

---

6 CROSSREFERENCES TO STRUCTURE DATABANKS

PIR/MIPS code

Brookhaven code

---

7 LITERATURE REFERENCES

[1] Sharma, C.B., Lehle, L., Tanner, W.: Eur. J. Biochem.,126,319–325 (1982)
[2] Kaushal, G.P., Elbein, A.D.: Arch. Biochem. Biophys.,250,38–47 (1986)
[3] Kaushal, G.P., Elbein, A.D.: Biochemistry,26,7953–7960 (1987)

1 NOMENCLATURE

EC number
2.4.1.143

Systematic name
UDP-N-acetyl-D-glucosamine:glycoprotein (N-acetyl-D-glucosamine to alpha-D-mannosyl-1,6-(R_1)-beta-D-mannosyl-R_2) beta-1,2-N-acetyl-D-glucosaminyl-transferase

Recommended name
alpha-1,6-Mannosyl-glycoprotein beta-1,2-N-acetylglucosaminyltransferase

Synonyms
N-Glycosyl-oligosaccharide-glycoprotein N-acetylglucosaminyltransferase II
N-Acetylglucosaminyltransferase II
Acetylglucosaminyltransferase II
Acetylglucosaminyltransferase, uridine diphosphoacetylglucosamine-mannoside alpha1→6-
Acetylglucosaminyltransferase, uridine diphosphoacetylglucosamine-alpha-1,6-mannosylglycoprotein beta-1–2-N-
Acetylglucosaminyltransferase, uridine diphosphoacetylglucosamine-alpha-D-mannoside beta1–2-
UDP-GlcNAc:mannoside alpha1–6 acetylglucosaminyltransferase
More (cf. EC 2.4.1.101, EC 2.4.1.144, EC 2.4.1.145, EC 2.4.1.155, EC 2.4.1.201)

CAS Reg. No.
91755-74-7; 91847-11-9

---

2 REACTION AND SPECIFICITY

Catalysed reaction
UDP-N-acetyl-D-glucosamine + alpha-D-mannosyl-1,6-(N-ace-tyl-beta-D-glucosaminyl-1,2-alpha-D-mannosyl-1,3)-beta-D-mannosyl-R →
→ UDP + N-acetyl-beta-D-glucosaminyl 1,2-alpha-D-mannosyl-1,6-(N-ace-tyl-beta-D-glucosaminyl-1,2-alpha-D-mannosyl-1,3)-beta-D-mannosyl-R

Reaction type
Hexosyl group transfer

Natural substrates
UDP-N-acetyl-D-glucosamine + alpha-D-mannosyl-1,6-(N-acetyl-D-gluco-saminyl-1,2-alpha-D-mannosyl-1,3)-beta-D-mannosyl-1,4-N-acetylglucos-aminyl-R (essential for biosynthesis of complex N-linked oligosaccharides) [1]

Substrate spectrum

1 UDP-N-acetyl-D-glucosamine + alpha-D-mannosyl-1,6-(N-acetyl-beta-
   D-glucosaminyl-1,2-alpha-D-mannosyl-1,3)-beta-D-mannosyl-1,4-N-ace-
   tylglucosaminyl-R (R represents the remainder of the N-oligosaccharide
   core (± fucose), R: H or 1,4-(+/-)-fucose-1,6-N-acetylglucosaminyl-1-N-as-
   paragine [1], or 1,4-(+/-)-fucose-1,6-N-acetylglucosamine-pyridinylamine
   (with or without xylose in 1,2-beta-linkage to the beta-mannosyl residue)
   [8], best acceptor, high specificity [3]. Acceptors are free and protein-matrix-bound glycans [7], UDP-N-acetyl-D-glucosamine cannot be replaced
   by UDP-N-acetylgalactosamine, UDPgalactose, UDPxylose, UDPglucose
   or GDPfucose [3], no acceptor substrates: overview [3]) [1–8]

Product spectrum

1 UDP + N-acetyl-beta-D-glucosaminyl 1,2-alpha-D-mannosyl-1,6-(N-ace-
   tyl-beta-D-glucosaminyl-1,2-alpha-D-mannosyl-1,3)-beta-D-mannosyl-
   1,4-N-acetylglucosaminyl-R [1, 3–6]

Inhibitor(s)

5-Hg-UDP (reversible) [1]; 5-Hg-UDP-N-acetylglucosamine (reversible) [1];
EDTA [5]

Cofactor(s)/prosthetic group(s)/activating agents

Triton X-100 (activation, 0.1%) [3]

Metal compounds/salts

Mn^{2+} (requirement [3, 4], 2–3 mM [3], 10–15 mM [4]) [3, 4]; Cd^{2+} (activation,
can replace Mn^{2+} to some extent, 0.5–1 mM) [3]; More (Ca^{2+}, Co^{2+}, Cu^{2+},
Hg^{2+}, Mg^{2+} or Zn^{2+} cannot replace Mn^{2+}, at 1 or 2.5 mM each) [3]

Turnover number (min^{-1})

Specific activity (U/mg)

0.0093 [3]; 27.5 (rat) [1, 4]

K_m-value (mM)

More (kinetic data of free and protein-matrix-bound acceptor substrates) [7];
0.0166 (alpha-D-mannosyl-1,6-(N-acetyl-D-glucosaminyl-1,2-alpha-D-man-
nosyl-1,3)-beta-D-mannosyl-1,4-N-acetylglucosaminyl-R) [3]; 0.018 (UDP-
N-acetylglucosamine) [3]; 0.1 (alpha-D-mannosyl-1,6-(N-acetyl-D-glucos-
aminyl-1,2-alpha-D-mannosyl-1,3)-beta-D-mannosyl-1,4-N-acetylglucos-
aminyl-R) [6]; 0.19 (alpha-D-mannosyl-1,6-(N-acetyl-D-glucosaminyl-1,2-
alpha-D-mannosyl-1,3)-beta-D-mannosyl-1,4-N-acetylglucosaminyl-R) [4];
0.96 (UDP-N-acetylglucosamine) [4]

pH-optimum

6–6.5 [4, 5]; 6.5–7 [3]

pH-range

alpha-1,6-Mannosyl-glycoprotein
beta-1,2-N-acetylglucosaminyltransferase
2.4.1.143

Temperature optimum (°C)
   37 (assay at) [1, 4–7]

Temperature range (°C)

---

3 ENZYME STRUCTURE

Molecular weight
   More (rat enzyme in crude extract exists in 2 forms, high-molecular weight
   form: above MW 200000, low-molecular weight form: MW 43000–48000, gel
   filtration) [1]

Subunits
   ? (x × 43000, x × 48000, rat, SDS-PAGE, one or both of these bands repre-
   sent the enzyme) [1]

Glycoprotein/Lipoprotein
   –

---

4 ISOLATION/PREPARATION

Source organism
   Bovine [2, 6]; Hamster [4]; Chicken (hen) [4]; Rat [1, 7]; Pig [5]; Acer pseudoplatanus (sycamore) [8]; Phaseolus aureus (mung bean) [3]

Source tissue
   Colostrum [2, 6]; Kidney (baby hamster) [4]; Liver [1, 5, 7]; Ovary cells
   (chinese hamster) [4]; Oviduct (hen) [4]; Trachea mucosa [5]; Cell suspension culture [8]; Seedling [3]

Localization in source
   Membrane-bound [1, 3, 4, 7]; Microsomes [3, 5]; Golgi apparatus [7, 8]

Purification
   Rat (solubilized with Triton X-100/NaCl, enzyme in crude extract exists in
   2 molecular weight forms: purification of low-molecular weight form, affinity
   chromatography on Hg-UDP-N-acetylglucosamine-thiopropyl-Sepharose)
   [1]; Phaseolus aureus (solubilized with Triton X-100) [3]

Crystallization
   –

Cloned
   –

Renatured
   –

---

5 STABILITY

pH

Temperature (°C)

Oxidation

Organic solvent

General stability information
    Triton X-100 stabilizes [1]; Bovine serum albumin stabilizes during purification [1, 4]; Acetone precipitation, unstable to [5]

Storage
    –20°C, several weeks in 20% glycerol, 0.0005 mM DTT and 0.1% Triton X-100 [3]; 0°C, at least 1 week [3]; 4°C, at least 6 months in 20% glycerol, 10 mM EDTA and 0.1% Triton X-100 [1]

---

6 CROSSREFERENCES TO STRUCTURE DATABANKS

PIR/MIPS code

Brookhaven code

---

7 LITERATURE REFERENCES

[1] Bendiak, B., Schachter, H.: J. Biol. Chem.,262,5775–5783 (1987)
[2] Rogers, G.N., Paulsen, J.C., Daniels, R.S.. Skehel, J.J., Wilson, I.A., Wiley, D.C.: Nature,304,76–78 (1983)
[3] Szumilo, T., Kaushal, G.P., Elbein, A.D.: Biochemistry,26,5498–5505 (1987)
[4] Schachter, H., Brockhausen, I., Hull, E.: Methods Enzymol.,179,351–397 (1989) (Review)
[5] Oppenheimer, C.L., Eckhardt, A.E., Hill, R.L.: J. Biol. Chem.,256,11477–11482 (1981)
[6] Harpaz, N., Schachter, H.: J. Biol. Chem.,255,4885–4893 (1980)
[7] Shao, M.-C., Wold, F.: J. Biol. Chem.,263,5771–5774 (1988)
[8] Tezuka, K., Hayashi, M., Ishihara, H., Akazawa, T., Takahashi, N.: Eur. J. Biochem., 203,401–413 (1992)

1 NOMENCLATURE

EC number
2.4.1.144

Systematic name
UDP-N-acetyl-D-glucosamine:glycoprotein (N-acetyl-D-glucosamine to beta-D-mannosyl of $R_1(R_2)$-beta-D-mannosyl-1,4-N-acetyl-beta-D-glucosaminyl-R_3) beta-1,4-N-acetyl-D-glucosaminyltransferase

Recommended name
beta-1,4-Mannosyl-glycoprotein beta-1,4-N-acetylglucosaminyltransferase

Synonyms
N-Glycosyl-oligosaccharide-glycoprotein N-acetylglucosaminyltransferase III
N-Acetylglucosaminyltransferase III
Acetylglucosaminyltransferase, uridine diphosphoacetylglucosamine-glyco-peptide beta4-, III
More (cf. EC 2.4.1.101, EC 2.4.1.143, EC 2.4.1.145, EC 2.4.1.155, EC 2.4.1.201)

CAS Reg. No.
83744-93-8

---

2 REACTION AND SPECIFICITY

Catalysed reaction
UDP-N-acetyl-D-glucosamine + (N-acetyl-beta-D-glucosaminyl-1,2-alpha-D-mannosyl-1,3)-(N-acetyl-beta-D-glucosaminyl-1,2-alpha-D-mannosyl-1,6)-beta-D-mannosyl-1,4-N-acetyl-beta-D-glucosaminyl-R →
→ UDP + N-acetyl-beta-D-glucosaminyl-1,2-alpha-D-mannosyl-1,3-(N-acetyl-beta-D-glucosaminyl-1,2-alpha-D-mannosyl-1,6)-(N-acetyl-beta-D-glucosaminyl-1,4)-beta-D-mannosyl-1,4-N-acetyl-beta-D-glucosaminyl-R

Reaction type
Hexosyl group transfer

Natural substrates
UDP-N-acetyl-D-glucosamine + (N-acetyl-beta-D-glucosaminyl-1,2-alpha-D-mannosyl-1,3)-(N-acetyl-beta-D-glucosaminyl-1,2-alpha-D-mannosyl-1,6)-beta-D-mannosyl-1,4-N-acetyl-beta-D-glucosaminyl-R (R represents the remainder of the N-oligosaccharide core (+/- fucose), essential for bio-synthesis of complex N-linked oligosaccharides) [3]

beta-1,4-Mannosyl-glycoprotein
beta-1,4-N-acetylglucosaminyltransferase
2.4.1.144

Substrate spectrum

1 UDP-N-acetyl-D-glucosamine + (N-acetyl-beta-D-glucosaminyl-1,2-
    alpha-D-mannosyl-1,3)-(N-acetyl-beta-D-glucosaminyl-1,2-alpha-D-man-
    nosyl-1,6)-beta-D-mannosyl-1,4-N-acetyl-beta-D-glucosaminyl-R (R repre-
    sents the remainder of the N-oligosaccharide core (+/- fucose), favourite
    substrate [3], incorporates a N-acetylglucosamine residue in beta-1,4-link-
    age to beta-linked mannosyl (i.e. bisecting N-acetylglucosamine), both
    terminal beta-1,2-linked N-acetylglucosamine residues required for maxi-
    mum activity, removal of one or both of them reduces activity by 85 or
    93%, respectively [1], galactosylation of N-acetylglucosaminyl-1,2-alpha-
    mannosyl-1,3-beta-mannosyl moiety prevents transferase III action [3],
    substrate specificity study [1, 3], pyridylaminated biantennary sugar
    chain [4]) [1–4]

Product spectrum

1 UDP + N-acetyl-beta-D-glucosaminyl-1,2-alpha-D-mannosyl-1,3-(N-ace-
    tyl-beta-D-glucosaminyl-1,2-alpha-D-mannosyl-1,6)-(N-acetyl-beta-D-
    glucosaminyl-1,4)-beta-D-mannosyl-1,4-N-acetyl-beta-D-glucosaminyl-R
    [1, 4]

Inhibitor(s)

Cofactor(s)/prosthetic group(s)/activating agents

Triton X-100 (activation, 0.13–1.3% v/v) [1–3]

Metal compounds/salts

Mn^{2+} (requirement, 12 mM) [1, 3, 4]

Turnover number (min^{-1})

Specific activity (U/mg)

0.000083 [1]; 5.52 [2]

K_m-value (mM)

0.19 (pyridylaminated acceptor substrate) [4]; 0.23 ((N-acetyl-beta-D-glucos-
aminyl-1,2-alpha-D-mannosyl-1,3)-(N-acetyl-beta-D-glucosaminyl-1,2-
alpha-D-mannosyl-1,6)-beta-D-mannosyl-1,4-N-acetyl-beta-D-glucosaminyl-R)
[1, 3]; 1.1 (UDP-N-acetylglucosamine) [1, 3]; 3.1 (UDP-N-acetylglucosamine
(+ pyridylaminated acceptor substrate)) [4]

pH-optimum

6.3 [4]; 6–7 [3]; 7 [1]

pH-range

Temperature optimum (°C)

37 (assay at) [1, 3, 4]

Temperature range (°C)

---

3 ENZYME STRUCTURE

Molecular weight
More (rat, nucleotide sequence) [2]

Subunits

Glycoprotein/Lipoprotein   .

–

---

4 ISOLATION/PREPARATION

Source organism
Chicken (hen, White Leghorn) [1]; Rat (male Donryu [2]) [2–4]; Chinese Hamster [3]; Human [3]

Source tissue
Brain [4]; Oviduct [1]; Kidney [2, 4]; Liver [4]; Ovary cell line LEC10 (ricin resistant mutant, hamster) [3]; Serum [4]; Hepatoma (rat) [3, 4]; Lymphoid (human) [3]; More (tissue distribution) [4]

Localization in source
Membrane-bound [1–3]; Microsomes [1, 3]

Purification
Rat (primary structure) [2]

Crystallization

–

Cloned
(rat, expression in COS-1 or HeLa-cells) [2]

Renatured

–

---

5 STABILITY

pH

Temperature (°C)

Oxidation

Organic solvent

General stability information

Storage

−20°C, chicken oviduct microsomal preparation, several months in detergent-free buffer [3]

6 CROSSREFERENCES TO STRUCTURE DATABANKS

PIR/MIPS code

Brookhaven code

7 LITERATURE REFERENCES

[1] Narasimhan, S.: J. Biol. Chem.,257,10235–10242 (1982)
[2] Nashikawa, A., Ihara, Y., Hatakeyama, M., Kangawa, K., Taniguchi, N.: J. Biol. Chem.,
267,18199–18204 (1992)
[3] Schachter, H., Brockhausen, I., Hull, E.: Methods Enzymol.,179,351–397 (1989)
(Review)
[4] Taniguchi, N., Nishikawa, A., Fujii, S., Gu, J.: Methods Enzymol.,179,397–408 (1989)
(Review)

1 NOMENCLATURE

EC number
2.4.1.145

Systematic name
UDP-N-acetyl-D-glucosamine:glycoprotein (N-acetyl-D-glucosamine to alpha-D-mannosyl-1,3- of R_1-alpha-D-mannosyl-1,3-(R_2)-beta-D-mannosyl-R_3) beta-1,4-N-acetyl-D-glucosaminyltransferase

Recommended name
alpha-1,3-Mannosylglycoprotein beta-1,4-N-acetylglucosaminyltransferase

Synonyms
N-Glycosyl-oligosaccharide-glycoprotein N-acetylglucosaminyltransferase IV
N-Acetylglucosaminyltransferase IV
beta-Acetylglucosaminyltransferase IV
Acetylglucosaminyltransferase, uridine diphosphoacetylglucosamine-glyco-peptide beta4-, IV
More (cf. EC 2.4.1.101, EC 2.4.1.143, EC 2.4.1.144, EC 2.4.1.155)

CAS Reg. No.
86498-16-0

---

2 REACTION AND SPECIFICITY

Catalysed reaction
UDP-N-acetyl-D-glucosamine + (N-acetyl-D-glucosaminyl-1,2)-alpha-D-man-nosyl-1,3-(beta-N-acetyl-D-glucosaminyl-1,2-alpha-D-mannosyl-1,6)-beta-D-mannosyl-R →
→ UDP + N-acetyl-D-glucosaminyl-1,4-(N-acetyl-beta-D-glucosaminyl-1,2)-alpha-D-mannosyl-1,3-(beta-N-acetyl-D-glucosaminyl-1,2-alpha-D-mannosyl-1,6)-beta-D-mannosyl-R

Reaction type
Hexosyl group transfer

Natural substrates
UDP-N-acetyl-D-glucosamine + (N-acetyl-D-glucosaminyl-1,2)-alpha-D-man-nosyl-1,3-(beta-N-acetyl-D-glucosaminyl-1,2-alpha-D-mannosyl-1,6)-beta-D-mannosyl-R (involved in N-glycan biosynthesis) [1]

Substrate spectrum

1 UDP-N-acetyl-D-glucosamine + (N-acetyl-D-glucosaminyl-1,2)-alpha-D-mannosyl-1,3-(beta-N-acetyl-D-glucosaminyl-1,2-alpha-D-mannosyl-1,6)-beta-D-mannosyl-R (R represents the remainder of the N-oligosaccharide core (+/- fucose), adds N-acetylglucosamine in beta-1,4-linkage to the alpha-1,3-linked mannosyl residues of the trimannosyl core of N-glyco-syloligosaccharides, maximal activity requires the presence of both termi-nal beta-1,2-linked N-acetylglucosamine residues in the substrate [1], acceptor specificity [1, 2], pyridylaminated sugar chain [3]) [1–3]

Product spectrum

1 UDP + N-acetyl-D-glucosaminyl-1,4-(N-acetyl-D-glucosaminyl-1,2)-alpha-D-mannosyl-1,3-(beta-N-acetyl-D-glucosaminyl-1,2-alpha-D-man-nosyl-1,6)-beta-D-mannosyl-R (triantennary structure) [1–3]

Inhibitor(s)

Cofactor(s)/prosthetic group(s)/activating agents
Triton X-100 (activation, 0.125% v/v) [1]

Metal compounds/salts
Mn^{2+} (activation, 7.5 mM [3], 12 mM [2]) [2, 3]

Turnover number (min^{-1})

Specific activity (U/mg)

K_m-value (mM)
3.4 (pyridylaminated acceptor substrate) [3]; 8.3 (UDP-N-acetylglucosamine (+ pyridylaminated acceptor substrate)) [3]

pH-optimum
6.5–7.5 (broad, at optimum Triton X-100 concentration) [1]; 7 (sharp, with 0.5–0.625% Triton X-100) [1, 2]; 7.3 [3]

pH-range

Temperature optimum (°C)
37 (assay at) [2, 3]

Temperature range (°C)

3 ENZYME STRUCTURE

Molecular weight

Subunits

Glycoprotein/Lipoprotein

–

4 ISOLATION/PREPARATION

Source organism
Chicken (hen, White Leghorn) [1, 2]; Rat [3]

Source tissue
Oviduct [1, 2]; Spleen [3]; Small intestine [3]

Localization in source
Membrane-bound [1, 2]

Purification
Chicken [1, 2]

Crystallization
−

Cloned
−

Renatured
−

5 STABILITY

pH

Temperature (°C)

Oxidation

Organic solvent

General stability information

Storage
−20°C, crude microsomal preparation, several months in detergent free
buffer [2]

6 CROSSREFERENCES TO STRUCTURE DATABANKS

PIR/MIPS code

Brookhaven code

7 LITERATURE REFERENCES

[1] Gleeson, P.A., Schachter, H.: J. Biol. Chem.,258,6162–6173 (1983)
[2] Schachter, H., Brockhausen, I., Hull, E.: Methods Enzymol.,179,351–397 (1989)
    (Review)
[3] Taniguchi, N., Nishikawa, A., Fujii, S., Gu, J.: Methods Enzymol.,179,397–408 (1989)
    (Review)

1 NOMENCLATURE

EC number
    2.4.1.146

Systematic name
    UDP-N-acetyl-D-glucosamine:O-glycosyl-glycoprotein (N-acetyl-D-glucos-
    amine to D-galactose of beta-D-galactosyl-1,3-(N-acetyl-D-glucosaminyl-
    1,6)-N-acetyl-D-galactosaminyl-R) beta-1,3-N-acetyl-D-glucosaminyltrans-
    ferase

Recommended name
    beta-1,3-Galactosyl-O-glycosyl-glycoprotein beta-1,3-N-acetylglucosaminyl-
    transferase

Synonyms
    O-Glycosyl-oligosaccharide-glycoprotein N-acetylglucosaminyltransferase II
    Acetylglucosaminyltransferase, uridine diphosphoacetylglucosamine-mucin
    beta(1→3)- (elongating)
    Elongation 3beta-GalNAc-transferase [2]
    More (cf. EC 2.4.1.102, EC 2.4.1.147 and EC 2.4.1.148)

CAS Reg. No.
    87927-99-9

---

2 REACTION AND SPECIFICITY

Catalysed reaction
    UDP-N-acetyl-D-glucosamine + beta-D-galactosyl-1,3-(N-ace-
    tyl-D-glucosaminyl-1,6)-N-acetyl-D-galactosaminyl-R →
    → UDP + N-acetyl-beta-D-glucosaminyl-1,3-beta-D-galactosyl-1,3-(N-acetyl-
    beta-D-glucosaminyl-1,6)-N-acetyl-D-galactosaminyl-R

Reaction type
    Hexosyl group transfer

Natural substrates
    UDP-N-acetyl-D-glucosamine + beta-D-galactosyl-1,3-(N-acetyl-D-glucos-
    aminyl-1,6)-N-acetyl-D-galactosaminyl-R (involved in elongation of O-glycan
    cores during mucin biosynthesis) [1]

Substrate spectrum
    1 UDP-N-acetyl-D-glucosamine + beta-D-galactosyl-1,3-(N-acetyl-D-
    glucosaminyl-1,6)-N-acetyl-D-galactosaminyl-R (i.e. core class 2, R: ben-
    zyl, o-nitrophenyl [1, 2], polypeptide from mucin or antifreeze glycoprotein
    [1]) [1, 2]

Product spectrum

1 UDP + N-acetyl-beta-D-glucosaminyl-1,3-beta-D-galactosyl-1,3-(N-ace-
tyl-D-glucosaminyl-1,6)-N-acetyl-D-galactosaminyl-R [1, 2]

Inhibitor(s)

EDTA [1]

Cofactor(s)/prosthetic group(s)/activating agents

Triton X-100 (stimulation, 0.1% v/v) [1]

Metal compounds/salts

Mn^{2+} (activation, 10–40 mM) [1]; Co^{2+} (activation) [1]; More (no activation by
Ca^{2+}, Mg^{2+}, Zn^{2+}) [1]

Turnover number (min⁻¹)

Specific activity (U/mg)

K_m-value (mM)

0.9 (beta-D-galactosyl-1,3-(N-acetyl-D-glucosaminyl-1,6)-N-acetyl-D-galac-
tosaminyl-o-nitrophenyl) [1]; 1.2 (beta-D-galactosyl-1,3-(N-acetyl-D-
glucosaminyl-1,6)-N-acetyl-D-galactosaminyl-benzyl) [1]; 1.6 (UDP-N-acetyl-D-glucosamine) [1]

pH-optimum

7 [1]

pH-range

Temperature optimum (°C)

Temperature range (°C)

3 ENZYME STRUCTURE

Molecular weight

Subunits

Glycoprotein/Lipoprotein

–

4 ISOLATION/PREPARATION

Source organism

Pig [1, 2]; Rat [1, 2]

Source tissue
  Stomach (pig) [1, 2]; Colon [1, 2]; More (low activity or absent in rat liver,
  stomach and submaxillary glands, dog and pig submaxillary glands) [1]

Localization in source
  Membrane-bound (pig stomach) [1]

Purification

Crystallization
  –

Cloned
  –

Renatured
  –

---

5 STABILITY

pH

Temperature (°C)

Oxidation

Organic solvent

General stability information

Storage
  –70°C, microsomal pig stomach enzyme preparation, detergent-free 0.25 M
  sucrose suspension, several years [1]

---

6 CROSSREFERENCES TO STRUCTURE DATABANKS

PIR/MIPS code

Brookhaven code

---

7 LITERATURE REFERENCES

[1] Schachter, H., Brockhausen, I., Hull, E.: Methods Enzymol.,179,351–397 (1989)
    (Review)
[2] Brockhausen, I., Rachaman, E.S., Matta, K.L., Schachter, H.: Carbohydr. Res.,
    120,3–16 (1983)

1 NOMENCLATURE

EC number
  2.4.1.147

Systematic name
  UDP-N-acetyl-D-glucosamine:O-glycosyl-glycoprotein (N-acetyl-D-glucosamine to N-acetyl-D-galactosaminyl-R) beta-1,3-N-acetyl-D-glucosaminyl-
  transferase

Recommended name
  Acetylgalactosaminyl-O-glycosyl-glycoprotein beta-1,3-N-acetylglucosaminyl-
  transferase

Synonyms
  O-Glycosyl-oligosaccharide-glycoprotein N-acetylglucosaminyltransferase III
  Acetylglucosaminyltransferase, uridine diphosphoacetylglucosamine-mucin
  beta(1→3)-
  Mucin core 3 beta3-GlcNAc-transferase [1]
  Core 3beta-GlcNAc-transferase [2]
  More (cf. EC 2.4.1.102, EC 2.4.1.146 and EC 2.4.1.148)

CAS Reg. No.
  87927-96-6

---

2 REACTION AND SPECIFICITY

Catalysed reaction
  UDP-N-acetyl-D-glucosamine + N-acetyl-beta-D-galactosaminyl-R →
  → UDP + N-acetyl-beta-D-glucosaminyl-1,3-N-acetyl-beta-D-galac-
  tosaminyl-R

Reaction type
  Hexosyl group transfer

Natural substrates
  UDP-N-acetyl-D-glucosamine + N-acetyl-beta-D-galactosaminyl-alpha-R
  (R: polypeptide derived from mucin, involved in mucin oligosaccharide
  biosynthesis) [1, 2]

Substrate spectrum
  1 UDP-N-acetyl-D-glucosamine + N-acetyl-beta-D-galactosaminyl-alpha-R
    (R: polypeptide derived from mucin [1], phenyl or benzyl [1–3], no accep-
    tor substrate: native ovine submaxillary mucin (containing sialyl-2,6-
    N-acetyl-D-galactosaminyl-groups) [1, 3]) [1–3]

Product spectrum
1 UDP + N-acetyl-beta-D-glucosaminyl-1,3-N-acetyl-beta-D-galactosaminyl-R
[1–3]

Inhibitor(s)
EDTA [1]; Triton X-100 (above 0.5% v/v) [3]; Co^{2+} (in the presence of Mn^{2+})
[3]; More (no inhibition by 2-mercaptoethanol or N-acetyl-D-glucosamine) [3]

Cofactor(s)/prosthetic group(s)/activating agents
Triton X-100 (stimulation, 0.1% v/v, inhibits at concentrations above 0.5% v/v,
pig colon) [1, 3]

Metal compounds/salts
Mn^{2+} (requirement, 10–20 mM, pig colon, cannot be replaced by Co^{2+}, Ca^{2+},
Mg^{2+} or Zn^{2+} [3]) [1, 3]

Turnover number (min^{-1})

Specific activity (U/mg)

K_m-value (mM)
2 (N-acetyl-D-galactosaminyl-alpha-benzyl, pig) [1, 3]; 3.2 (N-acetyl-D-ga-
lactosaminyl-alpha-ovine submaxillary mucin, pig) [1, 3]; 5 (N-acetyl-D-ga-
lactosaminyl-alpha-phenyl, pig) [1, 3]

pH-optimum
6.5 [1, 3]

pH-range

Temperature optimum (°C)
37 (assay at) [1, 3]

Temperature range (°C)

3 ENZYME STRUCTURE

Molecular weight

Subunits

Glycoprotein/Lipoprotein
–

4 ISOLATION/PREPARATION

Source organism
Pig [1, 3]; Rat [1–3]; Human [1, 3]; Monkey [1, 3]; Sheep [1, 3]

Source tissue
   Colon mucosa [1–3]; More (low activity in rat small intestine [1], rat, pig,
   monkey or sheep stomach [1, 3], absent from rat, pig or dog submaxillary
   glands [1, 3]) [1, 3]

Localization in source
   Microsomes [3]

Purification

Crystallization
   –

Cloned
   –

Renatured
   –

---

5 STABILITY

pH

Temperature (°C)

Oxidation

Organic solvent

General stability information

Storage

---

6 CROSSREFERENCES TO STRUCTURE DATABANKS

PIR/MIPS code

Brookhaven code

---

7 LITERATURE REFERENCES

[1] Schachter, H., Brockhausen, I., Hull, E.: Methods Enzymol.,179,351–397 (1989)
    (Review)
[2] Brockhausen, I., Rachaman, E.S., Matta, K.L., Schachter, H.: Carbohydr. Res.,
    120,3–16 (1983)
[3] Brockhausen, I., Matta, K.L., Orr, J., Schachter, H.: Biochemistry,24,1866–1874
    (1985)

1 NOMENCLATURE

EC number
2.4.1.148

Systematic name
UDP-N-acetyl-D-glucosamine:O-oligosaccharide-glycoprotein (N-acetyl-D-glucosamine to N-acetyl-D-galactosamine of N-acetyl-beta-D-glucosaminyl-1,3-N-acetyl-D-galactosaminyl-R) beta-1,6-N-acetyl-D-glucosaminyltransferase

Recommended name
Acetylgalactosaminyl-O-glycosyl-glycoprotein beta-1,6-N-acetylglucosaminyltransferase

Synonyms
O-Glycosyl-oligosaccharide-glycoprotein N-acetylglucosaminyltransferase IV
Acetylglucosaminyltransferase, uridine diphosphoacetylglucosamine-mucin beta(1→6)-, B
Core 4 beta6-GalNAc-transferase [1]
Core 6beta-GalNAc-transferase B [2]
More (may be identical with EC 2.4.1.102 [3], cf. EC 2.4.1.146 and EC 2.4.1.147)

CAS Reg. No.
87927-98-8

2 REACTION AND SPECIFICITY

Catalysed reaction
UDP-N-acetyl-D-glucosamine + N-acetyl-beta-D-glucosaminyl-1,3-N-acetyl-D-galactosaminyl-R →
→ UDP + N-acetyl-beta-D-glucosaminyl-1,6-(N-acetyl-beta-D-glucosaminyl-1,3)-N-acetyl-D-galactosaminyl-R

Reaction type
Hexosyl group transfer

Natural substrates
UDP-N-acetyl-D-glucosamine + N-acetyl-beta-D-glucosaminyl-1,3-N-acetyl-D-galactosaminyl-alpha-R (involved in mucin oligosaccharide biosynthesis) [1–3]

Substrate spectrum

1 UDP-N-acetyl-D-glucosamine + N-acetyl-beta-D-glucosaminyl-1,3-N-ace-
   tyl-D-galactosaminyl-alpha-R (i.e. core 3 [1], R: polypeptide derived from
   mucin [1], phenyl [1, 2] or benzyl [1–3], not H [1]) [1–3]

Product spectrum

1 UDP + N-acetyl-beta-D-glucosaminyl-1,6-(N-acetyl-beta-D-glucosaminyl-
   1,3)-N-acetyl-D-galactosaminyl-alpha-R (i.e. core 4 [1]) [1–3]

Inhibitor(s)

Zn^{2+} (inactivation) [3]; Co^{2+} [3]; Ca^{2+} (weak) [3]; Mg^{2+} (weak) [3]; EDTA
(weak) [1, 3]; 2-Mercaptoethanol (weak) [3]

Cofactor(s)/prosthetic group(s)/activating agents

Triton X-100 (slight stimulation, 0.1% v/v, rat colon) [1, 3]; AMP (slight stimulation) [3]; ATP (slight stimulation) [3]; N-Acetyl-D-glucosamine (slight stimulation) [3]

Metal compounds/salts

More (no Mn^{2+}-requirement, rat colon) [1, 3]

Turnover number (min^{-1})

Specific activity (U/mg)

K_m-value (mM)

0.6 (N-acetyl-beta-D-glucosaminyl-1,3-N-acetyl-D-galactosaminyl-alpha-ben-
zyl, rat colon) [1, 3] 1.8 (N-acetyl-beta-D-glucosaminyl-1,3-N-acetyl-D-galac-
tosaminyl-alpha-benzyl, pig stomach) [1]

pH-optimum

6.5 (rat colon) [1, 3]

pH-range

Temperature optimum (°C)

37 (assay at) [3]

Temperature range (°C)

3 ENZYME STRUCTURE

Molecular weight

Subunits

Glycoprotein/Lipoprotein

–

4 ISOLATION/PREPARATION

Source organism
Rat [1–3]; Pig [1, 3]; Dog [1]; Monkey [1, 3]; Human [1, 3]; Sheep [1, 3]

Source tissue
Colon mucosa (rat, pig, monkey, human [1]) [1–3]; Stomach (sheep, pig, monkey, rat [1]) [1, 3]; Small intestine (rat) [1]; Submaxillary glands (dog) [1, 3]; More (not in rat liver or pig submaxillary glands) [1]

Localization in source
Microsomes [3]

Purification

Crystallization
–

Cloned
–

Renatured
–

5 STABILITY

pH

Temperature (°C)

Oxidation

Organic solvent

General stability information

Storage

6 CROSSREFERENCES TO STRUCTURE DATABANKS

PIR/MIPS code

Brookhaven code

7 LITERATURE REFERENCES

[1] Schachter, H., Brockhausen, I., Hull, E.: Methods Enzymol.,179,351–397 (1989) (Review)
[2] Brockhausen, I., Rachaman, E.S., Matta, K.L., Schachter, H.: Carbohydr. Res., 120,3–16 (1983)
[3] Brockhausen, I., Matta, K.L., Orr, J., Schachter, H.: Biochemistry,24,1866–1874 (1985)

1 NOMENCLATURE

EC number
2.4.1.149

Systematic name
UDP-N-acetyl-D-glucosamine:beta-D-galactosyl-1,4-N-acetyl-D-glucosamine beta-1,3-acetyl-D-glucosaminyltransferase

Recommended name
N-Acetyllactosaminide beta-1,3-N-acetylglucosaminyltransferase

Synonyms
Acetylglucosaminyltransferase, uridine diphosphoacetylglucosamine-acetyllactosaminide beta1→3-
Poly-N-acetyllactosamine extension enzyme
Galbeta1→4GlcNAc-R beta1→3 N-acetylglucosaminyltransferase [1]
UDP-GlcNAc:GalR, beta-D-3-N-acetylglucosaminyltransferase [2]
N-Acetyllactosamine beta(1–3)N-acetylglucosaminyltransferase [3]
UDP-GlcNAc:Galbeta1→4GlcNAcbeta-Rbeta1→3-N-acetylglucosaminyltransferase [5]

CAS Reg. No.
85638-39-7

---

2 REACTION AND SPECIFICITY

Catalysed reaction
UDP-N-acetyl-D-glucosamine + beta-D-galactosyl-1,4-N-acetyl-D-glucosaminyl-R →
→ UDP + N-acetyl-beta-D-glucosaminyl-1,3-beta-D-galactosyl-1,4-N-acetyl-D-glucosaminyl-R

Reaction type
Hexosyl group transfer

Natural substrates
More (might function in the biosynthesis of cell surface polylactosaminoglycans on Novikoff cells, enzyme controls the synthesis of linear chain types by adding a N-acetylglucosaminyl residue to a Galbeta(1–4)GlcNAc-R "primer" structure present on glycoprotein or glycolipid yielding a GlcNAcbeta(1–3)Galbeta(1–4)GlcNAc-R sequence [1], the enzyme together with N-acetyllactosamine synthase (EC 2.4.1.90) catalyzes formation of linear

glycans containing alternating beta-3-O-substituted residues of galactose and beta-4-O-substituted residues of N-acetyl-D-glucosamine, structures are present among others in glycosphingolipids or H-II type, polyglycosylceramides and polyglycosylpeptides [2]) [1, 2]

Substrate spectrum

1 UDP-N-acetyl-D-glucosamine + beta-D-galactosyl-1,4-N-acetyl-D-glucosaminyl-R (acts on beta-galactosyl-1,4-N-acetylglucosaminyl termini on asialo-alpha$_1$-acid glycoproteins and other glycoproteins and oligosaccharides [1, 2], highly specific for acceptor oligosaccharides and glycoproteins carrying a terminal Galbeta(1–4)GlcNAcbeta1-R unit, catalyzes the formation of GlcNAcbeta(1–3)Galbeta(1–4)GlcNAcbeta-R sequence, Galbeta(1–4)GlcNAcbeta(1–2)[Galbeta(1–4)GlcNAcbeta(1–6)]Man pentasaccharide in the acceptor structure is a requirement for optimal activity [5], branches of this pentasaccharide structure, when contained in tri- and tetraantennary oligosaccharides are highly preferred over other branches for attachment of the 1st and 2nd mol of GlcNAc into the acceptor molecule, enzyme also shows activity towards oligosaccharides related to blood group I- and i-active polylactosaminoglycans [5]) [1–5]

2 UDP-N-acetylglucosamine + N-acetyllactosamine [2, 3]

3 UDP-N-acetylglucosamine + lactose [2, 3]

4 More (a similar enzyme from human colonic adenocarcinoma cell line SW403 catalyzes transfer of GlcNAc to lactosylceramides) [4]

Product spectrum

1 UDP + N-acetyl-beta-D-glucosaminyl-1,3-beta-D-galactosyl-1,4-N-acetyl-D-glucosaminyl-R [1–5]

2 UDP + GlcNAcbeta(1–3)Galbeta(1–4)Glc (GlcNAcbeta(1–3)Galbeta(1–4)GlcNAc [3]) [2]

3 ?

4 ?

Inhibitor(s)

NiCl$_2$ [5]; CaCl$_2$ [1, 5]; ZnCl$_2$ [1, 5]; EDTA [1, 5]; FeCl$_2$ [5]

Cofactor(s)/prosthetic group(s)/activating agents

Metal compounds/salts

More (absolute requirement for divalent cations) [5]; Mn^{2+} (essential for activity, neither Mg^{2+} nor Ca^{2+} can substitute for Mn^{2+} [3], MnCl$_2$: 10–25 mM required [1], stimulates [5]) [1, 3, 5]; CdCl$_2$ (stimulates) [1]; MgCl$_2$ (stimulates [1], no effect [5]) [1]

Turnover number (min^{-1})

Specific activity (U/mg)
  More [5]

K_m-value (mM)
  More [5]; 0.13 (UDP-GlcNAc (+ lactose)) [3]; 0.6 (asialo-alpha$_1$-acid glycoprotein) [5]; 1.0 (Galbeta(1–4)GlcNAcbeta(1–3)Galbeta(1–4)GlcNAc) [5];
  2.2 (Galbeta(1–4)GlcNAc) [5]; 5.2 (Galbeta(1–4)Glc) [5]; 7.0 (N-acetyllactosamine) [3]; 29.8 (lactose) [3]

pH-optimum
  6.5–8.0 [1]; 6.8–7.2 [5]

pH-range

Temperature optimum (°C)
  37 (assay at) [1]

Temperature range (°C)

---

3 ENZYME STRUCTURE

Molecular weight

Subunits

Glycoprotein/Lipoprotein
  –

---

4 ISOLATION/PREPARATION

Source organism
  Rat (ascites fluid of Novikoff tumor cells maintained in rat) [1, 5]; Human
  [2–4]

Source tissue
  Ascites fluid of Novikoff tumor cells [1, 5]; Serum [2, 3]; Colonic adenocarcinoma cell line SW403 [4]

Localization in source

Purification
  Rat (ascites fluid of Novikoff tumor cells maintained in rat, partial) [5]

Crystallization
  –

Cloned

–

Renatured

–

---

5 STABILITY

pH

Temperature (°C)

Oxidation

Organic solvent

General stability information

Storage

---

6 CROSSREFERENCES TO STRUCTURE DATABANKS

PIR/MIPS code

Brookhaven code

---

7 LITERATURE REFERENCES

[1] van den Eijnden, D.H., Winterwerp, H., Smeeman, P., Schiphorst, W.E.C.M.: J. Biol. Chem.,258,3435–3437 (1983)
[2] Zielenski, J., Koscielak, J.: FEBS Lett.,158,164–168 (1983)
[3] Hosomi, O., Takeya, A., Kogure, T.: J. Biochem.,95,1655–1659 (1984)
[4] Holmes, E.H.: Arch. Biochem. Biophys.,260,461–468 (1988)
[5] van den Eijnden, D.H., Koenderman, A.H.L., Schiphorst, W.E.C.M.: J. Biol. Chem., 263,12461–12471 (1988)

1 NOMENCLATURE

EC number

2.4.1.150

Systematic name

UDP-N-acetyl-D-glucosamine:beta-D-galactosyl-1,4-N-acetyl-D-glucos-aminide beta-1,6-N-acetyl-D-glucosaminyltransferase

Recommended name

N-Acetyllactosaminide beta-1,6-N-acetylglucosaminyltransferase

Synonyms

N-Acetylglucosaminyltransferase

Acetylglucosaminyltransferase, uridine diphosphoacetylglucosamine-ace-tyllactosaminide beta1→6-

Acetylglucosaminyltransferase, uridine diphosphoacetylglucosamine-ace-tyltransaminide beta1→6-

Galbeta1→4GlcNAc-R beta1→6 N-acetylglucosaminyltransferase [1]

UDP-GlcNAc:Gal-R, beta-D-6-N-acetylglucosaminyltransferase [3]

CAS Reg. No.

85638-40-0

---

2 REACTION AND SPECIFICITY

Catalysed reaction

UDP-N-acetyl-D-glucosamine + beta-D-galactosyl-1,4-N-acetyl-D-glucosaminyl-R →

→ UDP + N-acetyl-beta-D-glucosaminyl-1,6-beta-D-galactosyl-1,4-N-acetyl-D-glucosaminyl-R

Reaction type

Hexosyl group transfer

Natural substrates

More (might function in biosynthesis of cell surface polylactosaminoglycans on Novikoff cells, formation of the GlcNAcbeta(1–3)[GlcNAcbeta(1–6)]Gal-R branching points in the branched type of polylactosylaminoglycans [1], involved in midchain branching of oligo-(N-acetyllactosaminoglycans) [2]) [1–3]

Substrate spectrum
1 UDP-N-acetyl-D-glucosamine + beta-D-galactosyl-1,4-N-acetyl-D-glucos-
   aminyl-R (acts on beta-galactosyl-1,4-N-acetylglucosaminyl-termini on
   asialo-alpha$_1$-acid glycoproteins [1, 3]) [1–3]
2 UDPGlcNAc + GlcNAcbeta(1–3)Galbeta(1–4)GlcNAcbeta(1–3)Gal-
   beta(1–4)GlcNAc [2]
3 UDP-GlcNAc + N-acetyllactosamine [3]
4 UDP-GlcNAc + lactose [3]

Product spectrum
1 UDP + N-acetyl-beta-D-glucosaminyl-1,6-beta-D-galactosyl-1,4-N-acetyl-
   D-glucosaminyl-R [1–3]
2 UDP + GlcNAcbeta(1–3)Galbeta(1–4)GlcNAcbeta(1–3)[GlcNAc-
   beta(1–6)]Galbeta(1–4)GlcNAc [2]
3 UDP + GlcNAcbeta(1–6)Galbeta(1–4)Glc [3]
4 ?

Inhibitor(s)
   $CaCl_2$ [1]; $ZnCl_2$ [1]; EDTA [1]

Cofactor(s)/prosthetic group(s)/activating agents

Metal compounds/salts
   $MnCl_2$ (10–25 mM required) [1]; $CdCl_2$ (stimulates) [1]; $MgCl_2$ (stimulates)
   [1]

Turnover number (min^{-1})

Specific activity (U/mg)

K_m-value (mM)

pH-optimum
   6.5–8.0 [1]

pH-range

Temperature optimum (°C)
   37 (assay at) [1]

Temperature range (°C)

---

3 ENZYME STRUCTURE

Molecular weight

Subunits

Glycoprotein/Lipoprotein
   –

4 ISOLATION/PREPARATION

Source organism
Rat (ascites fluid of Novikoff tumor cells maintained in rat) [1]; Human [2, 3]

Source tissue
Ascites fluid of Novikoff tumor cells [1]; Serum [2, 3]

Localization in source

Purification

Crystallization
−

Cloned
−

Renatured
−

5 STABILITY

pH

Temperature (°C)

Oxidation

Organic solvent

General stability information

Storage

6 CROSSREFERENCES TO STRUCTURE DATABANKS

PIR/MIPS code

Brookhaven code

7 LITERATURE REFERENCES

[1] van den Eijnden, D.H., Winterwerp, H., Smeeman, P., Schiphorst, W.E.C.M.: J. Biol. Chem.,258,3435–3437 (1983)
[2] Leppänen, A., Penttilä, L., Niemelä, R., Helin, J., Seppo, A., Lusa, S., Renkonen, O.: Biochemistry,30,9287–9296 (1991)
[3] Zielenski, J., Koscielak, J.: FEBS Lett.,158,164–168 (1983)

1 NOMENCLATURE

EC number
2.4.1.151

Systematic name
UDPgalactose:beta-D-galactosyl-1,4-acetyl-D-glucosaminide alpha-1,3-D-galactosyltransferase

Recommended name
N-Acetyllactosaminide alpha-1,3-galactosyltransferase

Synonyms
UDP-Gal:beta-D-Gal(1,4)-D-GlcNAc alpha(1,3)-galactosyltransferase
Galactosyltransferase
Galactosyltransferase, uridine diphosphogalactose-acetyllactosamine alpha1→3-
Galactosyltransferase, uridine diphosphogalactose-acetyllactosamine
UDP-galactose-acetyllactosamine alpha-D-galactosyltransferase
UDP-Gal:N-acetyllactosaminide alpha-1,3-D-galactosyltransferase [3]
UDP-Gal:Galbeta1→4GlcNAc-R alpha1→3-galactosyltransferase [5]
UDP-Gal:N-acetyllactosaminide alpha(1,3)-galactosyltransferase [10]

CAS Reg. No.
96477-57-5; 78642-28-1

---

2 REACTION AND SPECIFICITY

Catalysed reaction
UDPgalactose + beta-D-galactosyl-1,4-N-acetyl-D-glucosaminyl-R →
→ UDP + alpha-D-galactosyl-1,3-beta-D-galactosyl-1,4-N-acetyl-D-glucosaminyl-R

Reaction type
Hexosyl group transfer

Natural substrates
More (functions in the biosynthesis of cell surface polylactosaminoglycans on Novikoff cells, attachs a galactose to the N-acetylglucosaminyl residue introduced by EC 2.4.1.149 [1], functions in the biosynthesis of calf thymocyte cell-surface glycoconjugates including glycoproteins [3], involved in the biosynthesis of alpha-D-galactosyl-terminated poly-N-acetyllactosamine glycans that occur on the surface of Ehrlich ascites tumor cells [4]) [1, 3, 4]

Substrate spectrum

1 UDPgalactose + beta-D-galactosyl-1,4-N-acetyl-D-glucosaminyl-R (the
most active acceptors have the structure beta-D-Gal-(1–4)-beta-D-
GlcNAc(1-) at their nonreducing termini [2], enzyme introduces galactosyl
residues in alpha anomeric configuration to the Galbeta(1–4)GlcNAc
units on the acceptor substrate [3], highly active with glycoproteins,
oligosaccharides and glycolipids having a terminal Galbeta(1–4)GlcNAc-
beta(1–2)Manalpha(1–3)Manbeta(1–4)GlcNAc and paragloboside [5],
acts preferentially on the alpha-D-Man(1,6) arm, this branch is preferred
2.5times in bi-, 5.5–8.5times in tri-, and 12.7times in tetraantennary structures over the alpha-D-Man(1–3) arm, within the alpha-D-Man(1,6) branch
there is a 1.3–1.9fold consistently higher frequency of galactosylation of
the beta-D-GlcNAc(1–2) as compared to the beta-D-GlcNAc(1–6) antenna
[7], very poor acceptor: lactose, beta-D-Gal(1–3)-D-GlcNAc [10], beta-D-
Gal(1–4)-[alpha-L-Fuc(1–3)]-D-GlcNAc [10], not: alpha-L-Fuc(1–2)-beta-
D-Gal(1–4)-D-GlcNAc [10]. Oligosaccharides, glycoproteins and glycos-
aminoglycans containing the terminal nonreducing N-acetyllactosamine
unit all serve as acceptors [10]) [1–10]
2 UDPgalactose + N-acetyllactosamine [2–4]

Product spectrum

1 UDP + alpha-D-galactosyl-1,3-beta-D-galactosyl-1,4-N-acetyl-D-glucos-
aminyl-R [1–10]
2 UDP + alpha-D-Gal-(1–3)-beta-D-Gal-(1–4)-D-GlcNAc [2]

Inhibitor(s)

EDTA (weak) [3]

Cofactor(s)/prosthetic group(s)/activating agents

Triton X-100 (activates [3], at 0.8% v/v 4fold stimulation [3], absolute requirement, maximal activity at 0.5–1.0% v/v [10]) [3, 10]

Metal compounds/salts

Mn^{2+} (required [1, 10], K_m 6.1 mM [3]) [1, 3, 10]; Mg^{2+} (slight activation) [3]

Turnover number (min^{-1})

Specific activity (U/mg)

4.3 [5]

K_m-value (mM)

0.10 (beta-D-Gal(1–4)-D-GlcNAc(1–2)-[beta-D-Gal(1–4)-beta-D-GlcNAc-(1–6)]-
D-Man) [10]; 0.2 (UDPgalactose) [3]; 0.25 (beta-D-Gal-(1–4)-beta-D-
GlcNAc-(1–2)-D-Man) [10]; 0.31 (beta-D-Gal-(1–4)-beta-D-GlcNAc-(1–6)-
D-Man) [10]; 0.57 (Galbeta(1–4)GlcNAcbeta(1–2)Man(1–3)Man-

beta(1–4)GlcNAc) [5]; 1.15 (beta-D-Gal-(1–4)-D-GlcNAc) [10]; 1.25 (asia-lo-alpha$_1$-acid glycoprotein) [5]; 1.39 (Galbeta(1–4)GlcNAc) [5]; 2.7 (N-ace-tyllactosamine) [3]; 3.7 (asialo-alpha$_1$-acid glycoprotein) [3]; 9.0 (Gal-beta(1–4)Glc) [5]; 12.6 (UDPgalactose) [10]; 16.0 (Galbeta(1–3)GlcNAc-beta(1–3)Galbeta(1–4)Glc) [5]

pH-optimum
5.5–7.0 (cacodylate buffer) [3]; 6.0 (acetate buffer) [3]; 6.2 [10]

pH-range
5.3–7.0 (5.3: about 45% of activity maximum, 7.0: about 30% of activity maximum) [10]

Temperature optimum (°C)
37 (assay at) [3]

Temperature range (°C)

---

3 ENZYME STRUCTURE

Molecular weight
43000 (calf, gel filtration) [5]

Subunits

Glycoprotein/Lipoprotein
–

---

4 ISOLATION/PREPARATION

Source organism
Rat (ascites fluid of Novikoff tumor cells maintained in rat) [1]; Mouse [2, 4, 7, 8, 10]; Bovine (calf [3, 5]) [3, 5, 6, 9]

Source tissue
Ehrlich ascites tumor cells [2, 4, 7, 10]; Novikoff ascites tumor cells [1]; Thymus [3, 5]; Spleen [6]; Liver (fetal) [6]

Localization in source
Membrane (bound [3]) [2–4]

Purification
Bovine (calf) [5]; Mouse (Ehrlich ascites tumor cells) [10]

Crystallization
–

Cloned
    (characterization of cDNA [8, 9], a single gene locus specifies 4 isoforms
    [9]) [8, 9]

Renatured
    –

---

5 STABILITY

pH

Temperature (°C)

Oxidation

Organic solvent

General stability information
    Detergent: rapid and irreversible inactivation in absence of detergent, more
    than 50% loss of activity after 4–6 weeks [10]

Storage
    –20°C, 20 mM MES, pH 6.2, 50 mM NaCl, stable for several months [10]

---

6 CROSSREFERENCES TO STRUCTURE DATABANKS

PIR/MIPS code

Brookhaven code

---

7 LITERATURE REFERENCES

[1] van den Eijnden, D.H., Winterwerp, H., Smeeman, P., Schiphorst, W.E.C.M.: J. Biol.
    Chem.,258,3435–3437 (1983)
[2] Blake, D.A., Goldstein, I.J.: J. Biol. Chem.,256,5387–5393 (1981)
[3] van den Eijnden, D.H., Blanken, W.M., Winterwerp, H., Schiphorst, W.E.C.M.: Eur. J.
    Biochem.,134,523–530 (1983)
[4] Elices, M.J., Goldstein, I.J.: Arch. Biochem. Biophys.,254,329–341 (1987)
[5] Blanken, W.M., van den Eijnden, D.H.: J. Biol. Chem.,260,12927–12934 (1985)
[6] Beyer, T.A., Sadler, J.E., Rearick, J.I., Paulson, J.C., Hill, R.L.: Adv. Enzymol. Relat.
    Areas Mol. Biol.,52,23–175 (1981) (Review)
[7] Elices, M.J., Goldstein, I.J.: J. Biol. Chem.,264,1375–1380 (1989)
[8] Joziasse, D.H., Shaper, N.L., Kim, D., van den Eijnden, D.H., Shaper, J.H.: J. Biol.
    Chem.,267,5534–5541 (1992)
[9] Joziasse, D.H., Shaper, J.H., van den Eijnden, D.H., van Tunen, A.J., Shaper, N.L.:
    J. Biol. Chem.,264,14290–14297 (1989)
[10] Elices, M.J., Blake, D.A., Goldstein, I.J.: J. Biol. Chem.,261,6064–6072 (1986)

1 NOMENCLATURE

EC number
2.4.1.152

Systematic name
GDP-L-fucose:1,4-beta-D-galactosyl-N-acetyl-D-glucosaminyl-R 3-L-fucosyl-transferase

Recommended name
Galactoside 3-fucosyltransferase

Synonyms
Lewis-negative alpha-3-fucosyltransferase
Plasma alpha-3-fucosyltransferase [3]
Fucosyltransferase, guanosine diphosphofucose-glucoside alpha1→3-
More (cf. EC 2.4.1.65 and EC 2.4.1.69)

CAS Reg. No.
111310-38-4

---

2 REACTION AND SPECIFICITY

Catalysed reaction
GDP-L-fucose + 1,4-beta-D-galactosyl-N-acetyl-D-glucosaminyl-R →
→ GDP + 1,4-beta-D-galactosyl-(alpha-1,3-L-fucosyl)-N-acetyl-D-glucosaminyl-R

Reaction type
Hexosyl group transfer

Natural substrates

Substrate spectrum
1 GDP-L-fucose + 1,4-beta-D-galactosyl-N-acetyl-D-glucosaminyl-R (unlike EC 2.4.1.65, EC 2.4.1.152 has no action on the corresponding 1,3-galactosyl derivative [3]) [1–3]
2 GDP-L-fucose + Galbeta(1–4)GlcNAc [2]
3 GDP-L-fucose + NeuAcalpha(2–3)Galbeta(1–4)GlcNAc [2]
4 GDP-L-fucose + Galbeta(1–3)GlcNAc [2]
5 GDP-L-fucose + NeuAcalpha(2–6)Galbeta(1–4)GlcNAc (poor substrate) [2]
6 GDP-L-fucose + Galbeta(1–4)glucal (lactal, poor substrate) [2]
7 GDP-L-fucose + Galbeta(1–4) (5-thioGlc) [2]

---

8 GDP-L-fucose + Galbeta(1–4)GlcNAcbetaOallyl [2]
9 GDP-L-fucose + Fucalpha(1–2)Galbeta(1–4)GlcNAcbeta-OR [3]
10 GDP-L-fucose + NeuAcalpha(2–3)Galbeta(1–4)GlcNAcbeta-OR [3]

Product spectrum
1 GDP + 1,4-beta-D-galactosyl-(alpha-1,3-L-fucosyl)-N-acetyl-D-
glucosaminyl-R [1–3]
2 ?
3 ?
4 ?
5 ?
6 ?
7 ?
8 ?
9 ?
10 ?

Inhibitor(s)
N-Ethylmaleimide [3]; Galbeta(1–4) (3-deoxy)GlcNAcbetaOallyl (weak) [2];
Galbeta(1–4)deoxynojirimycin [2]; Galbeta(1–3)GalNAc [2]; GDP [2]

Cofactor(s)/prosthetic group(s)/activating agents

Metal compounds/salts

Turnover number (min^{-1})

Specific activity (U/mg)

K_m-value (mM)
12 (Galbeta(1–4) (5-thioGlc)) [2]; 16 (Galbeta(1–4)GlcNAcbetaOallyl) [2];
34 (Galbeta(1–4)glucal) [2]; 35 (Galbeta(1–4)GlcNAc) [2]; 64 (NeuAc-
alpha(2–3)Galbeta(1–4)glucan) [2]; 70 (NeuAcalpha(2–6)Gal-
beta(1–4)GlcNAc) [2]; 100 (NeuAcalpha(2–3)Galbeta(1–4)GlcNAc) [2];
More [3]

pH-optimum
7.2–8.0 (liver) [3]

pH-range

Temperature optimum (°C)

Temperature range (°C)

2

3 ENZYME STRUCTURE

Molecular weight

Subunits

Glycoprotein/Lipoprotein

–

4 ISOLATION/PREPARATION

Source organism
  Human (recombinant [2]) [1–4]

Source tissue
  Saliva [1]; Plasma [3]; Hepatocytes [3]

Localization in source

Purification

Crystallization

–

Cloned
  [4]

Renatured

–

5 STABILITY

pH

Temperature (°C)
  50 ($t_{1/2}$: 4 min, plasma enzyme, 3 min, liver enzyme) [3]

Oxidation

Organic solvent

General stability information

Storage

Enzyme Handbook © Springer-Verlag Berlin Heidelberg 1996
Duplication, reproduction and storage in data banks are only
allowed with the prior permission of the publishers

6 CROSSREFERENCES TO STRUCTURE DATABANKS

PIR/MIPS code

Brookhaven code

---

7 LITERATURE REFERENCES

[1] Johnson, P.H., Yates, A.D., Watkins, W.M.: Biochem. Biophys. Res. Commun.,100, 1611–1618 (1981)
[2] Wong, C.-H., Dumas, D.P., Ichikawa, Y., Koseki, K., Danishefsky, S.J., Weston, B.W., Lowe, J.B.: J. Am. Chem. Soc.,114,7321–7322 (1992)
[3] Mollicone, R., Gibaud, A., Francois, A., Ratcliffe, M., Oriol, R.: Eur. J. Biochem., 191,169–176 (1990)
[4] Weston, B.W., Nair, R.P., Larsen, R.D., Lowe, J.B.: J. Biol. Chem.,267,4152–4160 (1992)

1 NOMENCLATURE

EC number
2.4.1.153

Systematic name
UDP-N-acetyl-D-glucosamine:dolichyl-phosphate alpha-N-acetyl-D-glucosaminyltransferase

Recommended name
Dolichyl-phosphate alpha-N-acetylglucosaminyltransferase

Synonyms
Acetylglucosaminyltransferase, uridine diphosphoacetylglucosamine-dolichol phosphate
Dolichyl phosphate acetylglucosaminyltransferase
Dolichyl phosphate N-acetylglucosaminyltransferase
UDP-N-acetylglucosamine-dolichol phosphate N-acetylglucosaminyltransferase

CAS Reg. No.
63363-73-5

---

2 REACTION AND SPECIFICITY

Catalysed reaction
UDP-N-acetyl-D-glucosamine + dolichyl phosphate →
→ UDP + dolichyl N-acetyl-alpha-D-glucosaminyl phosphate

Reaction type
Hexosyl group transfer

Natural substrates
More (formation of glycoproteins through the participation of a sugar lipid intermediate) [3]

Substrate spectrum
1 UDP-N-acetyl-D-glucosamine + dolichyl phosphate (lipid-phosphate [1, 3]) [1–5]

Product spectrum
1 UDP + dolichyl N-acetyl-alpha-D-glucosaminyl phosphate (reaction in microsomal preparation: UDP-GlcNAc + dolichyl-P→ UMP + dolichol-P-P-GlcNAc [2], → N-acetylglucosamine-pyrophosphoryl-dolichol + N,N-diacetyl-chitobiosyl-pyrophosphoryl-dolichol [4]) [2, 4]

Inhibitor(s)
UDP [1, 4]; UMP [1, 4]; Tunicamycin [4, 5]; EDTA [5]

Cofactor(s)/prosthetic group(s)/activating agents
Detergent (required for maximal activity, optimum concentration for Triton X-100: 0.01%-0.015%) [4]

Metal compounds/salts
Mg^{2+} (divalent cation required for maximal activity: Mg^{2+} > Mn^{2+} >> Ca^{2+} > Co^{2+} > Zn^{2+}, K_m for Mg^{2+}: 1.33 mM [4], divalent cations stimulate, Mn^{2+} most effective, Mg^{2+} and Co^{2+} partly effective [5]) [4, 5]; Mn^{2+} (divalent cation required for maximal activity: Mg^{2+} > Mn^{2+} >> Ca^{2+} > Co^{2+} > Zn^{2+} [4], divalent cations stimulate, Mn^{2+} most effective, Mg^{2+} and Co^{2+} partly effective [5]) [4, 5]; Ca^{2+} (divalent cation required for maximal activity: Mg^{2+} > Mn^{2+} >> Ca^{2+} > Co^{2+} > Zn^{2+}) [4]; Co^{2+} (divalent cation required for maximal activity: Mg^{2+} > Mn^{2+} >> Ca^{2+} > Co^{2+} > Zn^{2+}) [4]; Zn^{2+} (divalent cation required for maximal activity: Mg^{2+} > Mn^{2+} >> Ca^{2+} > Co^{2+} > Zn^{2+}) [4]

Turnover number (min^{-1})

Specific activity (U/mg)
More [5]

K_m-value (mM)
0.0115 (UDP-N-acetyl-D-glucosamine) [4]; 0.014 (UDP-N-acetyl-D-glucosamine) [5]; More (dolichol-phosphate: 0.0017 mg/ml) [4]

pH-optimum
7.5 [4]; 8.0 [5]

pH-range

Temperature optimum (°C)
37 (assay at) [5]

Temperature range (°C)

3 ENZYME STRUCTURE

Molecular weight

Subunits

Glycoprotein/Lipoprotein
−

4 ISOLATION/PREPARATION

Source organism
Rat [1, 2, 5]; Rabbit [1, 3]; Zea mays (L. inbred A 636) [4]

Source tissue
Liver [1–3, 5]; Endosperm culture [4]

Localization in source
Microsomes [1–3, 5]; Endoplasmic reticulum [4]

Purification

Crystallization
–

Cloned
–

Renatured
–

5 STABILITY

pH

Temperature (°C)

Oxidation

Organic solvent

General stability information
When sealed microsomal vesicles are incubated with trypsin for 30 min in absence of detergent, the activity is substantially reduced [4]

Storage

6 CROSSREFERENCES TO STRUCTURE DATABANKS

PIR/MIPS code

Brookhaven code

7 LITERATURE REFERENCES

[1] Tetas, M., Chao, H., Molnar, J.: Arch. Biochem. Biophys.,138,135–146 (1970)
[2] Leloir, L.F., Staneloni, R.J., Carminatti, H., Behrens, N.H.: Biochem. Biophys. Res. Commun.,52,1285–1292 (1973)
[3] Molnar, J., Chao, H., Ikehara, Y.: Biochim. Biophys. Acta,239,401–410 (1971)
[4] Riedell, W.E., Miernyk, J.A.: Plant Physiol.,87,420–426 (1988)
[5] Arakawa, H., Mookerjea, S.: Eur. J. Biochem.,140,297–302 (1984)

Enzyme Handbook © Springer-Verlag Berlin Heidelberg 1996
Duplication, reproduction and storage in data banks are only
allowed with the prior permission of the publishers

1 NOMENCLATURE

EC number
   2.4.1.154

Systematic name
   UDP-N-acetyl-D-galactosamine-globotriosylceramide beta-1,6-N-acetyl-galactosaminyltransferase

Recommended name
   Globotriosylceramide beta-1,6-N-acetylgalactosaminyltransferase

Synonyms
   Acetylgalactosaminyltransferase, uridine diphosphoacetylgalactosamine-glycosphingolipid
   Globoside N-acetylgalactosaminyltransferase
   Glycosphingolipid beta-N-acetylgalactosaminyltransferase [1]
   GalNAc transferase [1]

CAS Reg. No.
   83215-90-1

---

2 REACTION AND SPECIFICITY

Catalysed reaction
   UDP-N-acetyl-D-galactosamine + globotriosylceramide →
   → UDP + globotetraosylceramide

Reaction type
   Hexosyl group transfer

Natural substrates
   UDP-N-acetyl-D-galactosamine + globotriosylceramide (glycosphingolipid metabolism) [1]

Substrate spectrum
   1 UDP-N-acetyl-D-galactosamine + globotriosylceramide (highly acceptor specific, does not act on globotetraosylceramide or lactosylceramide) [1]

Product spectrum
   1 UDP + globotetraosylceramide (the product has a terminal beta-N-acetylgalactosamine residue) [1]

Inhibitor(s)
   Globotetraosylceramide [1]; II^3-alpha-N-Acetylneuraminyl-lactosylceramide [1]; More (no effect: lactosylceramide) [1]

Cofactor(s)/prosthetic group(s)/activating agents
  Detergents (greatest activation by sodium taurodeoxycholate, 1 mg/ml) [1]

Metal compounds/salts
  Mn^{2+} (required for maximal activity, 4 mM) [1]; More (Mg^{2+} cannot replace Mn^{2+}) [1]

Turnover number (min^{-1})

Specific activity (U/mg)

K_m-value (mM)
  0.14 (UDP-N-acetyl-D-galactosamine) [1]; 0.42 (globotriosylceramide) [1]

pH-optimum
  4.5–8.0 (maximal activity in 2-(N-morpholino)ethanesulfonic acid) [1]

pH-range

Temperature optimum (°C)
  37 (assay at) [1]

Temperature range (°C)

---

3 ENZYME STRUCTURE

Molecular weight

Subunits

Glycoprotein/Lipoprotein
  –

---

4 ISOLATION/PREPARATION

Source organism
  Hamster [1]

Source tissue
  Fibroblasts (cultured cells, NIL-8) [1]

Localization in source

Purification

Crystallization
  –

Cloned
  –

Renatured

–

---

5 STABILITY

pH

Temperature (°C)

Oxidation

Organic solvent

General stability information

Storage

---

6 CROSSREFERENCES TO STRUCTURE DATABANKS

PIR/MIPS code

Brookhaven code

---

7 LITERATURE REFERENCES

[1] Lockney, M.W., Sweely, C.S.: Biochim. Biophys. Acta,712,234–241 (1982)

---

1 NOMENCLATURE

EC number
    2.4.1.155

Systematic name
    UDP-N-acetyl-D-glucosamine:glycoprotein (N-acetyl-D-glucosamine to
    alpha-D-mannosyl-1,3(6)-N-acetyl-beta-D-glucosaminyl-R) beta-1,6-N-ace-
    tylglucosaminyltransferase

Recommended name
    alpha-1,3(6)-Mannosylglycoprotein beta-1,6-N-acetylglucosaminyltrans-
    ferase

Synonyms
    alpha-Mannoside beta-1,6-N-acetylglucosaminyltransferase
    Acetylglucosaminyltransferase, uridine diphosphoacetylglucosamine-
    alpha-mannoside beta1→6-
    UDP-N-acetylglucosamine:alpha-mannoside-beta1,6 N-acetylglucosaminyl-
    transferase [1]
    More (cf. EC 2.4.1.101 and EC 2.4.1.143–5)

CAS Reg. No.
    83588-90-3

2 REACTION AND SPECIFICITY

Catalysed reaction
    UDP-N-acetyl-D-glucosamine + N-acetyl-beta-D-glucosaminyl-1,2-alpha-
    D-mannosyl-1,3(6)-(N-acetyl-beta-D-glucosaminyl-1,2-alpha-D-mannosyl-
    1,6(3))-beta-D-mannosyl-1,4-N-acetyl-beta-D-glucosaminyl-R →
    → UDP + N-acetyl-beta-D-glucosaminyl-1,2-(N-acetyl-beta-D-glucosaminyl-
    1,6)-1,2-alpha-D-mannosyl-1,3(6)-(N-acetyl-beta-D-glucosaminyl-1,2-alpha-
    D-mannosyl-1,6(3))-beta-D-mannosyl-1,4-N-acetyl-beta-D-glucosaminyl-R
    (mechanism [4])

Reaction type
    Hexosyl group transfer

Natural substrates
    UDP-N-acetyl-D-glucosamine + N-acetyl-D-glucosaminyl-1,2-alpha-D-man-
    nosyl-1,3(6)-(N-acetyl-D-glucosaminyl-1,2-alpha-D-mannosyl-1,6(3))-beta-
    D-mannosyl-1,4-N-acetyl-D-glucosaminyl-R (involved in biosynthesis of bran-
    ched N-glycopeptides [1–3], enzyme requires prior action of N-ace-
    tylglucosaminyltransferase II which requires prior action of N-acetylglucos-
    aminyltransferase I [8]) [1–3, 8]

Substrate spectrum

1 UDP-N-acetyl-D-glucosamine + N-acetyl-D-glucosaminyl-1,2-alpha-D-man-
nosyl-1,3(6)-(N-acetyl-D-glucosaminyl-1,2-alpha-D-mannosyl-1,6(3))-
beta-D-mannosyl-1,4-N-acetyl-D-glucosaminyl-R (transfers N-acetylglucosamine to C-6 of alpha-linked mannose residue with inversion of configuration at its anomeric center [1], acceptor substrates are only branched
mannose glycopeptides with non-reducing N-acetylglucosamine terminal
residues [1, 2], acceptors are bi-antennary Asn-linked asialo- or aga-
lacto-oligosaccharides containing N-acetylglucosamine at the non-reducing terminal [1], best substrates are bi- and tri-antennary acceptors [2],
specificity study with various pyridylaminated sugar chains [2], synthetic
acceptor substrates [3, 7]. No substrates are bisected N-glycans [8],
glycopeptides with sialic acid [1] or galactose [1, 2] at non-reducing terminal) [1–8]

2 UDP-N-acetyl-D-glucosamine + N-acetyl-D-glucosaminyl-1,2-alpha-man-
nosyl-1,6-beta-glucosyl-O-R (R: hydrophobic group, e.g. $(CH_2)_7$-CH_3 [4, 5]
or $(CH_2)_8$-$COOCH_3$ [8], best substrate [4]) [4, 5, 8]

3 UDP-N-acetyl-D-glucosamine + N-acetyl-D-glucosaminyl-1,2-alpha-4-
deoxy-mannosyl-1,6-beta-glucosyl-O-$(CH_2)_7$-CH_3 (good substrate) [4]

4 UDP-N-acetyl-D-glucosamine + N-acetyl-D-glucosaminyl-1,2-alpha-man-
nosyl-1,6-beta-mannosyl-O-R (R: hydrophobic group, e.g.
$(CH_2)_8$-$COOCH_3$) [8]

Product spectrum

1 UDP + N-acetyl-D-glucosaminyl-1,2-(N-acetyl-D-glucosaminyl-1,6)-
1,2-alpha-D-mannosyl-1,3(6)-(N-acetyl-D-glucosaminyl-1,2-alpha-D-man-
nosyl-1,6(3))-beta-D-mannosyl-1,4-N-acetyl-D-glucosaminyl-R [7, 8]

2 UDP + N-acetyl-D-glucosaminyl-1,6-(N-acetyl-D-glucosaminyl-1,2)-alpha-
mannosyl-1,6-beta-D-glucosyl-O-R [4, 5, 8]

3 UDP + N-acetyl-D-glucosaminyl-1,6-(N-acetyl-D-glucosaminyl-1,2)-alpha-
4-deoxy-mannosyl-1,6-beta-D-mannosyl-O-$(CH_2)_7$-CH_3 [4]

4 UDP + N-acetyl-D-glucosaminyl-1,6-(N-acetyl-D-glucosaminyl-1,2)-alpha-
mannosyl-1,6-beta-D-mannosyl-O-R [8]

Inhibitor(s)

N-Acetyl-D-glucosaminyl-1,2-alpha-6-deoxy-mannosyl-1,6-beta-D-gluco-
syl-O-$(CH_2)_7$-CH_3 (kinetics [4]) [3–5]; N-Acetyl-D-glucosaminyl-1,2-alpha-
6-deoxy-mannosyl-1,6-beta-D-glucosyl-O-(CH_2)8-$COOCH_3$ [6]; N-Ace-
tyl-D-glucosaminyl-1,2-alpha-4-O-methyl-mannosyl-1,6-beta-D-gluco-
syl-O-$(CH_2)_7$-CH_3 [4]; More (no inhibitor: EDTA) [1, 6]

Cofactor(s)/prosthetic group(s)/activating agents

Triton X-100 (activation, 1.0–1.5%) [6]; Immunoglobulin G (activation and
stabilization, 0.5 mg/ml) [6]; Glycerol (activation and stabilization, 20%) [6]

Metal compounds/salts

NaCl (activation, 0.2 M) [6]; More (no requirement for Mn^{2+} [1–3, 6–8], Mg^{2+}, Ca^{2+} [6]) [1–3, 6–8]

Turnover number (min^{-1})

Specific activity (U/mg)

More [1, 3]; 0.139 [2]; 1.22 [6]

K_m-value (mM)

More (kinetic study) [4]; 0.023 (N-acetyl-D-glucosaminyl-1,2-alpha-man-nosyl-1,6-beta-glucosyl-O-(CH_2)7-CH_3) [4]; 0.074 (N-acetyl-D-glucos-aminyl-1,2-alpha-4-deoxy-mannosyl-1,6-beta-D-glucosyl-O-$(CH_2)_7$-CH_3) [4]; 0.133 (pyridylaminated biantennary acceptor substrates) [2]; 0.18 (N-ace-tyl-D-glucosaminyl-1,2-alpha-mannosyl-1,6-beta-mannosyl-O-(CH_2)-$COOCH_3$, hamster) [8]; 0.25 (N-acetyl-D-glucosamine, hamster) [8]; 0.27 (pyridylami-nated acceptor substrate) [7]; 0.66–0.7 (UDP-N-acetylgucosamine (+ N-ace-tyl-D-glucosaminyl-1,2-alpha-6-deoxy- (or –4-O-methyl-) mannosyl-1,6-beta-D-glucosyl-O-$(CH_2)_7$-CH_3)) [4]; 1.1–1.2 (UDP-N-acetylglucosamine (+ N-ace-tyl-D-glucosaminyl-1,2-alpha-mannosyl-1,6-beta-glucosyl-O-$(CH_2)_7$-CH_3 [4])) [3, 4]; 1.4 (UDP-N-acetylgucosamine (+ N-acetyl-D-glucosaminyl-1,2-alpha-4-deoxy-mannosyl-1,6-beta-D-glucosyl-O-$(CH_2)_7$-CH_3)) [4]; 7.6 (UDP-N-acetylglucosamine (+ pyridylaminated acceptor substrate)) [7]; 11 (UDP-N-acetylglucosamine) [6]

pH-optimum

6 (around, hamster) [8]; 6.3 [7]; 6.5–7 [6]

pH-range

6.1–8 (about half-maximal activity at pH 6.1 and 65% of maximal activity at pH 8) [6]

Temperature optimum (°C)

37 (assay at) [1, 3–5, 7, 8]

Temperature range (°C)

---

3 ENZYME STRUCTURE

Molecular weight

73000 (human, HPLC-gel filtration) [2]

Subunits

Monomer (1 × 73000, human, SDS-PAGE) [2]
More (rat kidney enzyme appears as doublet of MW 69000 and 75000 on SDS-PAGE) [6]

Glycoprotein/Lipoprotein

–

---

4 ISOLATION/PREPARATION

Source organism

Hamster [3–5, 8]; Chicken [8]; Human [2, 8]; Mouse [1, 8]; Rat [6, 7]

Source tissue

Lymphoma BW 5147 cell line [1, 8]; QC small lung cancer cell line [2]; Kidney (native BHK 21/c13 cells, cells transformed with Rous sarcoma virus (i.e. RS-BHK-cells) and lectin-resistant cell-lines LP 1.6 and 3.3 [3]) [3–5, 8]; Myoloid cell lines (human) [8]; Tumor cell lines (mouse) [8]; Brain [7]; Lung [7]; Oviduct (hen) [8]; Small intestine [7]; Spleen [7]; Testis [7]

Localization in source

Membrane-bound [1, 3–6]; Soluble [2]; Microsomes [3]

Purification

Human (partial) [2]; Rat [6]

Crystallization

–

Cloned

–

Renatured

–

---

5 STABILITY

pH

Temperature (°C)

Oxidation

Organic solvent

General stability information

Glycerol, 20%, stabilizes and enhances activity [6]; Immunoglobulin G, 0.5 mg/ml, stabilizes and enhances activity [6]; NaCl, 0.2 M, stabilizes and enhances activity [6]

Storage

4°C, several months in 20% glycerol [6]

6 CROSSREFERENCES TO STRUCTURE DATABANKS

PIR/MIPS code

Brookhaven code

7 LITERATURE REFERENCES

[1] Cummings, R.D., Trowbridge, I.S., Kornfeld, S.: J. Biol. Chem.,257,13421–13427 (1982)
[2] Gu, J., Nishikawa, A., Tsuruoka, N., Ohno, M., Yamaguchi, N., Kangawa, K., Taniguchi, N.: J. Biochem.,113,614–619 (1993)
[3] Palcic, M.M., Ripka, J., Kaur, K.J., Shoreibah, M., Hindsgaul, O., Pierce, M.: J. Biol. Chem.,265,6759–6769 (1990)
[4] Khan, S.H., Crawley, S.C., Kanie, O., Hindsgaul, O.: J. Biol. Chem.,268,2468–2473 (1993)
[5] Hindsgaul, O., Kaur, K.J., Srivastava, G., Blaszcyk-Thurin, M., Crawley, S.C., Heerze, L.D., Palcic, M.M.: J. Biol. Chem.,266,17858–17862 (1991)
[6] Shoreibah, M., Hindsgaul, O., Pierce, M.: J. Biol. Chem.,267,2920–2927 (1992)
[7] Taniguchi, N., Nishikawa, A., Fujii, S., Gu, J.: Methods Enzymol.,179,397–408 (1989) (Review)
[8] Schachter, H., Brockhausen, I., Hull, E.: Methods Enzymol.,179,351–397 (1989) (Review)

1 NOMENCLATURE

EC number
2.4.1.156

Systematic name
UDPgalactose:indol-3-ylacetyl-myo-inositol 5-O-D-galactosyltransferase

Recommended name
Indolylacetyl-myo-inositol galactosyltransferase

Synonyms
Galactosyltransferase, uridine diphosphogalactose-indolylacetylinositol
Indol-3-ylacetyl-myo-inositol galactoside synthase [1]

CAS Reg. No.
85537-80-0

---

2 REACTION AND SPECIFICITY

Catalysed reaction
UDPgalactose + indol-3-ylacetyl-myo-inositol →
→ UDP + 5-O-(indol-3-ylacetyl-myo-inositol) D-galactoside

Reaction type
Hexosyl group transfer

Natural substrates
UDPgalactose + indol-3-ylacetyl-myo-inositol [1] (involved in metabolism of
indol-3-ylacetic acid, a plant growth hormone) [1]

Substrate spectrum
1 UDPgalactose + indol-3-ylacetyl-myo-inositol (not UDPglucose) [1]

Product spectrum
1 UDP + 5-O-(indol-3-ylacetyl-myo-inositol) D-galactoside [1]

Inhibitor(s)
Zn^{2+} [1]; Mn^{2+} [1]; Mg^{2+} (weak) [1]; EDTA [1]

Cofactor(s)/prosthetic group(s)/activating agents

Metal compounds/salts
Ca^{2+} (slight activation) [1]

Turnover number (min^{-1})

Specific activity (U/mg)

K_m-value (mM)

pH-optimum
    7.6 (assay at) [1]

pH-range

Temperature optimum (°C)
    37 (assay at) [1]

Temperature range (°C)

---

3 ENZYME STRUCTURE

Molecular weight

Subunits

Glycoprotein/Lipoprotein
    –

---

4 ISOLATION/PREPARATION

Source organism
    Zea mays (sweet corn) [1]

Source tissue
    Immature kernels [1]

Localization in source

Purification
    Zea mays (partial) [1]

Crystallization
    –

Cloned
    –

Renatured
    –

5 STABILITY

pH

Temperature (°C)

Oxidation

Organic solvent

General stability information
2-Mercaptoethanol stabilizes during ammonium sulfate fractionation and dialysis [1]

Storage

6 CROSSREFERENCES TO STRUCTURE DATABANKS

PIR/MIPS code

Brookhaven code

7 LITERATURE REFERENCES

[1] Corcuera, L.J., Michalczuk, L., Bandurski, R.S.: Biochem. J.,207,283–290 (1982)

1 NOMENCLATURE

EC number
2.4.1.157

Systematic name
UDPglucose:1,2-diacylglycerol 3-D-glucosyltransferase

Recommended name
1,2-Diacylglycerol 3-glucosyltransferase

Synonyms
UDPglucose:diacylglycerol glucosyltransferase [1]
UDP-glucose:1,2-diacylglycerol glucosyltransferase [2]
Glucosyltransferase, uridine diphosphoglucose-diacylglycerol
UDP-glucose-diacylglycerol glucosyltransferase

CAS Reg. No.
83744-96-1

2 REACTION AND SPECIFICITY

Catalysed reaction
UDPglucose + 1,2-diacylglycerol →
→ UDP + 3-D-glucosyl-1,2-diacylglycerol

Reaction type
Hexosyl group transfer

Natural substrates

Substrate spectrum
1 UDPglucose + 1,2-diacylglycerol [1]
2 UDPglucose + 1,2-dipalmitoylglycerol [1, 2]
3 UDPglucose + 1-stearoyl-2-palmitoylglycerol [1]
4 UDPglucose + 1-oleoyl-2-palmitoylglycerol [1]

Product spectrum
1 UDP + 3-D-glucosyl-1,2-diacylglycerol [1]
2 UDP + 1,2-dipalmitoyl-3-D-glucosylglycerol
3 UDP + 1-stearoyl-2-palmitoyl-3-D-glucosylglycerol
4 UDP + 1-oleoyl-2-palmitoyl-3-D-glucosylglycerol

Inhibitor(s)
EDTA [1]; $CaCl_2$ [2]

Cofactor(s)/prosthetic group(s)/activating agents

Metal compounds/salts
Mg^{2+} (required, maximal activity at 20 mM $MgCl_2$) [1]

Turnover number (min^{-1})

Specific activity (U/mg)

K_m-value (mM)
0.045 (UDPglucose) [1]

pH-optimum
7.0 [1]

pH-range
5.0–8.7 (about 50% of activity maximum at pH 5.0 and 8.7) [1]

Temperature optimum (°C)
45 [1]; 38 (assay at) [2]

Temperature range (°C)
28–55 (28°C: 50% of activity maximum, 55°C: 35% of activity maximum) [1]

---

3 ENZYME STRUCTURE

Molecular weight

Subunits

Glycoprotein/Lipoprotein
–

---

4 ISOLATION/PREPARATION

Source organism
Anabaena variabilis [1]; Anacystis nidulans [2]

Source tissue
Cell [1]

Localization in source
Membrane (plasma membranes, thylakoid membranes, not in outer membrane [2]) [1, 2]

Purification

Crystallization
–

Cloned
–

Renatured
–

---

5 STABILITY

pH

Temperature (°C)

Oxidation

Organic solvent

General stability information

Storage
–80°C, stable for at least 2 months, crude enzyme extract [1]

---

6 CROSSREFERENCES TO STRUCTURE DATABANKS

PIR/MIPS code

Brookhaven code

---

7 LITERATURE REFERENCES

[1] Sato, N., Murata, N.: Plant Cell Physiol.,23,1115–1120 (1982)
[2] Omata, T., Murata, N.: Plant Cell Physiol.,27,485–490 (1986)

---

1 NOMENCLATURE

EC number
2.4.1.158

Systematic name
UDPglucose:13-hydroxydocosanoate 13-beta-D-glucosyltransferase

Recommended name
13-Hydroxydocosanoate 13-beta-glucosyltransferase

Synonyms
13-Glucosyloxydocosanoate 2'-beta-glucosyltransferase
UDP-glucose:13-hydroxydocosanoic acid glucosyltransferase [1]
Glucosyltransferase, uridine diphosphoglucose-hydroxydocosanoate
UDP-glucose-13-hydroxydocosanoate glucosyltransferase

CAS Reg. No.
70457-13-5

---

2 REACTION AND SPECIFICITY

Catalysed reaction
UDPglucose + 13-hydroxydocosanoate →
→ UDP + 13-beta-D-glucosyloxydocosanoate

Reaction type
Hexosyl group transfer

Natural substrates

Substrate spectrum
1 UDPglucose + 13-hydroxydocosanoic acid (glucosyltransferase I) [1]
2 UDPglucose + 13-(beta-D-glucopyranosyl)oxydocosanoic acid methyl
  ester (glucosyltransferase II) [1]
3 More (enzyme seems to be a single multifunctional protein catalyzing
  both transferase reactions) [1]

Product spectrum
1 UDP + 13-(beta-D-glucopyranosyl)oxydocosanoic acid [1]
2 UDP + 13-[(2'-O-beta-D-glucopyranosyl-beta-D-glucopyranosyl)oxy]do-
  cosanoic acid methyl ester [1]
3 ?

Inhibitor(s)
NaCl (0.2 mM: 50% inhibition of glucosyltransferase I, 0.25 mM: 50% inhibition of glucosyltransferase II) [1]

Cofactor(s)/prosthetic group(s)/activating agents

Metal compounds/salts

Turnover number (min^{-1})

Specific activity (U/mg)
0.0222 (glucosyltransferase I) [1]; 0.0242 (glucosyltransferase II) [1]

K_m-value (mM)
0.04 (UDPglucose, glucosyltransferase I and II) [1]

pH-optimum
6.5–8.0 (glucosyltransferase I, within this range activity of glucosyltransferase II increases) [1]

pH-range

Temperature optimum (°C)
25 (assay at) [1]; 30–37 [1]

Temperature range (°C)
5–37 (30°C-37°C: 100% of activity, at 5°C: 10% of maximal activity) [1]

3 ENZYME STRUCTURE

Molecular weight
50000 (Candida bogoriensis, gel filtration) [1]

Subunits
Monomer (1 × 52000, Candida bogoriensis, SDS-PAGE) [1]

Glycoprotein/Lipoprotein
–

4 ISOLATION/PREPARATION

Source organism
Candida bogoriensis [1]

Source tissue
Cell [1]

Localization in source

Purification
Candida bogoriensis (partial) [1]

Crystallization
–

Cloned
–

Renatured
–

5 STABILITY

pH
5.9–8.1 (glucosyltransferase I and II : stable) [1]

Temperature (°C)

Oxidation

Organic solvent

General stability information
Glycerol, 20%, stabilizes [1]

Storage
4°C, 20% glycerol: half-life 35 days [1]; 4°C, without glycerol: half-life of
5 days [1]

6 CROSSREFERENCES TO STRUCTURE DATABANKS

PIR/MIPS code

Brookhaven code

7 LITERATURE REFERENCES

[1] Breithaupt, T.B., Light, R.J.: J. Biol. Chem.,257,9622–9628 (1982)

1 NOMENCLATURE

EC number
2.4.1.159

Systematic name
UDP-L-rhamnose:flavonol-3-O-D-glucoside L-rhamnosyltransferase

Recommended name
Flavonol-3-O-glucoside L-rhamnosyltransferase

Synonyms
Rhamnosyltransferase, uridine diphosphorhamnose-flavonol 3-O-glucoside
UDP-rhamnose:flavonol 3-O-glucoside rhamnosyltransferase

CAS Reg. No.
83380-89-6

2 REACTION AND SPECIFICITY

Catalysed reaction
UDP-L-rhamnose + flavonol 3-O-D-glucoside →
→ UDP + flavonol 3-O-L-rhamnosylglucoside

Reaction type
Hexosyl group transfer

Natural substrates
UDPrhamnose + flavonol 3-O-glucoside (involved in flavonoid metabolism,
pathway of flavonol 3-O-triglycoside biosynthesis) [1]

Substrate spectrum
1 UDPrhamnose + flavonol 3-O-glucosides (acceptor substrates are
3-O-glucosides of quercetin, isorhamnetin or kaempferol, rutin, less efficient acceptors: quercetin 3-O-galactoside or quercetin 3-O-rhamnoside,
kaempferol 3-O-rhamnoside or myrecetin 3-O-rhamnoside, no substrates:
aglycones) [1]
2 UDPrhamnose + flavonol 3-O-diglycosides [1]

Product spectrum
1 UDP + flavonol 3-O-rhamnosylglucosides (i.e. 3-O-rutinosides) [1]
2 UDP + flavonol 3-O-triglycosides [1]

Inhibitor(s)
PCMB (weak) [1]

Cofactor(s)/prosthetic group(s)/activating agents

Dithioerythritol (slight stimulation) [1]; 2-Mercaptoethanol (slight stimulation) [1]; Glutathione (slight stimulation) [1]; More (no stimulation by sucrose or bovine serum albumin) [1]

Metal compounds/salts

Mg^{2+} (slight stimulation) [1]; More (no activation by Ca^{2+}, NH_4^+, Mn^{2+}, Cl^- or SO_4^{2-}) [1]

Turnover number (min^{-1})

Specific activity (U/mg)

K_m-value (mM)

pH-optimum
8.5–9 [1]

pH-range

Temperature optimum (°C)

Temperature range (°C)

3 ENZYME STRUCTURE

Molecular weight
40000 (Tulipa, gel filtration) [1]

Subunits

Glycoprotein/Lipoprotein
–

4 ISOLATION/PREPARATION

Source organism
Tulipa (tulip, cv. Apeldoorn) [1]

Source tissue
Anthers (tapetum) [1]

Localization in source

Purification
Tulipa (partial) [1]

Crystallization
–

Cloned

–

Renatured

–

5 STABILITY

pH

Temperature (°C)

Oxidation

Organic solvent

General stability information

Storage

6 CROSSREFERENCES TO STRUCTURE DATABANKS

PIR/MIPS code

Brookhaven code

7 LITERATURE REFERENCES

[1] Kleinehollenhorst, G., Behrens, H., Pegels, G., Srunk, N., Wiermann, R.: Z. Natur-forsch.,37c,587–599 (1982)

1 NOMENCLATURE

EC number

2.4.1.160

Systematic name

UDPglucose:pyridoxine 5'-O-beta-D-glucosyltransferase

Recommended name

Pyridoxine 5'-O-beta-D-glucosyltransferase

Synonyms

UDP-glucose:pyridoxine 5'-O-beta-glucosyltransferase [1]

Glucosyltransferase, uridine diphosphoglucose-pyridoxine 5'-beta-

UDP-glucose-pyridoxine glucosyltransferase

CAS Reg. No.

83744-97-2

---

2 REACTION AND SPECIFICITY

Catalysed reaction

UDPglucose + pyridoxine →

→ UDP + 5'-O-beta-D-glucosylpyridoxine

Reaction type

Hexosyl group transfer

Natural substrates

Substrate spectrum

1 UDPglucose + pyridoxine (high specificity for UDPglucose) [1]

2 UDPglucose + pyridoxamine [1]

3 UDPglucose + 4'-deoxypyridoxine [1]

Product spectrum

1 UDP + 5'-O-beta-D-glucosylpyridoxine [1]

2 ?

3 ?

Inhibitor(s)

Pyridoxine (competitive inhibitor of pyridoxamine glucosylation) [1];

Pyridoxamine (competitive inhibitor of pyridoxine glucosylation) [1];

4'-Deoxypyridoxine (competitive inhibitor of pyridoxamine and pyridoxine

glucosylation) [1]

Cofactor(s)/prosthetic group(s)/activating agents

Metal compounds/salts
Mg^{2+} (required, K_m: 1.0 mM) [1]; Mn^{2+} (can replace Mg^{2+}, 40% as effective as Mg^{2+}) [1]

Turnover number (min^{-1})

Specific activity (U/mg)

K_m-value (mM)
0.25 (pyridoxine, 4'-deoxypyridoxine) [1]; 0.26 (pyridoxamine) [1]; 0.67 (UDPglucose (+ pyridoxine)) [1]

pH-optimum
7.8–8.8 (pyridoxine + UDPglucose) [1]

pH-range
7–9.5 (about 50% of activity maximum at pH 7 and 9.5) [1]

Temperature optimum (°C)
30 (assay at) [1]

Temperature range (°C)

3 ENZYME STRUCTURE

Molecular weight

Subunits

Glycoprotein/Lipoprotein
–

4 ISOLATION/PREPARATION

Source organism
Pisum sativum (podded pea, L. cv. Kinusaya) [1]

Source tissue
Seedlings [1]

Localization in source

Purification

Crystallization
–

Cloned
–

Renatured

–

---

5 STABILITY

pH

Temperature (°C)

Oxidation

Organic solvent

General stability information

Storage

---

6 CROSSREFERENCES TO STRUCTURE DATABANKS

PIR/MIPS code

Brookhaven code

---

7 LITERATURE REFERENCES

[1] Tadera, K., Yagi, F., Kobayashi, A.: J. Nutr. Sci. Vitaminol.,28,359–366 (1982)

1 NOMENCLATURE

EC number
2.4.1.161

Systematic name
1,4-alpha-D-Glucan:1,4-alpha-D-glucan 4-alpha-D-glucosyltransferase

Recommended name
Oligosaccharide 4-alpha-D-glucosyltransferase

Synonyms
Amylase III
1,4-alpha-Glucan:1,4-alpha-glucan 4-alpha-glucosyltransferase [2]

CAS Reg. No.

2 REACTION AND SPECIFICITY

Catalysed reaction
Transfers the non-reducing terminal alpha-D-glucose residue from a
1,4-alpha-D-glucan to the 4-position of an alpha-D-glucan, thus bringing
about the hydrolysis of oligosaccharides

Reaction type
Hexosyl group transfer

Natural substrates
More (transfers the nonreducing terminal alpha-D-glucose residue from a
1,4-alpha-D-glucan to the 4-position of an alpha-D-glucan, thus bringinig
about the hydrolysis of oligosaccharides, acts on amylose, amylopectin,
glycogen and maltooligosaccharides [2], enzyme may be biologically rele-
vant in the environment of the intestinal tract of man [1]) [1, 2]

Substrate spectrum
1 Transfers the nonreducing terminal alpha-D-glucose residue from a
1,4-alpha-D-glucan to the 4-position of an alpha-D-glucan, thus bringinig
about the hydrolysis of oligosaccharides (acts on amylose, amylopectin,
glycogen and maltooligosaccharides (4-nitrophenol-alpha-glucoside [1, 2]
to 4-nitrophenylmaltoheptaoside [2]) not maltose [2]) [1, 2]

Product spectrum
1 ? (hydrolysis of oligosaccharides, no detectable free glucose is formed)
[2]

Enzyme Handbook © Springer-Verlag Berlin Heidelberg 1996

Inhibitor(s)
  1-Deoxynojirimycin (Bay h 5595) [2]; Acarbose (Bay g 5421) [2]

Cofactor(s)/prosthetic group(s)/activating agents

Metal compounds/salts

Turnover number (min^{-1})

Specific activity (U/mg)

K_m-value (mM)

pH-optimum
  6.0 [1]

pH-range

Temperature optimum (°C)
  33 [1]

Temperature range (°C)

3 ENZYME STRUCTURE

Molecular weight
  47000 (Entamoeba histolytica, gel filtration) [1]

Subunits

Glycoprotein/Lipoprotein
  −

4 ISOLATION/PREPARATION

Source organism
  Entamoeba histolytica [1, 2]

Source tissue
  Trophozoites [1]

Localization in source

Purification
  Entamoeba histolytica [1]

Crystallization
  −

2

Cloned

–

Renatured

–

5 STABILITY

pH

Temperature (°C)

Oxidation

Organic solvent

General stability information

Storage

6 CROSSREFERENCES TO STRUCTURE DATABANKS

PIR/MIPS code

Brookhaven code

7 LITERATURE REFERENCES

[1] Nebinger, P.: Biol. Chem. Hoppe-Seyler,367,161–167 (1986)
[2] Nebinger, P.: Biol. Chem. Hoppe-Seyler,367,169–176 (1986)

1 NOMENCLATURE

EC number
2.4.1.162

Systematic name
alpha-D-Aldosyl-beta-D-fructoside:aldose 1-beta-D-fructosyltransferase

Recommended name
Aldose beta-D-fructosyltransferase

Synonyms
More (cf. EC 2.4.1.10)

CAS Reg. No.

---

2 REACTION AND SPECIFICITY

Catalysed reaction
alpha-D-Aldosyl[1] beta-D-fructoside + D-aldose[2] →
→ D-aldose[1] + alpha-D-aldosyl[2] beta-D-fructoside

Reaction type
Hexosyl group transfer

Natural substrates

Substrate spectrum
1 Fructose donor + fructose acceptor (The donor must be a beta-fructosyl
  ring attached to the anomeric carbon of an aldose by an 1→2 link, e.g.
  sucrose, raffinose, stachyose. Acceptors can be monomeric or oligomeric
  hexoses or pentoses, e.g. ribose, sucrose, lyxose, D-arabinose, xylose,
  galactose, 3-O-methylglucose, 4-chlorogalactose, D-arabinose, methyl-
  alpha-D-glucopyranoside, 1-thio-D-glucose, sorbose, mannose, glucose
  6-phosphate, 6-deoxyglucose, 6-O-methylglucose, 6-O-methylgalactose,
  rhamnose, mellibiose, lactose, isomaltose, cellobiose, glucose 6-acetate,
  glucose 6-benzyl ether, glucose 6-alkyl ether, 4-chloro-4-deoxyglucose
  6-acetate, maltotriose, 3-O-methyl alpha-D-glucose, maltose, maltopen-
  taose, maltohexaose, 6-chloro-6-deoxyglucose, L-arabinose, L-fucose,
  D-glycero-D-galactoheptose, lyxose, gluconic acid, xylitol, glycerol, ethanol, galactose 6-acetate, mannose, no acceptors are inosital, 2-deoxyglu-
  cose, glucosamine) [1]
2 Sucrose + xylose [1]
3 Sucrose + D-galactose [1]
4 Glucose 6-acetate + sucrose [1]

1

5 6-Deoxy-D-glucose + sucrose [1]
6 D-Glucose 6-benzoate + sucrose [1]
7 Raffinose + xylose [1]
8 Raffinose + galactose [1]

Product spectrum
1 D-aldose + D-aldosyl beta-D-fructoside (the products can act as acceptors for further reactions, leading to oligosaccharides or polysaccharides) [1]
2 beta-D-Fructofuranosyl(2→1)alpha-D-xylopyranoside + D-glucose [1]
3 beta-D-Fructofuranosyl(2→1)D-galactopyranoside + D-glucose [1]
4 Sucrose 6-acetate + D-glucose [1]
5 6-Deoxysucrose + D-glucose [1]
6 Sucrose 6-benzoate + D-glucose [1]
7 beta-D-Fructofuranosyl(2→1)alpha-D-xylopyranoside + alpha-D-galactopyranosyl-alpha-D-glucopyranoside [1]
8 beta-D-Fructofuranosyl(2→1) D-galactopyranoside + alpha-D-galactopyranosyl-alpha-D-glucopyranoside [1]

Inhibitor(s)

Cofactor(s)/prosthetic group(s)/activating agents

Metal compounds/salts

Turnover number (min^{-1})

Specific activity (U/mg)

K_m-value (mM)
200 (sucrose, Bacillus subtilis NCIB 11871) [1]

pH-optimum
5.4–6.0 [1]

pH-range

Temperature optimum (°C)
30 [1]

Temperature range (°C)

3 ENZYME STRUCTURE

Molecular weight

Subunits

Glycoprotein/Lipoprotein
–

4 ISOLATION/PREPARATION

Source organism
Bacillus subtilis (NCIB 11871, 11872, 11873) [1]; Erwinia sp. (previously Aerobacter levanicum) [1]

Source tissue

Localization in source
Exocellular [1]

Purification
Bacillus subtilis [1]

Crystallization
–

Cloned
–

Renatured
–

5 STABILITY

pH

Temperature (°C)
45 (at least 20 min stable) [1]

Oxidation

Organic solvent

General stability information

Storage

6 CROSSREFERENCES TO STRUCTURE DATABANKS

PIR/MIPS code

Brookhaven code

7 LITERATURE REFERENCES

[1] Rathbone, E.B., Hacking, A.J., Cheetham, P.S.J.: UK Patent Application, GB2145080 A (1985)

1 NOMENCLATURE

EC number
2.4.1.163

Systematic name
UDP-N-acetyl-D-glucosamine:beta-D-galactosyl-1,4-N-acetyl-beta-D-glu-
cosaminyl-1,3-beta-D-galactosyl-1,4-beta-D-glucosylceramide beta-1,3-
acetylglucosaminyltransferase

Recommended name
beta-Galactosyl-N-acetylglucosaminylgalactosylglucosylceramide beta-1,3-
acetylglucosaminyltransferase

Synonyms
Acetylglucosaminyltransferase, uridine diphosphoacetylglucosamine-ace-
tyllactosaminide beta1→3-poly-N-acetyllactosamine extension enzyme

CAS Reg. No.
85638-39-7

2 REACTION AND SPECIFICITY

Catalysed reaction
UDP-N-acetyl-D-glucosamine + beta-D-galactosyl-1,4-N-acetyl-beta-D-
glucosaminyl-1,3-beta-D-galactosyl-1,4-beta-D-glucosylceramide →
→ UDP + N-acetyl-D-glucosaminyl-1,3-beta-D-galactosyl-1,4-N-ace-
tyl-beta-D-glucosaminyl-1,3-beta-D-galactosyl-1,4-beta-D-glucosylceramide

Reaction type
Hexosyl group transfer

Natural substrates
UDP-N-acetyl-D-glucosamine + beta-D-galactosyl-1,4-N-acetyl-beta-D-
glucosaminyl-1,3-beta-D-galactosyl-1,4-beta-D-glucosylceramide (biosynthesis of Ii core glycosphingolipids) [1]

Substrate spectrum
1 UDP-N-acetyl-D-glucosamine + beta-D-galactosyl-1,4-N-acetyl-beta-
  D-glucosaminyl-1,3-beta-D-galactosyl-1,4-beta-D-glucosylceramide (i.e.
  neolactotetraosylceramide) [1]
2 UDP-N-acetyl-D-glucosamine + lactosylceramide (15% of the activity with
  neolactotetraosylceramide) [1]
3 UDP-N-acetyl-D-glucosamine + sialic acid depleted alpha$_1$-acid glycoprotein (30% of the activity with neolactotetraosylceramide) [1]

Product spectrum
1 UDP + N-acetyl-D-glucosaminyl-1,3-beta-D-galactosyl-1,4-N-acetyl-beta-
D-glucosaminyl-1,3-beta-D-galactosyl-1,4-beta-D-glucosylceramide
2 ?
3 ?

Inhibitor(s)
EDTA [1]

Cofactor(s)/prosthetic group(s)/activating agents
Detergent (reaction rate is optimal at detergent:protein ratio of 1.6 in Triton
CF-54 or T-100, 1:3 in Nonidet P-40) [1]

Metal compounds/salts
Mn^{2+} (required, K_m: 1.25 mM) [1]; Ca^{2+} (48% as effective as Mn^{2+}) [1]; Co^{2+}
(52% as effective as Mn^{2+}) [1]; More (Mn^{2+} cannot be replaced by Mg^{2+},
Zn^{2+}, Cu^{2+} or Cd^{2+}) [1]

Turnover number (min^{-1})

Specific activity (U/mg)

K_m-value (mM)
0.09 (neolactotetraosylceramide) [1]; 0.33 (UDP-N-acetyl-D-glucosamine,
solubilized enzyme) [1]; 2.5 (UDP-N-acetylglucosamine, membrane-bound
enzyme) [1]

pH-optimum
7.0–8.0 [1]

pH-range

Temperature optimum (°C)

Temperature range (°C)

3 ENZYME STRUCTURE

Molecular weight

Subunits

Glycoprotein/Lipoprotein
–

4 ISOLATION/PREPARATION

Source organism
Mouse [1]

Source tissue
  Lymphoma P-1798 cell line [1]

Localization in source
  Membrane [1]

Purification

Crystallization
  –

Cloned
  –

Renatured
  –

---

5 STABILITY

pH

Temperature (°C)

Oxidation

Organic solvent

General stability information

Storage

---

6 CROSSREFERENCES TO STRUCTURE DATABANKS

PIR/MIPS code

Brookhaven code

---

7 LITERATURE REFERENCES

[1] Basu, M., Basu, S.: J. Biol. Chem.,259,12557–12562 (1984)

1 NOMENCLATURE

EC number
2.4.1.164

Systematic name
UDP-N-acetyl-D-glucosamine:D-galactosyl-1,4-N-acetyl-beta-D-glucosaminyl-
1,3-beta-D-galactosyl-1,4-beta-D-glucosylceramide beta-1,6-N-acetyl-
glucosaminyltransferase

Recommended name
Galactosyl-N-acetylglucosaminylgalactosylglucosylceramide beta-1,6-N-
acetylglucosaminyltransferase

Synonyms
Acetylglucosaminyltransferase, uridine diphosphoacetylglucosamine-ace-
tyllactosaminide beta1→6-

CAS Reg. No.
85638-40-0

---

2 REACTION AND SPECIFICITY

Catalysed reaction
UDP-N-acetyl-D-glucosamine + D-galactosyl-1,4-N-acetyl-beta-D-glucos-
aminyl-1,3-beta-D-galactosyl-1,4-beta-D-glucosylceramide →
→ UDP + N-acetyl-D-glucosaminyl-1,6-beta-D-galactosyl-1,4-N-acetyl-
beta-D-glucosaminyl-1,3-beta-D-galactosyl-1,4-beta-D-glucosylceramide

Reaction type
Hexosyl group transfer

Natural substrates
UDP-N-acetyl-D-glucosamine + beta-D-galactosyl-1,4-N-acetyl-beta-D-glu-
cosaminyl-1,3-beta-D-galactosyl-1,4-beta-D-glucosylceramide (biosynthesis
of li core glycosphingolipids) [1]

Substrate spectrum
1 UDP-N-acetyl-D-glucosamine + beta-D-galactosyl-1,4-N-acetyl-beta-
   D-glucosaminyl-1,3-beta-D-galactosyl-1,4-beta-D-glucosylceramide (i.e.
   neolactotetraosylceramide) [1]
2 UDP-N-acetyl-D-glucosamine + lactosylceramide (15% of the activity with
   neolactotetraosylceramide) [1]
3 UDP-N-acetyl-D-glucosamine + sialic acid depleted alpha₁-acid glycoprotein (30% of the activity with neolactotetraosylceramide) [1]

Product spectrum

1 UDP + N-acetyl-D-glucosaminyl-1,6-beta-D-galactosyl-1,4-N-acetyl-
beta-D-glucosaminyl-1,3-beta-D-galactosyl-1,4-beta-D-glucosylceramide
2 ?
3 ?

Inhibitor(s)
EDTA [1]

Cofactor(s)/prosthetic group(s)/activating agents
Detergent (reaction rate is optimal at detergent:protein ratio of 1.6 in Triton
CF-54 or T-100, 1:3 in Nonidet P-40) [1]

Metal compounds/salts
Mn^{2+} (required, K_m: 1.25 mM) [1]; Ca^{2+} (48% as effective as Mn^{2+}) [1]; Co^{2+}
(52% as effective as Mn^{2+}) [1]; More (Mn^{2+} cannot be replaced by: Mg^{2+},
Zn^{2+}, Cu^{2+} or Cd^{2+}) [1]

Turnover number (min^{-1})

Specific activity (U/mg)

K_m-value (mM)
0.09 (neolactotetraosylceramide) [1]; 0.33 (UDP-N-acetyl-D-glucosamine,
solubilized enzyme) [1]; 2.5 (UDP-N-acetylglucosamine, membrane-bound
enzyme) [1]

pH-optimum
7.0–8.0 [1]

pH-range

Temperature optimum (°C)

Temperature range (°C)

---

3 ENZYME STRUCTURE

Molecular weight

Subunits

Glycoprotein/Lipoprotein
—

4 ISOLATION/PREPARATION

Source organism
   Mouse [1]

Source tissue
   Lymphoma P-1798 cell line [1]

Localization in source
   Membrane [1]

Purification

Crystallization
   –

Cloned
   –

Renatured
   –

---

5 STABILITY

pH

Temperature (°C)

Oxidation

Organic solvent

General stability information

Storage

---

6 CROSSREFERENCES TO STRUCTURE DATABANKS

PIR/MIPS code

Brookhaven code

---

7 LITERATURE REFERENCES

[1] Basu, M., Basu, S.: J. Biol. Chem.,259,12557–12562 (1984)

1 NOMENCLATURE

EC number
2.4.1.165

Systematic name
UDP-N-acetyl-D-galactosamine:N-acetylneuraminyl-2,3-alpha-D-galactosyl-
1,4-beta-D-glucosylceramide beta-1,4-N-acetylgalactosaminyltransferase

Recommended name
N-Acetylneuraminylgalactosylglucosylceramide beta-1,4-N-acetylgalac-
tosaminyltransferase

Synonyms
Acetylgalactosaminyltransferase, uridine diphosphoacetylgalactosamine-
acetylneuraminyl(alpha2→3)galactosyl(beta1→4)glucosyl beta1→4-

CAS Reg. No.
109136-50-7

2 REACTION AND SPECIFICITY

Catalysed reaction
UDP-N-acetyl-D-galactosamine + N-acetylneuraminyl-2,3-alpha-D-galac-
tosyl-1,4-beta-D-glucosylceramide →
→ UDP + N-acetyl-D-galactosaminyl-1,4-beta-N-acetylneuraminyl-2,3-
alpha-D-galactosyl-1,4-beta-D-glucosylceramide

Reaction type
Hexosyl group transfer

Natural substrates
UDP-N-acetyl-D-galactosamine + N-acetylneuraminyl-2,3-alpha-D-galac-
tosyl-1,4-beta-D-glucosylceramide (responsible for addition of the immuno-
dominant sugar of the Sda-isto-blood-group determinant) [2]

Substrate spectrum
1 UDP-N-acetyl-D-galactosamine + N-acetylneuraminyl-2,3-alpha-D-galac-
tosyl-1,4-beta-D-glucosylceramide (i.e. 3'-sialyllactose, transfers N-acetylgalactosamine to position C-4 of the galactosyl-residue of sialyllactose
[1], in beta-linkage [1, 2], strictly requires the presence of sialic acid in
alpha-2,3-linkage to subterminal galactose [2], efficient acceptors are
fetuin [1, 2], Tamm-Horsfall glycoprotein from human urine, $alpha_1$-acid

glycoprotein [1], human chorionic gonadotropin, glycophorin [2], poor acceptors are sialosylparagloboside, glycophorin A [1], no substrates are 6'-sialyllactose, 2'-fucosyllactose, ganglioside GM3, globotriglycosylceramide, globoside, paragloboside, lactose [1] or human transferrin [2]) [1, 2]

Product spectrum
   1 UDP + N-acetyl-D-galactosaminyl-1,4-beta-N-acetylneuraminyl-2,3-alpha-D-galactosyl-1,4-beta-D-glucosylceramide [1]

Inhibitor(s)
   ATP (at increased concentrations) [1]; Triton X-100 (weak [1], activation [2]) [1]

Cofactor(s)/prosthetic group(s)/activating agents
   Triton X-100 (activation [2], weak inhibition [1]) [2]

Metal compounds/salts
   Mn^{2+} (requirement, K_m: 16 mM [1]) [1, 2]; More (no activation by Co^{2+}, Cd^{2+}, Ni^{2+}, Fe^{2+}, Mg^{2+}, Ca^{2+}, Sn^{2+}, Zn^{2+}) [1]

Turnover number (min^{-1})

Specific activity (U/mg)

K_m-value (mM)
   0.055 (UDP-N-acetylgalactosamine) [1]; 1.1 (3'-sialyllactose) [1]

pH-optimum
   7.5 [1]

pH-range
   5.8–9.5 (about half-maximal activity at pH 5.8 and 9.5) [1]

Temperature optimum (°C)

Temperature range (°C)

---

3 ENZYME STRUCTURE

Molecular weight

Subunits

Glycoprotein/Lipoprotein
   –

4 ISOLATION/PREPARATION

Source organism
  Human [1–3]

Source tissue
  Blood plasma (group O) [1]; Large intestine (colon mucosa or colon carci-
  noma cell line CaCo-2, not in cell lines SW-948, SW-948FL, SW-480, SW-48,
  SW-1417, COLO-205, LOVO or HT-29) [2]; Kidney [3]

Localization in source

Purification

Crystallization
  –

Cloned
  –

Renatured
  –

---

5 STABILITY

pH

Temperature (°C)

Oxidation

Organic solvent

General stability information

Storage

---

6 CROSSREFERENCES TO STRUCTURE DATABANKS

PIR/MIPS code

Brookhaven code

---

7 LITERATURE REFERENCES

[1] Takeya, A., Hosomi, O., Kogure, T.: J. Biochem.,101,251–259 (1987)
[2] Malagolini, N., Dall'Olio, F., Serafini-Cessi, F.: Biochem. Biophys. Res. Commun.,
    180,681–686 (1991)
[3] Morton, J.A., Pickles, M.M., Vanhegan, R.I.: Immunol. Invest.,17,217–224 (1988)

---

1 NOMENCLATURE

EC number
2.4.1.166

Systematic name
Raffinose:raffinose alpha-D-galactosyltransferase

Recommended name
Raffinose-raffinose alpha-galactotransferase

Synonyms
Galactosyltransferase, raffinose (raffinose donor)
Raffinose:raffinose alpha-galactosyltransferase [1]

CAS Reg. No.
93389-38-9

2 REACTION AND SPECIFICITY

Catalysed reaction
2 Raffinose →
→ 1^F-alpha-D-galactosylraffinose + sucrose

Reaction type
Hexosyl group transfer

Natural substrates
More (involved in the accumulation of the tetrasaccharides lychnose and isolychnose in the leaves of Cerastium arvense and other Caryophyllaceae during late autumn) [1]

Substrate spectrum
1 Raffinose + raffinose [1]
2 Raffinose + raffinose [1]

Product spectrum
1 Lychnose (i.e. 1^F-alpha-D-galactosylraffinose) + sucrose [1]
2 Isolychnose + sucrose (isolychose is 3^F-alpha-galactosylraffinose, enzyme can also be found in Cerastium arvense leaves, probably another raffinose:raffinose alpha-D-galactosyltransferase) [1]

Inhibitor(s)

Cofactor(s)/prosthetic group(s)/activating agents

Metal compounds/salts

Turnover number (min^{-1})

Specific activity (U/mg)
  0.00025 (lychnose) [1]; 0.000316 (isolychnose) [1]

K_m-value (mM)

pH-optimum
  6.0 (both transferases) [1]

pH-range

Temperature optimum (°C)
  38 (assay at) [1]

Temperature range (°C)

---

3 ENZYME STRUCTURE

Molecular weight

Subunits

Glycoprotein/Lipoprotein
  –

---

4 ISOLATION/PREPARATION

Source organism
  Cerastium arvense (enzyme can also be isolated from other species of the
  Caryophyllaceae) [1]

Source tissue
  Leaf [1]

Localization in source

Purification
  Cerastium arvense (partial) [1]

Crystallization
  –

Cloned
  –

Renatured
  –

5 STABILITY

pH

Temperature (°C)

Oxidation

Organic solvent

General stability information

Storage
  −20°C, 6 months [1]

---

6 CROSSREFERENCES TO STRUCTURE DATABANKS

PIR/MIPS code

Brookhaven code

---

7 LITERATURE REFERENCES

[1] Hopf, H., Gruber, G., Zinn, A., Kandler, O.: Planta,162,283–288 (1984)

1 NOMENCLATURE

EC number
2.4.1.167

Systematic name
UDPgalactose:sucrose 6F-alpha-D-galactosyltransferase

Recommended name
Sucrose 6F-alpha-galactotransferase

Synonyms
Galactosyltransferase, uridine diphosphogalactose-sucrose 6F-alpha-
UDPgalactose:sucrose 6fru-alpha-galactosyltransferase [1]

CAS Reg. No.
92480-04-1

2 REACTION AND SPECIFICITY

Catalysed reaction
UDPgalactose + sucrose →
→ UDP + 6F-alpha-D-galactosylsucrose

Reaction type
Hexosyl group transfer

Natural substrates
More (probably involved in the biosynthesis of planteose in seeds of
Sesamum indicum) [1]

Substrate spectrum
1 UDPgalactose + sucrose [1]

Product spectrum
1 UDP + 6F-alpha-galactosylsucrose (i.e. planteose) [1]

Inhibitor(s)
5'-UMP (10 mM: 55% inhibition, 1 mM: 34% inhibition) [1]; More (UDP, UDP-
glucose, glucose 1-phosphate, glucose, galactose 1-phosphate, galactose
are not inhibitory up to 1 mM) [1]

Cofactor(s)/prosthetic group(s)/activating agents
More (no requirement for a sulfhydryl reagent) [1]; DTT (10 mM: increase of
activity by about 30%) [1]

Metal compounds/salts
   More (bivalent metal ions, e.g. Ca^{2+}, Cu^{2+}, Fe^{2+}, Mg^{2+}, Zn^{2+} have no effect)
   [1]

Turnover number (min^{-1})

Specific activity (U/mg)

K_m-value (mM)
   0.2–0.5 (UDPgalactose) [1]; 3.6–6 (sucrose) [1]

pH-optimum
   6.2 [1]

pH-range
   5.0–7.0 (30% of maximal activity at pH 5.0, 70% of maximal activity at
   pH 7.0) [1]

Temperature optimum (°C)
   38 (assay at) [1]

Temperature range (°C)

3 ENZYME STRUCTURE

Molecular weight

Subunits

Glycoprotein/Lipoprotein
   –

4 ISOLATION/PREPARATION

Source organism
   Sesamum indicum [1]

Source tissue
   Seeds [1]

Localization in source

Purification
   Sesamum indicum [1]

Crystallization
   –

Cloned
   –

Renatured

–

---

5 STABILITY

pH

Temperature (°C)

Oxidation

Organic solvent

General stability information
Freezing and thawing have no effect [1]

Storage
–20°C, 6 months [1]

---

6 CROSSREFERENCES TO STRUCTURE DATABANKS

PIR/MIPS code

Brookhaven code

---

7 LITERATURE REFERENCES

[1] Hopf, H., Spanfelner, M., Kandler, O.: Z. Pflanzenphysiol.,114,485–492 (1984)

---

1 NOMENCLATURE

EC number
   2.4.1.168

Systematic name
   UDPglucose:xyloglucan 1,4-beta-D-glucosyltransferase

Recommended name
   Xyloglucan 4-glucosyltransferase

Synonyms
   Glucosyltransferase, uridine diphosphoglucose-xyloglucan 4beta-
   Xyloglucan 4beta-D-glucosyltransferase
   Xyloglucan glucosyltransferase
   More (not identical with EC 2.4.1.12)

CAS Reg. No.
   80237-91-8

---

2 REACTION AND SPECIFICITY

Catalysed reaction
   Transfers a beta-D-glucosyl residue from UDPglucose on to a glucose
   residue in xyloglucan, forming a beta-1,4-D-glucosyl-D-glucose linkage

Reaction type
   Hexosyl group transfer

Natural substrates
   UDP-D-glucose + xyloglucan (involved in biosynthesis of xyloglucan in
   higher plants [2], responsible for the formation of the xyloglucan backbone
   [1]) [1, 2]

Substrate spectrum
   1 UDP-D-glucose + (glucosyl)xyloglucan (transfers a beta-D-glucosyl residue from UDPglucose on to a glucose residue in xyloglucan, other gluco-
   syl-acceptors are beta-1,3-glucan and xylan [1], no acceptors are
   beta-1,4-glucan [4], cello-oligosaccharides and fragment oligosaccharides from endoglucanase digest [1]) [1–7]

Product spectrum
   1 UDP + (glucosyl-glucosyl)xyloglucan (forms a beta-1,4-D-glucosyl-D-glu-
   cose linkage) [1–7]

Enzyme Handbook © Springer-Verlag Berlin Heidelberg 1996
Duplication, reproduction and storage in data banks are only
allowed with the prior permission of the publishers   1

Inhibitor(s)
   GTPglucose (weak) [1, 4]; TDPglucose (weak) [1]; UTP [1]; UTPxylose
   (above 0.03 mM) [2]; More (no inhibition by tunicamycin, ATP or GTP) [1]

Cofactor(s)/prosthetic group(s)/activating agents
   UDPxylose (activation, 0.01–0.03 mM) [1, 3–5]; More (fusicoccin, indole
   acetic acid and D_2O activate in vivo, via proton extrusion, in the presence of
   O_2 and sugar substrate, mannitol, vanadate or nigrecin inhibits this activation, no direct activation in vitro) [7]

Metal compounds/salts
   Mn^{2+} (activation [3, 4, 6], 3–5 mM [6]) [3, 4, 6]; Mg^{2+} (activation [3, 6, 7], 3–5
   mM [6], can replace Mn^{2+} [6], to some extent [3]) [3, 6, 7]; More (no activation by Ca^{2+}) [6]

Turnover number (min^{-1})

Specific activity (U/mg)

K_m-value (mM)
   0.056 (UDPglucose) [4]

pH-optimum
   6.5–7.0 [1]; 7.0–8.0 [4]

pH-range

Temperature optimum (°C)
   35 [1]

Temperature range (°C)

---

3 ENZYME STRUCTURE

Molecular weight

Subunits

Glycoprotein/Lipoprotein
   –

---

4 ISOLATION/PREPARATION

Source organism
   Glycine max (soy bean) [1, 2, 4, 5]; Pisum sativum (pea, cv. Alaska [3, 6, 7]
   or Caprice [3]) [3, 6, 7]

Source tissue
Cell suspension culture [1, 4, 5]; Hypocotyls (elongation region) [2]; Seedlings [3, 7]

Localization in source
Membrane-bound [1, 4–6]; Golgi membranes [2, 3, 7]

Purification

Crystallization
–

Cloned
–

Renatured
–

---

5 STABILITY

pH

Temperature (°C)

Oxidation

Organic solvent

General stability information
DTT, 1 mM, EDTA, 1 mM, sucrose, 0.4 M, bovine serum albumin, 0.1%, stabilize [1]

Storage
0°C, crude membrane-bound enzyme preparation, 1 day [4]

---

6 CROSSREFERENCES TO STRUCTURE DATABANKS

PIR/MIPS code

Brookhaven code

---

7 LITERATURE REFERENCES

[1] Hayashi, T., Matsuda, K.: J. Biol.Chem.,256,11117–11122 (1981)
[2] Hayashi, T., Koyama, T., Matsuda, K.: Plant Physiol.,87,341–345 (1988)
[3] White, A.R., Xin, Y., Pezeshk, V.: Biochem. J.,294,231–238 (1993)
[4] Hayashi, T., Matsuda, K.: Plant Cell Physiol.,22,1571–1584 (1981)
[5] Hayashi, T., Nakajima, T., Matsuda, K.: Agric. Biol. Chem.,48,1023–1027 (1984)
[6] Gordon, R., MacLachlan, G.: Plant Physiol.,91,373–378 (1989)
[7] Ray, P.M.: Plant Physiol.,78,466–472 (1985)

---

1 NOMENCLATURE

EC number
2.4.1.169

Systematic name
UDP-D-xylose:xyloglucan 1,6-alpha-D-xylosyltransferase

Recommended name
Xyloglucan 6-xylosyltransferase

Synonyms
Xylosyltransferase, uridine diphosphoxylose-xyloglucan 6alpha-
Xyloglucan 6-alpha-D-xylosyltransferase

CAS Reg. No.
80238-01-3

---

2 REACTION AND SPECIFICITY

Catalysed reaction
Transfers an alpha-D-xylosyl residue from UDP-D-xylose to a glucose residue in xyloglucan, forming an alpha-1,6-D-xylosyl-D-glucose linkage

Reaction type
Hexosyl group transfer

Natural substrates
UDP-D-xylose + xyloglucan (involved in xyloglucan biosynthesis of higher plants [2], responsible for xyloglucan side-chain formation [1]) [1, 2]

Substrate spectrum
1 UDP-D-xylose + (glucosyl)xyloglucan (transfers an alpha-D-xylosyl residue from UDP-D-xylose to a glucose residue in xyloglucan [1–4], xylosyl-transfer is closely linked to glucosyl-transfer (EC 2.4.1.168) [4], GDP-D-glucose or GDP-D-mannose cannot replace UDPxylose [5], other xylosyl-acceptors are beta-1,3-glucan and xylan [1], no acceptors are cello-oligosaccharides and fragment oligosaccharides from endo-glucanase digest [1]) [1–5]

Product spectrum
1 UDP + (xylosyl-glucosyl)xyloglucan (forms an alpha-1,6-D-xylosyl-D-glucose linkage) [1–5]

Inhibitor(s)
EDTA (weak) [1]; Tris-HCl buffer [1]; Phosphate buffer [1]; Detergents (no solubilization possible due to this inhibition) [1]; GDPglucose (weak) [4]; Cu^{2+} [5]; More (no inhibition by tunicamycin, ATP or GTP) [1]

Cofactor(s)/prosthetic group(s)/activating agents
UDPglucose (activation, xylose is effectively incorporated in the presence of UDPglucose, the transfer must be preceded by elongation of the beta-1,4-glucan-backbone, because xylosyl residues constitute the side chains, in concentrations exceeding UDPxylose, no activation by CDPglucose or ADP-glucose [1], 2 mM [5]) [1, 2, 5]; TDPglucose (activation, as effective as UDPglucose, in concentrations exceeding UDPxylose) [1]; GDP-D-glucose (activation [1, 5], about half as effective as UDPglucose, in concentrations exceeding UDPxylose [1], protection [5]) [1, 5]; GDP-D-mannose (activation, protection) [5]

Metal compounds/salts
Mn^{2+} (activation [1, 3, 5], 10 mM [1]) [1, 3, 5]; Mg^{2+} (activation, can replace Mn^{2+} to some extent) [1, 5]; Ca^{2+} (activation, can replace Mn^{2+} to some extent) [5]; Co^{2+} (activation, can replace Mn^{2+} to some extent) [5]

Turnover number (min^{-1})

Specific activity (U/mg)

K_m-value (mM)
More (kinetic study) [1]

pH-optimum
6.0 (incorporation of xylosyl residues into polymeric acceptors) [1]; 6.5–7.0 (UDPglucose + UDPxylose) [1]; 7.0 (broad) [5]

pH-range

Temperature optimum (°C)
35 [1]

Temperature range (°C)

3 ENZYME STRUCTURE

Molecular weight

Subunits

Glycoprotein/Lipoprotein
–

4 ISOLATION/PREPARATION

Source organism
   Glycine max (soy bean) [1, 2, 4]; Pisum sativum (pea, cv. Alaska or Caprice)
   [3]; Phaseolus vulgaris (dwarf-french-bean, cv. Canadian Wonder) [5]

Source tissue
   Cell suspension culture [1, 4, 5]; Hypocotyls (elongation region) [2]; Seedlings [3]

Localization in source
   Membrane-bound [1, 4, 5]; Golgi membranes [2, 3]

Purification

Crystallization
   −

Cloned
   −

Renatured
   −

5 STABILITY

pH

Temperature (°C)

Oxidation

Organic solvent

General stability information
   DTT, 1 mM, EDTA, 1 mM, sucrose, 0.4 M, bovine serum albumin, 0.1%,
   stabilize [1]

Storage
   0°C, crude membrane-bound enzyme preparation, 1 day [4]

6 CROSSREFERENCES TO STRUCTURE DATABANKS

PIR/MIPS code

Brookhaven code

7 LITERATURE REFERENCES

[1] Hayashi, T., Matsuda, K.: J. Biol.Chem.,256,11117–11122 (1981)
[2] Hayashi, T., Koyama, T., Matsuda, K.: Plant Physiol.,87,341–345 (1988)
[3] White, A.R., Xin, Y., Pezeshk, V.: Biochem. J.,294,231–238 (1993)
[4] Hayashi, T., Matsuda, K.: Plant Cell Physiol.,22,1571–1584 (1981)
[5] Campbell, R.E., Brett, C.T., Hillman, J.R.: Biochem. J.,253,795–800 (1988)

1 NOMENCLATURE

EC number
2.4.1.170

Systematic name
UDPglucose:isoflavone 7-O-beta-D-glucosyltransferase

Recommended name
Isoflavone 7-O-glucosyltransferase

Synonyms
Glucosyltransferase, uridine diphosphoglucose-isoflavone 7-O-
UDPglucose-favonoid 7-O-glucosyltransferase
UDPglucose:isoflavone 7-O-glucosyltransferase

CAS Reg. No.
97089-62-8

2 REACTION AND SPECIFICITY

Catalysed reaction
UDPglucose + isoflavone →
→ UDP + isoflavone 7-O-beta-D-glucoside

Reaction type
Hexosyl group transfer

Natural substrates
UDPglucose + isoflavone [1]

Substrate spectrum
1 UDPglucose + biochanin A [1]
2 UDPglucose + formononetin [1]
3 UDPglucose + genistein (slowly) [1]
4 UDPglucose + daidzein (slowly) [1]
5 More (no reaction as acceptor: isoflavanones, flavones, flavanones, flavanols, coumarin) [1]

Product spectrum
1 UDP + isoflavone 7-O-beta-glucoside [1]
2 ?
3 ?
4 ?
5 ?

Inhibitor(s)

Cofactor(s)/prosthetic group(s)/activating agents

Metal compounds/salts

Turnover number (min^{-1})

Specific activity (U/mg)
    More [1]

K_m-value (mM)
    0.012 (biochanin A) [1]; 0.024 (formononetin) [1]; 0.2 (UDPglucose) [1]

pH-optimum
    8.5–9.0 [1]

pH-range

Temperature optimum (°C)
    30 (assay at) [1]

Temperature range (°C)

---

3 ENZYME STRUCTURE

Molecular weight
    42000 (Cicer arietinum, sucrose gradient ultracentrifugation) [1]
    50000 (Cicer arietinum, gel filtration) [1]

Subunits

Glycoprotein/Lipoprotein
    −

---

4 ISOLATION/PREPARATION

Source organism
    Cicer arietinum [1]

Source tissue
    Root [1]

Localization in source

Purification
    Cicer arietinum [1]

Crystallization
    −

Cloned

–

Renatured

–

5 STABILITY

pH
8.5–9.0 (optimal stability) [1]

Temperature (°C)

Oxidation

Organic solvent

General stability information
2-Mercaptoethanol, 40 mM, stabilizes during purification [1]; Dithioerythritol, 10 mM, effectively stabilizes during purification [1]

Storage
–20°C, 20% v/v glycerol, stable for several months [1]

6 CROSSREFERENCES TO STRUCTURE DATABANKS

PIR/MIPS code

Brookhaven code

7 LITERATURE REFERENCES

[1] Köster, J., Barz, W.: Arch. Biochem. Biophys.,212,98–104 (1981)

1 NOMENCLATURE

EC number
2.4.1.171

Systematic name
UDPglucose:methyl-ONN-azoxymethanol beta-D-glucosyltransferase

Recommended name
Methyl-ONN-azoxymethanol beta-D-glucosyltransferase

Synonyms
Cycasin synthase
Glucosyltransferase, uridine diphosphoglucose-methylazoxymethanol
UDPglucose-methylazoxymethanol glucosyltransferase [1]

CAS Reg. No.
99283-65-5

---

2 REACTION AND SPECIFICITY

Catalysed reaction
UDPglucose + CH_3-N(O)=N-CH_2OH →
→ UDP + cycasin

Reaction type
Hexosyl group transfer

Natural substrates

Substrate spectrum
1 UDPglucose + methylazoxymethanol [1]

Product spectrum
1 UDP + cycasin [1]

Inhibitor(s)

Cofactor(s)/prosthetic group(s)/activating agents

Metal compounds/salts

Turnover number (min^{-1})

Specific activity (U/mg)

K_m-value (mM)
0.30 (methylazoxymethanol) [1]; 1.58 (UDPglucose) [1]

---

pH-optimum
   8.3 [1]

pH-range

Temperature optimum (°C)
   30 (assay at) [1]

Temperature range (°C)

---

3 ENZYME STRUCTURE

Molecular weight

Subunits

Glycoprotein/Lipoprotein
   –

---

4 ISOLATION/PREPARATION

Source organism
   Cycas revoluta Thunb. (Japanese cycad) [1]

Source tissue
   Leaf [1]

Localization in source

Purification
   Cycas revoluta Thunb. [1]

Crystallization
   –

Cloned
   –

Renatured
   –

---

5 STABILITY

pH

Temperature (°C)

Oxidation

Organic solvent

General stability information

Storage

6 CROSSREFERENCES TO STRUCTURE DATABANKS

PIR/MIPS code

Brookhaven code

7 LITERATURE REFERENCES

[1] Tadera, K., Yagi, F., Arima, M., Kobayashi, A.: Agric. Biol. Chem.,49,2827–2828 (1985)

1 NOMENCLATURE

EC number
2.4.1.172

Systematic name
UDPglucose:salicyl-alcohol beta-D-glucosyltransferase

Recommended name
Salicyl-alcohol beta-D-glucosyltransferase

Synonyms
Glucosyltransferase, uridine diphosphoglucose-salicyl alcohol 2-
UDPglucose:salicyl alcohol phenyl-glucosyltransferase [2]

CAS Reg. No.
89400-32-8

---

2 REACTION AND SPECIFICITY

Catalysed reaction
UDPglucose + salicyl alcohol →
→ UDP + salicin

Reaction type
.Hexosyl group transfer

Natural substrates

Substrate spectrum
1 UDPglucose + salicyl alcohol [1, 2]

Product spectrum
1 UDP + salicin [1, 2]

Inhibitor(s)
Cu^{2+} (2 mM, 80% inhibition) [1]; Zn^{2+} (2 mM, 60% inhibition) [1]; Mn^{2+}
(2 mM, 20% inhibition) [1]; p-Chloromercuribenzoate (10 mM, 25% inhibition) [1]; N-Ethylmaleimide (10 mM, 35% inhibition) [1]

Cofactor(s)/prosthetic group(s)/activating agents

Metal compounds/salts

Turnover number (min^{-1})

Specific activity (U/mg)
   0.165 [1]

K$_m$-value (mM)
   0.031 (UDPglucose) [1]; 0.11 (salicyl alcohol) [1]; 0.53 (salicyl alcohol) [2];
   0.64 (UDPglucose) [2]

pH-optimum
   9.0 [2]; 9.0–9.5 [1]

pH-range
   7.5–10.2 (half maximal activity at pH 7.5 and 10.2) [2]; 6.0–10.5 (45% of
   maximal activity at pH 6.0, 10% of maximal activity at pH 10.5) [1]

Temperature optimum (°C)
   30 (assay at) [2]; 37 (assay at) [1]; 50 [2]

Temperature range (°C)
   30–55 (20% of maximal activity at 30°C, 35% of maximal activity at 55°C) [2]

---

3 ENZYME STRUCTURE

Molecular weight
   51000 (Gardenia jasminoides, gel filtration) [1]

Subunits

Glycoprotein/Lipoprotein
   –

---

4 ISOLATION/PREPARATION

Source organism
   Gardenia jasminoides Ellis [1, 2]

Source tissue
   Leaf (cell culture) [1, 2]

Localization in source

Purification
   Gardenia jasminoides (partial) [1, 2]

Crystallization
   –

Cloned

–

Renatured

–

5 STABILITY

pH
7.0–9.0 (stable) [1]

Temperature (°C)
50 (5 min, 10% remaining activity) [1]

Oxidation

Organic solvent

General stability information

Storage
4°C, 2 weeks, 40% loss of activity, or –20°C, 2 weeks, 40% loss of activity
[1]; –20°C, 1 week, 20% loss of activity, –20°C, 3 weeks, 50% loss of activity
[2]

6 CROSSREFERENCES TO STRUCTURE DATABANKS

PIR/MIPS code

Brookhaven code

7 LITERATURE REFERENCES

[1] Mizukami, H., Terao, T., Ohashi, H.: Planta Med.,1985,104–107 (1985)
[2] Terao, T., Ohashi, H., Mizukami, H.: Plant Sci. Lett.,33,47–52 (1984)

1 NOMENCLATURE

EC number
2.4.1.173

Systematic name
UDPglucose:sterol 3-O-beta-D-glucosyltransferase

Recommended name
Sterol 3beta-glucosyltransferase

Synonyms
UDPG:sterol glucosyltransferase
UDP-glucose-sterol beta-glucosyltransferase [5]
Sterol:UDPG glucosyltransferase [7]
UDPG-SGTase [15]
Glucosyltransferase, uridine diphosphoglucose-poriferasterol
Glucosyltransferase, uridine diphosphoglucose-sterol
Sterol glucosyltransferase
Sterol-beta-D-glucosyltransferase
UDP-glucose-sterol glucosyltransferase
Uridine diphosphoglucose-sterol glucosyltransferase
More (not identical with EC 2.4.1.192 or EC 2.4.1.193)

CAS Reg. No.
9075-00-7; 123940-38-5

2 REACTION AND SPECIFICITY

Catalysed reaction
UDPglucose + a sterol →
→ UDP + a sterol 3-beta-D-glucoside

Reaction type
Hexosyl group transfer

Natural substrates
UDPglucose + poriferasterol (synthesis of poriferasterol monoglucoside may
be an important process of differentiation) [4]

Substrate spectrum

1 UDPglucose + sitosterol (beta-sitosterol [14], sitosterol as well as poriferasterol are best substrates [4], best substrate [3], 87% of the activity with epiandrosterone [7]) [1–4, 7–10, 12, 14, 15, 17]

2 UDPglucose + poriferasterol (sitosterol as well as poriferasterol are best substrates [4]) [4, 8]

3 UDPglucose + epiandrosterone (epiandrosterone as well as stigmasterol are best substrates) [7]

4 UDPglucose + stigmasterol (14% of the activity with poriferasterol [4], 50% of the activity with sitosterol [3], epiandrosterone as well as stigmasterol are best substrates [7]) [3, 4, 7, 8, 14]

5 UDPglucose + campesterol (28% of the activity with sitosterol [3], 80% of the activity with poriferasterol [4]) [3, 4, 7]

6 UDPglucose + cholesterol (12% of the activity with poriferasterol [4], 91% of the activity with epiandrosterone [7], 12% of the activity with sitosterol [3]) [3, 4, 6–8, 11, 12, 14]

7 UDPglucose + alpha-spinasterol (31% of the activity with sitosterol) [3]

8 UDPglucose + 5alpha-cholestanol (8% of the activity with sitosterol [3]) [3, 8]

9 UDPglucose + 7-dehydrocholesterol (6% of the activity with sitosterol) [3]

10 UDPglucose + ergosterol (11% of the activity with sitosterol [3], 9% of the activity with poriferasterol [4]) [3, 4, 6, 8]

11 UDPglucose + pregnenolone (72% of the activity with epiandrosterone [7], reaction catalyzed with Physarum polycephalum enzyme, not with Sinapis alba enzyme [8]) [7, 8]

12 UDPglucose + testosterone (10.4% of the activity with epiandrosterone) [7]

13 More (Sinapis albe enzyme shows absolute specificity towards UDPglucose, Physarum polycephalum enzyme also utilizes CDPglucose and TDPglucose at an about six-times lower rate, overview: specificity of Sinapis alba and Physarum polycephalum enzyme [8], sterols possessing an alkyl group at C-24 are better substrates than C_{27} sterols [3], the presence of a DELTA^{22} double bond decreases affinity of the sterol for the enzyme [3], presence and localization of double bonds in the ring system exert a pronounced effect on the rate of glucosylation [3], DELTA^5-sterols are glucosylated at higher rate than DELTA^7-sterols [3], stanols and sterols with conjugated double bonds in ring B are poor substrates [3], not: 4-methylsterols [3], specific for UDPglucose [13, 14], only slight activity with ADPglucose, GDPglucose, CDPglucose, TDPglucose, UDPgalactose, UDPmannose [14]) [3, 8, 13, 14]

Product spectrum
  1 UDP + sitosterol 3-beta-D-monoglucoside [3]
  2 UDP + poriferasterol monoglucoside [4]
  3 Epiandrosterone 3beta-glucoside + UDP [7]
  4 ?
  5 ?
  6 ?
  7 ?
  8 ?
  9 ?
 10 ?
 11 ?
 12 ?
 13 ?

Inhibitor(s)
  Ca^{2+} (0.001–10 mM: stimulates [8], above 0.1 mM slight inhibition [3]) [3]; Mg^{2+} (0.001–10 mM: stimulation [8], above 0.1 mM slight inhibition [3]) [3]; Zn^{2+} (strong [3, 8]) [3, 7, 8, 10, 12, 13]; Hg^{2+} (strong [3, 8]) [3, 8, 10]; PCMB [3, 10]; UDP [3, 7, 10]; UTP [3, 10]; UMP [10]; Triton X-100 (stimulates [1, 2, 11, 15]) [3, 10]; Tween 20 (stimulation [11]) [3]; Tween 60 (stimulation [11]) [3]; Sodium deoxycholate [11]; Deoxycholate [3]; Sodium taurocholate [11]; Stigmasterol (in presence of poriferasterol) [4]; Mn^{2+} [7]; NH_4^+ [7]; Steryl glucosides [7]; Cu^{2+} [10]; EDTA (slight [12], slight stimulation [3]) [12]; EGTA (slight) [12]; More (partial loss of activity after treatment with phospholipase A, C or D, phosphatidylethanolamine completely restores activity after phospholipase A treatment, phosphatidylcholine and phosphatidylserine are without effect, after phospholipase C or D treatment each phospholipid brings about a partial recovery of activity but phosphatidylethanolamine is superior) [12]

Cofactor(s)/prosthetic group(s)/activating agents
  Dithiothreitol (stimulates) [3]; 2-Mercaptoethanol (slight stimulation [10], stimulation [3]) [3, 7, 10]; ATP (slight activation) [3]; Phospholipids (stimulated by negatively charged phospholipids [5, 9], reconstitution of enzyme activity into unilamellar lipid vesicles [9], phosphatidylcholine or phosphatidylglycerol decreases K_m of sterol and increases V_{max} [5], stimulation by phosphatidylethanolamine, phosphatidylcholine or phosphatidylserine [12], activity depends on phospholipids [15, 17], no enhancement of activity by lipids in presence and absence of Triton X-100 [11]) [5, 9, 12, 15, 17]; Triton X-100 (plasma-membrane bound enzyme is activated by Triton X-100 and other non-ionic detergents, either at low (ratio Triton X-100/protein of 0.2 w/w) or at high concentration (ratio Triton X-100/protein of 2), at intermediate concentration (ratio Triton X-100/protein of 0.4–0.8) this stimulation is abolished [15], stimulates [1, 2, 11], maximum stimulation at 0.25% [1], 6.5fold

stimulation at 0.1% [11], inhibition [3, 10]) [1, 2, 11, 15]; Lecithin (stimulates [3], stimulation by plant lecithin, egg lecithin has no effect [14]) [3, 14]; EDTA (slight stimulation [3], activates [7], slight inhibition [12]) [3, 7]; Ethanol (7–15%: stimulation, no inhibition up to 20%) [6]; Tween 20 (stimulates 2fold at 0.1% [11], inhibition [3]) [11]; Tween 40 (70% enhancement at 0.05%, 35% enhancement at 0.1%) [11]; Tween 60 (70% enhancement at 0.05%, 35% enhancement at 0.1% [11], inhibition [3]) [11]; More (hydroxyl ions or phospholipase A_2 can be used in place of detergents to reveal latent enzymatic sites) [16]

Metal compounds/salts

More (no metal ion requirement [12], no activation by divalent metals [3, 10]) [3, 10, 12]; Mg^{2+} (stimulates [12, 13], 0.001–10 mM: stimulation [8], no activation [10]) [8, 12, 13]; Ca^{2+} (0.001–10 mM stimulate [8], stimulates [12, 13]) [8, 12, 13]

Turnover number (min⁻¹)

Specific activity (U/mg)

0.0085 [14]; More (effect of phosphatidylcholine and phosphatidylglycerol) [5]

K_m-value (mM)

0.0048 (poriferasterol) [4]; 0.005 (sitosterol, Sinapis alba) [8]; 0.0056 (sitosterol) [3]; 0.08 (UDPglucose, Sinapis alba) [8]; 0.12 (UDPglucose) [4]; 0.18 (UDPglucose) [3]; More [17]

pH-optimum

7.0 [4]; 7.2 [3]; 7.5 [7]; 8 (near) [13]; 8.3 [10]

pH-range

Temperature optimum (°C)

30 (assay at [3, 7, 8, 15], activity maximum [10]) [3, 7, 8, 10, 15]

Temperature range (°C)

3 ENZYME STRUCTURE

Molecular weight

70000 (Physarum polycephalum, gel filtration) [3]
72000 (Physarum polycephalum, gel filtration) [4]
140000 (Sinapis alba, gel filtration) [8]

Subunits

Glycoprotein/Lipoprotein

—

4 ISOLATION/PREPARATION

Source organism
Cotton [14]; Avena sativa (oat) [1, 2]; Physarum polycephalum [3, 4, 8];
Maize [5, 9, 15–17]; Candida bogoriensis [6]; Digitalis purpurea [7]; Sinapis
alba [8, 11]; Asparagus plumosus [10]; Pisum sativum (pea) [12, 13]

Source tissue
Seedlings (located predominantly in axis tissue of etiolated seedlings [12])
[8, 12, 13]; Leaf [1, 2, 7]; Coleoptiles (etiolated [15–17]) [5, 9, 15–17]; Cell
culture [7]; Shoots [10]; Seeds [14]; Developing fibers [14]; More (no activity
in haploid cells, enzyme activity expressed after conjugation) [4]

Localization in source
Membrane (in membranous and soluble fractions [10], weakly bound [3],
associated with plasma membrane [4], bound to plasma membrane [15,
16], membrane vesicles [14]) [1–4, 8, 10, 12, 14–16]; Soluble (in membranous and soluble fractions) [10]; Microplasmodia [3]; Microsomes [17]

Purification
Maize (partial) [5]; Cotton (partial) [14]; Avena sativa [1]; Physarum
polycephalum (partial) [3, 4, 8]; Digitalis purpurea [7]; Sinapis alba (partial)
[8]; Pisum sativum [12]

Crystallization
–

Cloned
–

Renatured
(reconstitution of enzyme activity into unilamellar lipid vesicles) [9]

---

5 STABILITY

pH

Temperature (°C)

Oxidation

Organic solvent

General stability information

Storage

6 CROSSREFERENCES TO STRUCTURE DATABANKS

PIR/MIPS code

Brookhaven code

---

7 LITERATURE REFERENCES

[1] Kalinowska, M., Wojciechowski, Z.A.: Phytochemistry,25,2525–2529 (1986)
[2] Kalinowska, M., Wojciechowski, Z.A.: Phytochemistry,26,353–357 (1987)
[3] Wojciechowski, J., Zimowski, J., Tyski, S.: Phytochemistry,16,911–914 (1977)
[4] Murakami-Murofushi, K., Ohta, J.: Biochim. Biophys. Acta,992,412–415 (1989)
[5] Ullmann, P., Bouvier-Nave, P., Benveniste, P.: Plant Physiol.,85,51–55 (1987)
[6] Esders, T.W., Light, R.J.: J. Biol. Chem.,247,7494–7497 (1972)
[7] Yoshikawa, T., Furuya, T.: Phytochemistry,18,239–241 (1979)
[8] Wojciechowski, Z.A., Zimowski, J., Zimowski, J.G., Lyznik, A.: Biochim. Biophyş. Acta,570,363–370 (1979)
[9] Ury, A., Benveniste, P., Bouvier-Nave, P.: Plant Physiol.,91,567–573 (1989)
[10] Paczkowski, C., Zimowski, J., Krawczyk, D., Wojciechowski, A.: Phytochemistry, 29,63–70 (1990)
[11] Kalinowska, M., Wojciechowski, Z.A.: Phytochemistry,25,45–49 (1986)
[12] Fang, T.-Y., Baisted, D.J.: Phytochemistry,15,273–278 (1976)
[13] Staver, M.J., Glick, K., Baisted, D.J.: Biochem. J.,169,297–303 (1978)
[14] Forsee, W.T., Laine, R.A., Elbein, A.D.: Arch. Biochem. Biophys.,161,248–259 (1974)
[15] Bouvier-Nave, P., Ullmann, P., Rimmele, D., Benveniste, P.: Plant Sci. Lett.,36,19–27 (1984)
[16] Quantin-Martenot, E., Benveniste, P., Hartmann, M.A., Bouvier-Nave, P.: Plant Sci. Lett.,29,305–314 (1983)
[17] Ullmann, P., Rimmele, D., Benveniste, P., Bouvier-Nave, P.: Plant Sci. Lett.,36,29–36 (1984)

1 NOMENCLATURE

EC number
2.4.1.174

Systematic name
UDP-N-acetyl-D-galactosamine:D-glucuronyl-1,3-beta-D-galactosylproteogly-
can beta-1,4-N-acetylgalactosaminyltransferase

Recommended name
Glucuronylgalactosylproteoglycan beta-1,4-N-acetylgalactosaminyltrans-
ferase

Synonyms
N-Acetylgalactosaminyltransferase I
Acetylgalactosaminyltransferase, uridine diphosphoacetylgalactos-
amine-chondroitin, I

CAS Reg. No.
96189-39-8

2 REACTION AND SPECIFICITY

Catalysed reaction
UDP-N-acetyl-D-galactosamine + beta-D-glucuronyl-1,3-D-galac-
tosylproteoglycan →
→ UDP + N-acetyl-D-galactosaminyl-1,4-beta-D-glucuronyl-1,3-beta-D-galac-
tosylproteoglycan

Reaction type
Hexosyl group transfer

Natural substrates
UDP-N-acetyl-D-galactosamine + glucuronylgalactosyl glycosides (involved
in chondroitin sulfate biosynthesis: transfer of the first N-acetylgalactosamine residue) [1]

Substrate spectrum
1 UDP-N-acetyl-D-galactosamine + glucuronylgalactosyl glycosides (acceptors are tri- or disaccharides with terminal glucuronyl residue and a
galactosyl residue in penultimate position, transfers acetylgalactosamine
in beta-linkage, no substrates are glucuronic acid or disaccharides with
terminal galactosyl residues) [1]
2 UDP-N-acetyl-D-galactosamine + beta-D-glucuronyl-1,3-D-galactosyl-1,3-
D-galactose (best substrate) [1]

3 UDP-N-acetyl-D-galactosamine + beta-D-glucuronyl-1,3-D-galactosyl-1,4-glucose [1]

4 UDP-N-acetyl-D-galactosamine + beta-D-glucuronyl-1,3-D-galactose (minimum requirement for acceptor structure) [1]

Product spectrum
1 UDP + N-acetyl-D-galactosaminylglucuronylgalactosyl glycosides [1]
2 UDP + N-acetyl-D-galactosaminyl-beta-D-glucuronyl-1,3-D-galactosyl-1,3-D-galactose [1]
3 UDP + ?
4 UDP + ?

Inhibitor(s)

Cofactor(s)/prosthetic group(s)/activating agents

Metal compounds/salts
Mn^{2+} (requirement) [1]

Turnover number (min^{-1})

Specific activity (U/mg)
More (13.1 cpm/min/mg protein) [1]

K_m-value (mM)

pH-optimum
6.4 [1]

pH-range

Temperature optimum (°C)
30 (assay at) [1]

Temperature range (°C)

3 ENZYME STRUCTURE

Molecular weight

Subunits

Glycoprotein/Lipoprotein
–

4 ISOLATION/PREPARATION

Source organism
Bovine (calf) [1]

Source tissue
Arterial tissue (Aorta thoracica) [1]

Localization in source
Microsomes [1]

Purification
Bovine (partially solubilized with a buffer containing 2% Triton X-100, 20% glycerol and phospholipase A_2) [1]

Crystallization
–

Cloned
–

Renatured
–

---

5 STABILITY

pH

Temperature (°C)
50 (1 h, 10% loss of activity) [1]

Oxidation

Organic solvent

General stability information

Storage

---

6 CROSSREFERENCES TO STRUCTURE DATABANKS

PIR/MIPS code

Brookhaven code

---

7 LITERATURE REFERENCES

[1] Rohrmann, K., Niemann, R., Buddecke, E.: Eur. J. Biochem.,148,463–469 (1985)

1 NOMENCLATURE

EC number
  2.4.1.175

Systematic name
  UDP-N-acetyl-D-galactosamine:D-glucuronyl-N-acetyl-1,3-beta-D-galac-
  tosaminylproteoglycan beta-1,4-N-acetylgalactosaminyltransferase

Recommended name
  Glucuronyl-N-acetylgalactosaminylproteoglycan beta-1,4-N-acetylgalac-
  tosaminyltransferase

Synonyms
  N-Acetylgalactosaminyltransferase II
  Acetylgalactosaminyltransferase, uridine diphosphoacetylgalactosamine-
  chondroitin, II

CAS Reg. No.
  96189-40-1

2 REACTION AND SPECIFICITY

Catalysed reaction
  UDP-N-acetyl-D-galactosamine + D-glucuronyl-N-acetyl-1,3-beta-D-galac-
  tosaminylproteoglycan →
  → UDP + N-acetyl-D-galactosaminyl-1,4-beta-D-glucuronyl-N-acetyl-1,3-beta-
  D-galactosaminylproteoglycan

Reaction type
  Hexosyl group transfer

Natural substrates
  UDP-N-acetyl-D-galactosamine + chondroitin oligosaccharide (involved in
  chondroitin sulfate biosynthesis: elongation of growing chondroitin sulfate
  chain) [1]

Substrate spectrum
  1 UDP-N-acetyl-D-galactosamine + chondroitin oligosaccharide (acceptors are even-numbered chondroitin oligosaccharides with terminal glu-
  curonyl-N-acetyl-1,3-beta-D-galactosamine sequence at non-reducing
  end, transfers N-acetyl-galactosamine to non-reducing glucuronic acid
  residues in beta-glycosidic linkage, no substrates are glucuronyl-N-ace-
  tyl-1,3-beta-galactosamine or glucuronyl-N-acetyl-1,3-beta-D-galac-
  tosaminyl-N-acetyl-1,3-beta-D-galactosaminyl-N-acetyl-galactosaminol) [1]

2 UDP-N-acetyl-D-galactosamine + (glucuronyl-N-acetyl-1,3-beta-D-ga-
lactosamine)$_5$ (i.e. chondroitin decasaccharide, best substrate) [1]

3 UDP-N-acetyl-D-galactosamine + (glucuronyl-N-acetyl-1,3-beta-D-ga-
lactosamine)$_4$ (i.e. chondroitin octasaccharide, glycosylated at 75% the
rate of decamer glycosylation) [1]

4 UDP-N-acetyl-D-galactosamine + (glucuronyl-N-acetyl-1,3-beta-D-ga-
lactosamine)$_3$ (i.e. chondroitin hexasaccharide, glycosylated at 60% the
rate of decamer glycosylation) [1]

5 UDP-N-acetyl-D-galactosamine + (glucuronyl-N-acetyl-1,3-beta-D-ga-
lactosamine)$_6$ (i.e. chondroitin dodecasaccharide, glycosylated at about
60% the rate of decamer glycosylation) [1]

6 UDP-N-acetyl-D-galactosamine + (glucuronyl-N-acetyl-1,3-beta-D-ga-
lactosamine)$_7$ (i.e. chondroitin tetradecasaccharide, glycosylated at
about 45% the rate of decamer glycosylation) [1]

7 UDP-N-acetyl-D-galactosamine + (glucuronyl-N-acetyl-1,3-beta-D-ga-
lactosamine)$_2$ (i.e. chondroitin tetrasaccharide, glycosylated at 40% the
rate of decamer glycosylation) [1]

8 UDP-N-acetyl-D-galactosamine + (glucuronyl-N-acetyl-1,3-beta-D-ga-
lactosamine)$_8$ (i.e. chondroitin hexadecasaccharide, glycosylated at
15% the rate of decamer glycosylation) [1]

9 UDP-N-acetyl-D-galactosamine + (glucuronyl-N-acetyl-1,3-beta-D-ga-
lactosamine 4-sulfate)$_4$ (i.e. chondroitin sulfate octasaccharide,
glycosylated at 32% the rate of chondroitin octasaccharide) [1]

10 UDP-N-acetyl-D-galactosamine + (glucuronyl-N-acetyl-1,3-beta-D-glu-
cosamine)$_4$ (i.e. hyaluronate octasaccharide, glycosylated at 25% the
rate of chondroitin octasaccharide) [1]

Product spectrum

1 UDP + N-acetyl-D-galactosaminyl chondroitin oligosaccharide [1]

2 UDP + N-acetyl-D-galactosaminyl-glucuronyl-N-acetyl-1,3-beta-D-gala-
ctosaminyl-(glucuronyl-N-acetyl-1,3-beta-D-glucosamine)$_3$ [1]

3 UDP + ?

4 UDP + ?

5 UDP + ?

6 UDP + ?

7 UDP + ?

8 UDP + ?

9 UDP + ?

10 UDP + ?

Inhibitor(s)

Cofactor(s)/prosthetic group(s)/activating agents

Metal compounds/salts

Turnover number (min^{-1})

Specific activity (U/mg)
    More (34.6 cpm/min/mg protein) [1]

K_m-value (mM)

pH-optimum
    7.4 [1]

pH-range

Temperature optimum (°C)
    30 (assay at) [1]

Temperature range (°C)

---

3 ENZYME STRUCTURE

Molecular weight

Subunits

Glycoprotein/Lipoprotein
    –

---

4 ISOLATION/PREPARATION

Source organism
    Bovine (calf) [1]

Source tissue
    Arterial tissue (Aorta thoracica) [1]

Localization in source
    Microsomes [1]

Purification
    Bovine (partially solubilized with a buffer containing 2% Triton X-100, 20%
    glycerol and phospholipase A_2) [1]

Crystallization
    –

Cloned
    –

Renatured
    –

---

5 STABILITY

pH

Temperature (°C)
   50 ($t_{1/2}$: 5 min, 90% loss of activity within 20 min) [1]

Oxidation

Organic solvent

General stability information
   Unstable upon purification [1]; Rapid thermodenaturation [1]; Ion-exchange
   chromatography on DEAE-trisacryl inactivates [1]

Storage

6 CROSSREFERENCES TO STRUCTURE DATABANKS

PIR/MIPS code

Brookhaven code

7 LITERATURE REFERENCES

[1] Rohrmann, K., Niemann, R., Buddecke, E.: Eur. J. Biochem.,148,463–469 (1985)

4

1 NOMENCLATURE

EC number
2.4.1.176

Systematic name
UDPglucose:gibberellin 2-O-beta-D-glucosyltransferase

Recommended name
Gibberellin beta-D-glucosyltransferase

Synonyms
Glucosyltransferase, uridine diphosphoglucose-gibberellate 7-
Glucosyltransferase, uridine diphosphoglucose-gibberellate 3-O-
Giberellin beta-D-glucosyltransferase

CAS Reg. No.
99775-14-1; 94489-97-1

2 REACTION AND SPECIFICITY

Catalysed reaction
UDPglucose + gibberellin →
→ UDP + gibberellin 2-O-beta-D-glucoside

Reaction type
Hexosyl group transfer

Natural substrates
UDPglucose + gibberellin GA_3 (acts on the plant hormone gibberellin GA_3 and related compounds) [1]

Substrate spectrum
1 UDPglucose + gibberellin GA_3 (Phaseolus, not: Lycopersicon) [1]
2 UDPglucose + gibberellin GA_9 (Lycopersicon) [1]
3 UDPglucose + gibberellin GA_7 (Phaseolus and Lycopersicon) [1]
4 UDPglucose + gibberellin GA_{30} (Phaseolus) [1]
5 More (gibberellins not accepted as substrates by Phaseolus enzyme:
GA_1, GA_4, GA_5, GA_8, 3-epi-GA_3, iso-GA_3, 3-dehydro-GA_3, 3-OH⁻epi-all-
ogibberic acid, 1beta-OH-GA_5, gibberellins not accepted as substrates
by Lycopersicon enzyme: GA_1, GA_3, GA_4, GA_5, GA_6, GA_8, GA_{20}, GA_7-
methylester, GA_8-methylester) [1]

Product spectrum
1 UDP + gibberellin GA_3-O(3)-beta-D-glucopyranoside [1]
2 UDP + gibberellin GA_9-O(3)-beta-D-glucopyranoside [1]
3 ?
4 ?
5 ?

Inhibitor(s)

Cofactor(s)/prosthetic group(s)/activating agents

Metal compounds/salts

Turnover number (min^{-1})

Specific activity (U/mg)

K_m-value (mM)

pH-optimum
8.2 (assay at, Phaseolus) [1]; 8.4 (assay at, Lycopersicon) [1]

pH-range

Temperature optimum (°C)
25 (assay at, Phaseolus) [1]; 37 (assay at, Lycopersicon) [1]

Temperature range (°C)

---

3 ENZYME STRUCTURE

Molecular weight

Subunits

Glycoprotein/Lipoprotein
–

---

4 ISOLATION/PREPARATION

Source organism
Phaseolus coccineus [1]; Lycopersicon peruvianum [1]

Source tissue
Maturing fruits [1]; Suspension cultures [1]

Localization in source
Soluble [1]

Purification

Crystallization
 —

Cloned
 —

Renatured
 —

---

5 STABILITY

pH

Temperature (°C)

Oxidation

Organic solvent

General stability information

Storage
 Deep-frozen, stable for months [1]

---

6 CROSSREFERENCES TO STRUCTURE DATABANKS

PIR/MIPS code

Brookhaven code

---

7 LITERATURE REFERENCES

[1] Sembdner, G., Knöfel, H.-D., Schwarzkopf, E., Liebisch, H.W.: Biol. Plant.,27, 231–236 (1985)

1 NOMENCLATURE

EC number
2.4.1.177

Systematic name
UDPglucose:trans-cinnamate beta-D-glucosyltransferase

Recommended name
Cinnamate beta-D-glucosyltransferase

Synonyms
Glucosyltransferase, uridine diphosphoglucose-cinnamate
UDPG:t-Cinnamate glucosyltransferase

CAS Reg. No.
83744-95-0

---

2 REACTION AND SPECIFICITY

Catalysed reaction
UDPglucose + trans-cinnamate →
→ UDP + trans-cinnamoyl beta-D-glucoside

Reaction type
Hexosyl group transfer

Natural substrates
UDPglucose + cinnamate (involved in biosynthesis of chlorogenic acid in root of Ipomoea batatas [1], enzyme of benzoate metabolism [2, 3]) [1–3]

Substrate spectrum
1 UDPglucose + cinnamate (trans-cinnamate [1], best substrate [1], glucosylation at 8% the rate of 4-methoxycinnamate [2]) [1–4]
2 UDPglucose + p-coumarate (glucosylation at 57% the rate of cinnamate [1] and 43% the rate of 4-methoxycinnamate [2], the glucosyl-group is introduced exclusively into the carboxyl-group [1]) [1, 2]
3 UDPglucose + 4-methoxycinnamate (best substrate) [2, 3]
4 UDPglucose + 4-hydroxy-3-methoxycinnamate (glucosylation at 35% the rate of cinnamate) [1]
5 UDPglucose + 3,4-dimethoxycinnamate (glucosylation at 78% the rate of 4-methoxycinnamate) [2]
6 UDPglucose + 3,4,5-trimethoxycinnamate (glucosylation at 84% the rate of 4-methoxycinnamate) [2]

7 UDPglucose + 3-methoxycinnamate (glucosylation at 52% the rate of 4-methoxycinnamate) [2]
8 UDPglucose + isoferulic acid (glucosylation at 46% the rate of 4-methoxycinnamate) [2]
9 UDPglucose + o-coumarate (glucosylation at 52% the rate of cinnamate) [1]
10 UDPglucose + benzoate (glucosylation at 71% the rate of cinnamate) [1]
11 UDPglucose + feruloate (glucosylation at 27% the rate of cinnamate [1] and 26% the rate of 4-methoxycinnamate [2]) [1, 2]
12 UDPglucose + caffeate (glucosylation at 15% the rate of cinnamate) [1]
13 UDPglucose + sinapic acid (glucosylation at 6.5% the rate of 4-methoxycinnamate) [2]

Product spectrum
1 UDP + cinnamoyl beta-D-glucose (trans-isomer [1]) [1, 4]
2 UDP + p-coumaroyl-D-glucose [1]
3 UDP + 4-methoxycinnamoyl beta-D-glucose
4 UDP + 4-hydroxy-3-methoxycinnamoyl beta-D-glucose
5 UDP + 3,4-dimethoxycinnamoyl beta-D-glucose
6 UDP + 3,4,5-trimethoxycinnamoxy beta-D-glucose
7 UDP + 3-methoxycinnamoyl beta-D-glucose
8 UDP + isoferuloyl beta-D-glucose
9 UDP + o-coumaroyl beta-D-glucose
10 UDP + benzoyl beta-D-glucose
11 UDP + feruloyl beta-D-glucose
12 UDP + caffeoyl beta-D-glucose
13 UDP + sinapoyl beta-D-glucose

Inhibitor(s)
PCMB [1]; $HgCl_2$ (strong) [1]; More (no inhibition by $MgCl_2$, $CaCl_2$, $MnCl_2$, chlorogenic acid (1 mM each) [1], or hydroxycinnamic acid [4]) [1, 4]

Cofactor(s)/prosthetic group(s)/activating agents

Metal compounds/salts
More (no activation by $MgCl_2$, $CaCl_2$ or $MnCl_2$, 1 mM each) [1]

Turnover number (min^{-1})

Specific activity (U/mg)
0.0279 [2]; 70.1 [1]

K_m-value (mM)
0.1 (UDPglucose) [1]; 0.2 (trans-cinnamic acid) [1]; 0.4 (cinnamic acid) [4]

pH-optimum
    5.8 [1]

pH-range

Temperature optimum (°C)
    30 (assay at) [1, 2]

Temperature range (°C)

---

3 ENZYME STRUCTURE

Molecular weight
    45000 (Ipomoea batatas, gel filtration) [1]

Subunits

Glycoprotein/Lipoprotein
    –

---

4 ISOLATION/PREPARATION

Source organism
    Ipomoea batatas (sweet potato, Lam cv. Norin 1) [1]; Medicago sativa
    (alfalfa, cv. Apollo) [4]; Phaseolus vulgaris (bean, cv. Canadian Wonder) [4];
    Vanilla planifolia Andr. [2, 3]

Source tissue
    Cell suspension culture [2–4]; Root [1]

Localization in source
    Soluble [1]; Cytosol [4]

Purification
    Ipomoea batatas (partial) [1]; Medicago sativa (partial) [4]; Phaseolus
    vulgaris (partial) [4]; Vanilla planifolia Andr. (partial) [2]

Crystallization
    –

Cloned
    –

Renatured
    –

5 STABILITY

pH

Temperature (°C)

Oxidation

Organic solvent

General stability information
  PMSF and 2-mercaptoethanol stabilize during purification [1]; Freezing inactivates, sorbitol stabilizes [1]; Sorbitol, 10%, stabilizes during storage [1]

Storage
  −20°C, 10% sorbitol, about 10% loss of activity within 30 days [1]

---

6 CROSSREFERENCES TO STRUCTURE DATABANKS

PIR/MIPS code

Brookhaven code

---

7 LITERATURE REFERENCES

[1] Shimizu, T., Kojima, M.: J. Biochem.,95,205–212 (1984)
[2] Funk, C., Brodelius, P.E.: Plant Physiol.,94,102–108 (1990)
[3] Funk, C., Brodelius, P.E.: Plant Physiol.,99,256–262 (1992)
[4] Edwards, R., Mavandad, M., Dixon, R.A.: Phytochemistry,29,1867–1873 (1990)

1 NOMENCLATURE

EC number
2.4.1.178

Systematic name
UDPglucose:4-hydroxymandelonitrile glucosyltransferase

Recommended name
Hydroxymandelonitrile glucosyltransferase

Synonyms
Cyanohydrin glucosyltransferase
Glucosyltransferase, uridine diphosphoglucose-cyanohydrin

CAS Reg. No.
89287-39-8

---

2 REACTION AND SPECIFICITY

Catalysed reaction
UDPglucose + 4-hydroxymandelonitrile →
→ UDP + taxiphyllin

Reaction type
Hexosyl group transfer

Natural substrates
UDPglucose + 4-hydroxymandelonitrile (last step in biosynthesis of
cyanogenic glucosides) [1]

Substrate spectrum
1 UDPglucose + 4-hydroxymandelonitrile [1]
2 UDPglucose + 3,4-dihydroxymandelonitrile [1]

Product spectrum
1 UDP + taxiphyllin [1]
2 UDP + ? (properties of the product) [1]

Inhibitor(s)

Cofactor(s)/prosthetic group(s)/activating agents

Metal compounds/salts
High salt concentrations (e.g. NaCl, KCl, $(NH_4)_2SO_4$ stimulate) [1]

Turnover number (min^{-1})

Specific activity (U/mg)
    More [1]

K$_m$-value (mM)
    0.5 (UDPglucose + (4-hydroxymandelonitrile or 3,4-dihydroxymandeloni-
    trile)) [1]

pH-optimum
    6.5–8.5 [1]

pH-range
    6–9.5 (about 50% of activity maximum at pH 6 and 9.5) [1]

Temperature optimum (°C)
    30 [1]

Temperature range (°C)

---

3 ENZYME STRUCTURE

Molecular weight
    50000 (Triglochin maritima, gel filtration) [1]

Subunits
    Monomer (1 × 47000, Triglochin maritima, SDS-PAGE) [1]

Glycoprotein/Lipoprotein
    No glycoprotein [1]

---

4 ISOLATION/PREPARATION

Source organism
    Triglochin maritima [1]

Source tissue
    Seedlings [1]

Localization in source
    Soluble [1]

Purification
    Triglochin maritima [1]

Crystallization
    –

Cloned

–

Renatured

–

---

5 STABILITY

pH
   7.5 (unstable below, crude enzyme extract) [1]

Temperature (°C)

Oxidation

Organic solvent

General stability information

Storage
   –20°C, 10% glycerol, crude extract stable for several weeks [1]

---

6 CROSSREFERENCES TO STRUCTURE DATABANKS

PIR/MIPS code

Brookhaven code

---

7 LITERATURE REFERENCES

[1] Hösel, W., Schiel, O.: Arch. Biochem. Biophys.,229,177–186 (1984)

1 NOMENCLATURE

EC number
2.4.1.179

Systematic name
UDPgalactose:D-galactosyl-1,4-beta-D-glucosyl-R beta-1,3-galactosyltrans-
ferase

Recommended name
Lactosylceramide beta-1,3-galactosyltransferase

Synonyms
Galactosyltransferase, uridine diphosphogalactose-lactosylceramide
beta1→3-

CAS Reg. No.
106769-64-6

---

2 REACTION AND SPECIFICITY

Catalysed reaction
UDPgalactose + D-galactosyl-1,4-beta-D-glucosyl-R →
→ UDP + D-galactosyl-1,3-beta-D-galactosyl-1,4-beta-D-glucosyl-R

Reaction type
Hexosyl group transfer

Natural substrates
UDP + D-galactosyl-1,3-beta-D-galactosyl-1,4-beta-D-glucosyl-glycolipid (involved in elongation of oligosaccharide chains in glycolipids [1], associated
with biosynthesis of type 1 lactoseries core chain carbohydrate structure
[4, 5]) [1, 4, 5]

Substrate spectrum
1 UDPgalactose + D-galactosyl-1,4-beta-D-glucosyl-R (R may be an
oligosaccharide or a glycolipid, substrates are methyl-beta-D-galactoside
(poor [2]) [1], 3'-galactosyllactose, lacto-N-tetraose, lactulose, methyl-
alpha-galactoside [2], p-nitrophenyl-beta-glycosides of galactose (not [2])
or lactose are by far the best substrates among oligosaccharides [1],
substrate specificity [1, 5], hydrophobic substrates preferred [1]. Poor
substrates are alpha-galactosides [1], galactose [1, 2], 4'-galactosyllac-
tose, N-acetyllactosamine, stachyose, D-galactose-1,6-N-acetylglucosyl-
amine, p-nitrophenyl-alpha-N-acetylgalactosylamine, N-acetylgalactosyl-

amine [2]. No substrates are asialo-glycophorin A, asialo-fetuin, asialo-orosomucoid, galactoside GM2, globoside, D-galactose-1,3-beta-N-acetylgalactosylamine [1], melibiose, p-nitrophenyl-beta-galactoside, lactosylceramide or glucosylceramide [2]) [1–5]

2 UDPgalactose + lactose (best substrate [2]) [1–3]
3 UDPgalactose + lactosylceramide (best glycolipid substrate [1]) [1, 3]
4 UDPgalactose + lactotriaosylceramide [4, 5]

Product spectrum

1 UDP + D-galactosyl-1,3-beta-D-galactosyl-1,4-beta-D-glucosyl-R [1]
2 UDP + 3'-galactosyllactose [2]
3 ?
4 UDP + lactotetraosylceramide [4, 5]

Inhibitor(s)

EDTA [2, 3]; Salt-concentrations above 0.1 M [1]; More (no inhibition by N-acetylgalactosamine [1], bovine or tammar alpha-lactalbumin [2]) [1, 2]

Cofactor(s)/prosthetic group(s)/activating agents

Triton CF-54 (activation, 0.1%) [5]; Triton X-100 (activation, can replace CF-54 with 75% efficiency, less effective are deoxycholate, CHAPSO or taurodeoxycholate) [5]; Bovine liver non-specific transfer proteins (activation) [5]; Phosphatidylethanolamine (activation, less efficient than Triton CF-54) [5]; Glycerol (requirement, above 20% w/v) [1]; More (little or no activation by phosphatidylcholine, phosphatidylglycerol, phosphatidylserine, phosphatidylinositol or cardiolipin, Brij-58, Empigen BB or detergent G-3634-A) [5]

Metal compounds/salts

Mn^{2+} (requirement, 25 mM [5]) [2, 3, 5]; Co^{2+} (activation, can replace Mn^{2+} with 51% efficiency) [5]; Ca^{2+} (activation, can replace Mn^{2+} with 26% efficiency) [5]; Cu^{2+} (activation, can replace Mn^{2+} with 15% efficiency) [5]; Ni^{2+} (activation, can replace Mn^{2+} with 16% efficiency) [5]; Mg^{2+} (activation, can replace Mn^{2+} with 13% efficiency) [5]; More (no activation by Zn^{2+} or Cd^{2+}) [5]

Turnover number (min^{-1})

Specific activity (U/mg)

0.00767 [5]; 0.02–0.032 (from day 90 to day 225 post partum) [2]

K_m-value (mM)

0.013 (lactotriaosylceramide) [5]; 0.048 (UDPgalactose) [5]; 0.17 (UDPgalactose) [1]; 2.5 (lactosylceramide) [1]; 46 (lactose) [2]; 242 (lactose) [1]

pH-optimum

6 [1]; 6.5 [2]; 7 (HEPES buffer) [5]

pH-range
5–8 (about 75% of maximal activity at pH 5 and about half-maximal activity
at pH 8) [2]; 5.3–7.5 (about 80% of maximal activity at pH 5.3 and about
55% of maximal activity at pH 7.5) [1]; 5.8–8.2 (about half-maximal activity
at pH 5.8 and 8.2, cacodylate or Tris-HCl buffer) [5]

Temperature optimum (°C)
37 (assay at) [1–5]

Temperature range (°C)

---

3 ENZYME STRUCTURE

Molecular weight

Subunits

Glycoprotein/Lipoprotein
–

---

4 ISOLATION/PREPARATION

Source organism
Human [1, 3–5]; Macropus eugenii (female tammar wallaby) [2]; More (not
in lactating mammary gland of mouse) [2]

Source tissue
Kidney [1]; Mammary gland (lactating) [2]; Colonic adenocarcinoma (cell
lines Colo 205 [4, 5] and Colo SW403 [5]) [4, 5]; Cell suspension culture [5]

Localization in source
Microsomes [1, 3]; Membrane-bound [1, 3, 5]; Golgi apparatus [5]; Endoplasmic reticulum [5]

Purification
Human (partial [3, 5], Colo 205 cell line [5]) [3, 5]

Crystallization
–

Cloned
–

Renatured
–

5 STABILITY

pH

Temperature (°C)
36 ($t_{1/2}$: 20 min) [2]; 39 (30 min, complete inactivation, in the presence of Mn^{2+} 11% loss of activity within 60 min) [2]

Oxidation

Organic solvent

General stability information
Glycerol, 20% w/v and above, stabilizes [1]; Mn^{2+} enhances thermal stability [2]

Storage
4°C, in 20% glycerol, several weeks [3]

---

6 CROSSREFERENCES TO STRUCTURE DATABANKS

PIR/MIPS code

Brookhaven code

---

7 LITERATURE REFERENCES

[1] Bailly, P., Piller, F., Cartron, J.-P.: Eur. J. Biochem.,173,417–422 (1988)
[2] Messer, M., Nicholas, K.R.: Biochim. Biophys. Acta,1077,79–85 (1991)
[3] Bailly, P., Piller, F., Cartron, J.-P.: Biochem. Biophys. Res. Commun.,141,84–91 (1986)
[4] Holmes, E.H., Levery, S.B.: Arch. Biochem. Biophys.,274,14–25 (1989)
[5] Holmes, E.H.: Arch. Biochem. Biophys.,270,630–646 (1989)

1 NOMENCLATURE

EC number
2.4.1.180

Systematic name
UDP-N-acetyl-beta-D-mannosaminouronate:lipopolysaccharide N-acetyl-beta-D-mannosaminouronosyltransferase

Recommended name
Lipopolysaccharide N-acetylmannosaminouronosyltransferase

Synonyms
ManNAcA transferase [1]
Acetylmannosaminuronosyltransferase, uridine diphosphoacetyl-mannosaminuronate-acetylglucosaminylpyrophosphorylundecaprenol

CAS Reg. No.
113478-30-1

---

2 REACTION AND SPECIFICITY

Catalysed reaction
UDP-N-acetyl-beta-D-mannosaminouronate + lipopolysaccharide →
→ UDP + N-acetyl-beta-D-mannosaminouronosyl-1,4-lipopolysaccharide

Reaction type
Hexosyl group transfer

Natural substrates
More (involved in the biosynthesis of common antigen in Entero-bacteriaceae) [1]

Substrate spectrum
1 UDP-N-acetyl-D-mannosaminuronic acid + N-acetyl-D-glucosamine-pyro-phosphorylundecaprenol (i.e. lipid I) [1]

Product spectrum
1 UDP + ManNAcA-GlcNAc-pyrophosphorylundecaprenol (i.e. lipid II) [1]

Inhibitor(s)

Cofactor(s)/prosthetic group(s)/activating agents

Metal compounds/salts

Turnover number (min⁻¹)

Specific activity (U/mg)

K_m-value (mM)

pH-optimum

pH-range

Temperature optimum (°C)
    37 (assay at) [1]

Temperature range (°C)

---

3 ENZYME STRUCTURE

Molecular weight

Subunits

Glycoprotein/Lipoprotein
    –

---

4 ISOLATION/PREPARATION

Source organism
    E. coli [1]

Source tissue
    Cell [1]

Localization in source
    Membrane-associated [1]

Purification
    E. coli [1]

Crystallization
    –

Cloned
    –

Renatured
    –

5 STABILITY

pH

Temperature (°C)

Oxidation

Organic solvent

General stability information

Storage

6 CROSSREFERENCES TO STRUCTURE DATABANKS

PIR/MIPS code

Brookhaven code

7 LITERATURE REFERENCES

[1] Barr, K., Ward, S., Meier-Dieter, U., Mayer, H., Rick, P.D.: J. Bacteriol.,170,228–233 (1988)

1 NOMENCLATURE

EC number
2.4.1.181

Systematic name
UDPglucose:hydroxyanthraquinone O-glucosyltransferase

Recommended name
Hydroxyanthraquinone glucosyltransferase

Synonyms
Glucosyltransferase, uridine diphosphoglucose-anthraquinone
Anthraquinone-specific glucosyltransferase

CAS Reg. No.
112198-78-4

---

2 REACTION AND SPECIFICITY

Catalysed reaction
UDPglucose + a hydroxyanthraquinone →
→ UDP + a glucosyloxyanthraquinone

Reaction type
Hexosyl group transfer

Natural substrates

Substrate spectrum
1 UDPglucose + a hydroxyanthraquinone (a range of anthraquinones and
  some flavones can act as acceptors) [1]
2 UDPglucose + emodin [1]
3 UDPglucose + anthrapurpurin [1]
4 UDPglucose + quinizarin [1]
5 UDPglucose + 2,6-dihydroanthraquinone [1]
6 UDPglucose + 1,8-dihydroanthraquinone [1]

Product spectrum
1 UDP + a glucosyloxyanthraquinone [1]
2 ?
3 ?
4 ?
5 ?
6 ?

Inhibitor(s)
p-Chloromercuribenzoate [1]; Co^{2+} [1]; Cu^{2+} [1]; Zn^{2+} [1]; Mn^{2+} (weak) [1]; Iodoacetamide [1]; Iodoacetate [1]; N-Ethylmaleimide [1]; Phenylmercuric acetate [1]

Cofactor(s)/prosthetic group(s)/activating agents

Metal compounds/salts
No cation requirement [1]

Turnover number (min^{-1})

Specific activity (U/mg)
More [1]

K_m-value (mM)
0.01 (emodin, anthrapurpurin, quinizarin, 2,6-dihydroanthraquinone, 1,8-dihydroanthraquinone) [1]

pH-optimum
7 [1]

pH-range

Temperature optimum (°C)
30 (assay at) [1]

Temperature range (°C)

3 ENZYME STRUCTURE

Molecular weight
50000 (Cinchona succirubra, gel filtration) [1]

Subunits

Glycoprotein/Lipoprotein
–

4 ISOLATION/PREPARATION

Source organism
Cinchona succirubra (5 distinct glucosylating activities with different pI values and substrate specificities) [1]

Source tissue
Cell suspension culture [1]

Localization in source

Purification
    Cinchona succirubra [1]

Crystallization
    –

Cloned
    –

Renatured
    –

---

5 STABILITY

pH

Temperature (°C)

Oxidation

Organic solvent

General stability information
    In absence of SH-group protectors, 50% loss of activity after 48 h, in
    presence of 14 mM 2-mercaptoethanol, 50% loss of activity after 2 weeks
    [1]

Storage
    –20°C, 25 mM histidine/HCl buffer, pH 7.0, 10% glycerol,
    10 mM dithioerythritol, stable for 2 months [1]

---

6 CROSSREFERENCES TO STRUCTURE DATABANKS

PIR/MIPS code

Brookhaven code

---

7 LITERATURE REFERENCES

[1] Khouri, H.E., Ibrahim, R.K.: Phytochemistry,26,2531–2535 (1987)

1 NOMENCLATURE

EC number
2.4.1.182

Systematic name
UDP-2,3-bis(3-hydroxytetradecanoyl)glucosamine:2,3-bis(3-hydroxytetra-
decanoyl)-beta-D-glucosaminyl-1-phosphate 2,3-bis(3-hydroxytetradeca-
noyl)glucosaminyltransferase

Recommended name
Lipid-A-disaccharide synthase

Synonyms
Synthase, lipid A disaccharide
Lipid A disaccharide synthase

CAS Reg. No.
105843-81-0

2 REACTION AND SPECIFICITY

Catalysed reaction
UDP-2,3-bis(3-hydroxytetradecanoyl)glucosamine + 2,3-bis(3-hydroxytetra-
decanoyl)-beta-D-glucosaminyl 1-phosphate →
→ UDP + 2,3-bis(3-hydroxytetradecanoyl)-D-glucosaminyl-1,6-beta-D-2,3-
bis(3-hydroxytetradecanoyl)-beta-D-glucosaminyl 1-phosphate

Reaction type
Hexosyl group transfer

Natural substrates
2,3-Diacylglucosamine 1-phosphate + UDP-2,3-diacylglucosamine (involved
with EC 2.3.1.129 and EC 2.7.1.130 in the biosynthesis of the phosphoryla-
ted glycolipid, Lipid A, in the outer membrane of E. coli and other Gram-
negative bacteria) [1]

Substrate spectrum

1 2,3-Diacylglucosamine 1-phosphate + UDP-2,3-diacylglucosamine
   (2,3-diacylglucosamine 1-phosphate is lipid X [5]) [1–5]
2 2,3-Diacylglucosamine 1-phosphate + TDP-2,3-diacylglucosamine
   (33.3% of the activity with UDP-2,3-diacylglucosamine) [4]
3 2,3-Diacylglucosamine 1-phosphate + ADP-2,3-diacylglucosamine
   (2.66% of the activity with UDP-2,3-diacylglucosamine) [4]
4 2,3-Diacylglucosamine 1-phosphate + CDP-2,3-diacylglucosamine
   (0.70% of the activity with UDP-2,3-diacylglucosamine) [4]
5 2,3-Diacylglucosamine 1-phosphate + GDP-2,3-diacylglucosamine
   (0.11% of the activity with UDP-2,3-diacylglucosamine) [4]
6 2,3-Diacylglucosamine 1-phosphate + UDP-3-O-((R)-3-hydroxymyris-
   toyl)GlcNAc (0.44% of the activity with UDP-2,3-diacylglucosamine) [4]

Product spectrum

1 2,3-Diacylglucosamine-(beta-1,6)2,3diacylglucosamine 1-phosphate +
   UDP [1–5]
2 ?
3 ?
4 ?
5 ?
6 ?

Inhibitor(s)

Octyl-beta-D-glucoside [1]

Cofactor(s)/prosthetic group(s)/activating agents

No requirement for detergent [1, 4, 5]

Metal compounds/salts

No requirement for divalent cations [1]

Turnover number (min^{-1})

Specific activity (U/mg)

15.9 [4, 5]

K_m-value (mM)

0.11 (UDP-2,3-diacylglucosamine) [4]; 0.27 (2,3-diacylglucosamine 1-phos-
phate) [4]

pH-optimum

8 (assay at [5]) [1, 5]

pH-range

Temperature optimum (°C)

37 (assay at) [4]

Temperature range (°C)

3 ENZYME STRUCTURE

Molecular weight
   86000 (E. coli, gel filtration) [4]
   More (nucleotide sequence) [1]

Subunits
   Dimer (2 × 42000, E. coli, SDS-PAGE) [4]
   ? (x × 42339, E. coli) [5]

Glycoprotein/Lipoprotein
   –

4 ISOLATION/PREPARATION

Source organism
   E. coli (MC1061/pSR8 [4, 5]) [1–5]

Source tissue

Localization in source
   Cytosol [1]; Cytoplasm [5]

Purification
   E. coli [4, 5]

Crystallization
   –

Cloned
   [3]

Renatured
   –

5 STABILITY

pH

Temperature (°C)
   60 (30 min, inactivation) [1]

Oxidation

Organic solvent

General stability information

Storage
   –80°C, frozen in liquid N_2, stable for several years [5]

6 CROSSREFERENCES TO STRUCTURE DATABANKS

PIR/MIPS code

Brookhaven code

---

7 LITERATURE REFERENCES

[1] Ray, B.L., Painter, G., Raetz, C.R.H.: J. Biol. Chem.,259,4852–4859 (1984)
[2] Crowell, D.N., Reznikoff, W.S., Raetz, C.R.H.: J. Bacteriol.,169,5727–5734 (1987)
[3] Crowell, D.N., Anderson, M.S., Raetz, C.R.H.: J. Bacteriol.,168,152–159 (1986)
[4] Radika, K., Raetz, C.R.H.: J. Biol. Chem.,263,14859–14867 (1988)
[5] Raetz, C.R.H.: Methods Enzymol.,209,455–466 (1992) (Review)

1 NOMENCLATURE

EC number
2.4.1.183

Systematic name
UDPglucose:alpha-D-(1–3)-glucan 3-alpha-D-glucosyltransferase

Recommended name
alpha-1,3-Glucan synthase

Synonyms
Glucosyltransferase, uridine diphosphoglucose-1,3-alpha-glucan
1,3-alpha-D-Glucan synthase [1]

CAS Reg. No.
113478-38-9

---

2 REACTION AND SPECIFICITY

Catalysed reaction
UDPglucose + [alpha-D-glucosyl-(1–3)]$_n$ →
→ UDP + [alpha-D-glucosyl-(1–3)]$_{n+1}$

Reaction type
Hexosyl group transfer

Natural substrates

Substrate spectrum
1 UDPglucose + [alpha-D-glucosyl-(1–3)]$_n$ (enzyme catalyzes both the
hydrolysis of sucrose to glucose and fructose and the glucosyl transfer to
glucosyl polymers to yield water-insoluble glucan, the enzyme catalyzes
only sucrose hydrolysis however in the absence of 1,6-alpha-D-glucan as
acceptor [1], a glucan primer is needed to begin the reaction which
brings about elongation of the glucan chains) [1]

Product spectrum
1 UDP + [alpha-D-glucosyl-(1–3)]$_{n+1}$ [1]

Inhibitor(s)
Antiserum against the isolated 1,3-alpha-D-glucan synthase (glucosyl
transfer activity is completely inhibited, not sucrose hydrolysis activity) [1]

Cofactor(s)/prosthetic group(s)/activating agents

Metal compounds/salts

Turnover number (min^{-1})

Specific activity (U/mg)

K_m-value (mM)

pH-optimum
    6.0 (assay at) [1]

pH-range

Temperature optimum (°C)

Temperature range (°C)

3 ENZYME STRUCTURE

Molecular weight

Subunits

Glycoprotein/Lipoprotein
    −

4 ISOLATION/PREPARATION

Source organism
    Streptococcus mutans [1]

Source tissue

Localization in source

Purification
    Streptococcus mutans [1]

Crystallization
    −

Cloned
    −

Renatured
    −

5 STABILITY

pH

Temperature (°C)

Oxidation

Organic solvent

General stability information

Storage

6 CROSSREFERENCES TO STRUCTURE DATABANKS

PIR/MIPS code

Brookhaven code

7 LITERATURE REFERENCES

[1] Yamashita, Y., Hanada, N., Itoh-Andoh, M., Takehara, T.: FEBS Lett.,243,343–346 (1989)

1 NOMENCLATURE

EC number
2.4.1.184

Systematic name
Mono-beta-D-galactosyldiacylglycerol:mono-beta-D-galactosyldiacylglycerol
beta-D-galactosyltransferase

Recommended name
Galactolipid galactosyltransferase

Synonyms
Galactolipid:galactolipid galactosyltransferase
Galactosyltransferase, galactolipid-galactolipid
Interlipid galactosyltransferase
GGGT [4]

CAS Reg. No.
66676-74-2

2 REACTION AND SPECIFICITY

Catalysed reaction
2 Mono-beta-D-galactosyldiacylglycerol →
→ alpha-D-galactosyl-beta-D-galactosyldiacylglycerol + diacylglycerol

Reaction type
Hexosyl group transfer

Natural substrates
Mono-beta-D-galactosyldiacylglycerol + mono-beta-D-galactosyldiacylglycerol [2]
alpha-D-Galactosyl-beta-D-galactosyldiacylglycerol + diacylglycerol (mono-
galactosyldiacylglycerol synthesis, maintaining of a low concentration of
diacylglycerol in the chloroplast outer envelope membrane) [5]

Substrate spectrum
1 Mono-beta-D-galactosyldiacylglycerol + mono-beta-D-galactosyldiacylglycerol (r [5], digalactosyldiacylglycerol synthesis proceeds most rapidly
by galactosyl transfer to hexaene species of monogalactosyldiacylglyc-
erol [4]) [1–6]

Product spectrum
1 alpha-D-Galactosyl-beta-D-galactosyldiacylglycerol + diacylglycerol (by further transfer of galactosyl residues to the digalactosyldiacylglycerol, trigalactosyldiacylglycerol and tetragalactosyldiacylglycerol are also formed [2–4]) [1–6]

Inhibitor(s)
Zn^{2+} [3, 4]; 5-(Hydroxymercuri)benzoic acid [4]; N-Ethylmaleimide [4]; UDP (no effect [5], 8 mM, 50% inhibition [4]) [4]; UMP (10 mM, 50% inhibition) [4]; $CdCl_2$ (2 mM, 50% inhibition) [4]

Cofactor(s)/prosthetic group(s)/activating agents
alpha-Linolenic acid (stimulates in presence of $MnCl_2$, $CaCl_2$ or KCl) [6]

Metal compounds/salts
Mg^{2+} (stimulates forward reaction, no effect on reverse reaction [5], stimulation [3, 4], EDTA inhibits stimulation [3, 6], unsaturated 16- and 18-carbon fatty acids stimulate in presence of $MgCl_2$ [6], alpha-linolenic acid causes drastic increase in activity under limiting concentrations of $MgCl_2$, without affecting its maximum activity at higher $MgCl_2$ concentration, free alpha-linolenic acid alone does not affect the activity [6]) [3–6]; Mn^{2+} (alpha-linolenic acid stimulates activity in presence of $MnCl_2$ [6], stimulation [3, 4]) [3, 4, 6]; Li^+ (weak stimulation) [3]; Na^+ (weak stimulation) [3]; K^+ (weak stimulation [3], alpha-linolenic acid stimulates in presence of high concentrations of KCl [6]) [3, 6]; Ba^{2+} (stimulation) [4]; Ca^{2+} (stimulation [4], alpha-linolenic acid stimulates activity in presence of $CaCl_2$ [6]) [4, 6]

Turnover number (min^{-1})

Specific activity (U/mg)

K_m-value (mM)

pH-optimum
5.9–7.0 [3]; 6–7 (forward reaction) [5]; 6.5 (reverse reaction) [5]; 8 (assay at) [2]

pH-range
5.3–8.0 (50% of activity maximum at pH 5.3 and 8.0) [3]

Temperature optimum (°C)
30 (assay at) [4]

Temperature range (°C)
4 (still active at 4°C) [3]

3 ENZYME STRUCTURE

Molecular weight

Subunits

Glycoprotein/Lipoprotein

–

4 ISOLATION/PREPARATION

Source organism
  Spinacia oleracea [1–6]

Source tissue
  Leaf [3, 6]

Localization in source
  Chloroplast (envelope [3–6], outer envelope membrane [1, 2, 5], outer surface of [2]) [1–6]

Purification

Crystallization

–

Cloned

–

Renatured

–

5 STABILITY

pH

Temperature (°C)

Oxidation

Organic solvent

General stability information

Storage

6 CROSSREFERENCES TO STRUCTURE DATABANKS

PIR/MIPS code

Brookhaven code

---

7 LITERATURE REFERENCES

[1] Heemskerk, J.W.M., Wintermans, J.F.G.M., Joyard, J., Block, M.A., Dorne, A.-J., Douce, R.: Biochim. Biophys. Acta,877,281–289 (1986)

[2] Dorne, A.-J., Block, M.A., Joyard, J., Douce, R.: FEBS Lett.,145,30–34 (1982)

[3] Heemskerk, J.W., Bögemann, G., Wintermans, J.F.G.M.: Biochim. Biophys. Acta,754,181–189 (1983)

[4] Heemskerk, J.W.M., Jacobs, F.H.H., Scheijen, M.A.M., Helsper, J.P.F.G., Wintermans, J.F.G.M.: Biochim. Biophys. Acta,918,189–203 (1987)

[5] Heemskerk, J.W.M., Jacobs, F.H.H., Wintermans, J.F.G.M.: Biochim. Biophys. Acta, 961,38–47 (1988)

[6] Sakaki, T., Kondo, N., Yamada, M.: Plant Physiol.,94,781–787 (1990)

1 NOMENCLATURE

EC number
2.4.1.185

Systematic name
UDPglucose:flavanone 7-O-beta-D-glucosyltransferase

Recommended name
Flavanone 7-O-beta-glucosyltransferase

Synonyms
Glucosyltransferase, uridine diphosphoglucose-flavanone 7-O-
Naringenin 7-O-glucosyltransferase [1]
Hesperetin 7-O-glucosyl-transferase [1]

CAS Reg. No.
125752-73-0

2 REACTION AND SPECIFICITY

Catalysed reaction
UDPglucose + a flavanone →
→ UDP + a flavanone 7-O-beta-D-glucoside

Reaction type
Hexosyl group transfer

Natural substrates

Substrate spectrum
1 UDPglucose + naringenin (i.e. 4',5',7-trihydroxyflavanone) [1, 2]
2 UDPglucose + hesperetin (i.e. 3',5,7-trihydroxy-4-methoxyflavanone) [1, 2]
3 More (enzyme does not accept other flavanone or flavonol aglycones) [1]

Product spectrum
1 UDP + prunin (i.e. naringenin 7-O-glucoside) [1, 2]
2 UDP + hesperetin 7-O-glucoside [1, 2]
3 ?

Inhibitor(s)
ATP (10 mM, strong) [2]; UTP (10 mM, strong) [2]; ADP (10 mM, strong) [2];
TTP (10 mM, strong) [2]; UDP (10 mM, weak) [2]; NADPH (10 mM, weak) [2]

Cofactor(s)/prosthetic group(s)/activating agents

Metal compounds/salts
More ($CaCl_2$, $MnCl_2$, NaCl, $MgCl_2$ at 0.1 mM, 1.0 mM and 10 mM have no effect) [2]

Turnover number (min^{-1})

Specific activity (U/mg)

K_m-value (mM)
0.063 (naringenin) [1]; 0.08 (naringenin) [2]; 0.11 (hesperetin) [2]; 0.124 (hesperetin) [1]; 0.243 (UDPglucose) [1]

pH-optimum
6.5–7.5 [2]; 7.5 (naringenin) [1]; 7.5–8.0 (hesperetin) [1]

pH-range
5.5–7.5 (60% of maximal activity at pH 5.5 and pH 7.5) [1]

Temperature optimum (°C)
30 (assay at) [1]; 37 [2]

Temperature range (°C)

3 ENZYME STRUCTURE

Molecular weight
54900 (Citrus paradisi, gel filtration) [1]

Subunits

Glycoprotein/Lipoprotein
–

4 ISOLATION/PREPARATION

Source organism
Citrus paradisi (grapefruit) [1, 2]

Source tissue
Leaf [1]; Seedlings (highest activity in leaves, more than in stem and roots) [2]

Localization in source

Purification
Citrus paradisi (partial) [1, 2]

Crystallization
–

Cloned
–

Renatured
–

---

5 STABILITY

pH

Temperature (°C)

Oxidation

Organic solvent

General stability information

Storage
  4°C, 1% bovine serum albumin, stable for 2–3 days [2]

---

6 CROSSREFERENCES TO STRUCTURE DATABANKS

PIR/MIPS code

Brookhaven code

---

7 LITERATURE REFERENCES

[1] McIntosh, C.A., Latchinian, L., Mansell, R.L.: Arch. Biochem. Biophys.,282,50–57
    (1990)
[2] McIntosh, C.A., Mansell R.L.: Phytochemistry,29,1533–1538 (1990)

---

1 NOMENCLATURE

EC number
2.4.1.186

Systematic name
UDPglucose:glycogenin glucosyltransferase

Recommended name
Glycogenin glucosyltransferase

Synonyms
Glyogenin
Priming glucosyltransferase

CAS Reg. No.

2 REACTION AND SPECIFICITY

Catalysed reaction
UDPglucose + glycogenin →
→ UDP + glucosylglycogenin

Reaction type
Hexosyl group transfer

Natural substrates
UDPglucose + glycogenin (the glycogenin subunit of glycogen synthase
(EC 2.4.1.11) catalyzes this reaction, i.e. the enzyme catalyzes its own
glucosylation) [1]

Substrate spectrum
1 UDPglucose + glycogenin [1]

Product spectrum
1 Glucosylated glycogenin + UDP (glucosylation reaches a plateau, when
5 additional glucose residues have been added to glycogenin) [1]

Inhibitor(s)
More (not inhibitory: UDP-pyridoxal, LiBr) [1]

Cofactor(s)/prosthetic group(s)/activating agents

Metal compounds/salts
Mn^{2+} (dependent on divalent cations e.g. Mn^{2+}) [1]

Turnover number (min^{-1})

Specific activity (U/mg)

K$_m$-value (mM)
    0.002 (UDPglucose) [1]

pH-optimum

pH-range

Temperature optimum (°C)

Temperature range (°C)

---

3 ENZYME STRUCTURE

Molecular weight
    38000 (rabbit, SDS-PAGE, glycogenin glucosyltransferase represents the
    smaller subunit of glycogen synthase) [1, 2]

Subunits
    More (38000, rabbit, SDS-PAGE, glycogenin glucosyltransferase represents
    the smaller subunit of glycogen synthase, both enzyme form a complex of
    molar ratio 1:1) [2]

Glycoprotein/Lipoprotein
    –

---

4 ISOLATION/PREPARATION

Source organism
    Rabbit [1, 2]

Source tissue
    Skeletal muscle [1, 2]

Localization in source

Purification
    Rabbit (separation from glycogen synthase, EC 2.4.1.11) [2]

Crystallization
    –

Cloned
    –

Renatured
    –

5 STABILITY

pH

Temperature (°C)

Oxidation

Organic solvent

General stability information

Storage

6 CROSSREFERENCES TO STRUCTURE DATABANKS

PIR/MIPS code
  PIR2:A45094 (rabbit)

Brookhaven code

7 LITERATURE REFERENCES

[1] Pitcher, J., Smythe, C., Cohen, P.: Eur. J. Biochem.,176,391–395 (1988)
[2] Pitcher, J., Smythe, C., Campell, D.G., Cohen, P.: Eur. J. Biochem.,169,497–502
    (1987)

1 NOMENCLATURE

EC number
2.4.1.187

Systematic name
UDP-N-acetyl-D-mannosamine:N-acetyl-beta-D-glucosaminyldiphosphounde-caprenol beta-1,4-N-acetylmannosaminyltransferase

Recommended name
N-Acetylglucosaminyldiphosphoundecaprenol N-acetyl-beta-D-mannos-aminyltransferase

Synonyms
Acetylmannosaminyltransferase, uridine diphosphoacetyl-mannosamine-acetylglucosaminylpyrophosphorylundecaprenol
N-Acetylmannosaminyltransferase [1]
UDP-N-Acetylmannosamine:N-acetylglucosaminyl pyrophosphorylundeca-prenol N-acetylmannosaminyltransferase [1]

CAS Reg. No.
118731-82-1

2 REACTION AND SPECIFICITY

Catalysed reaction
UDP-N-acetyl-D-mannosamine + N-acetyl-D-glucosaminyldiphosphounde-caprenol →
→ UDP + N-acetyl-beta-D-mannosaminyl-1,4-N-acetyl-D-glucosaminyl-diphosphoundecaprenol

Reaction type
Hexosyl group transfer

Natural substrates
More (involved in the biosynthesis of teichoic acid linkage units in bacterial cell walls) [1]

Substrate spectrum
1 UDP-N-acetylmannosamine + N-acetylglucosaminyl pyrophosphoryl-un-decaprenol [1]

Product spectrum
1 UDP + N-acetylmannosamine(beta-1–4)-N-acetylglucosaminylpyro-phos-phorylundecaprenol [1]

Inhibitor(s)
UDP (2 mM, strong) [1]; UMP (weak) [1]; Triton X-100 (strong) [1]

Cofactor(s)/prosthetic group(s)/activating agents
Glycerol (25%: twofold increase of activity) [1]; Nonidet P-40 (0.1%: small stimulatory effect) [1]

Metal compounds/salts
$MgCl_2$ (10 mM, highest stimulatory effect) [1]; NaCl (0.3–0.5 M: stimulation) [1]; KCl (0.3 M: stimulation) [1]; $(NH_4)_2SO_4$ (0.3–0.5 M: stimulation) [1]; NH_4Cl (0.3–0.5 M: stimulation) [1]

Turnover number (min⁻¹)

Turnover number (min^{-1})

Specific activity (U/mg)
0.0000022 [1]

K_m-value (mM)
0.0044 (UDP-N-acetylmannosamine) [1]

pH-optimum
7.3 [1]

pH-range
6–9 (40% of maximal activity at pH 6, 45% of maximal activity at pH 9) [1]

Temperature optimum (°C)
25 (assay at) [1]

Temperature range (°C)

---

3 ENZYME STRUCTURE

Molecular weight

Subunits

Glycoprotein/Lipoprotein
–

---

4 ISOLATION/PREPARATION

Source organism
Bacillus subtilis AHU 1035 [1]

Source tissue
Cell [1]

Localization in source

Purification
Bacillus subtilis AHU 1035 (partial) [1]

Crystallization
–

Cloned
–

Renatured
–

---

5 STABILITY

pH

Temperature (°C)

Oxidation

Organic solvent

General stability information
Glycerol, 25%, stabilizes [1]; Nonidet P-40, 0.1–0.2%, stabilizes [1]

Storage
–18°C, several months, no significant loss of activity [1]

---

6 CROSSREFERENCES TO STRUCTURE DATABANKS

PIR/MIPS code

Brookhaven code

---

7 LITERATURE REFERENCES

[1] Murazumi, N., Kumita, K., Araki, Y., Ito, E.: J. Biochem.,104,980–984 (1988)

1 NOMENCLATURE

EC number
2.4.1.188

Systematic name
UDPglucose:N-acetyl-D-glucosaminyldiphosphoundecaprenol
4-beta-D-glucosyltransferase

Recommended name
N-Acetylglucosaminyldiphosphoundecaprenol glucosyltransferase

Synonyms
UDP-D-glucose:N-acetylglucosaminyl pyrophosphorylundecaprenol
glucosyltransferase [1]
Glucosyltransferase, uridine diphosphoglucose-acetylglucosaminylpyro-
phosphorylundecaprenol

CAS Reg. No.
118731-83-2

2 REACTION AND SPECIFICITY

Catalysed reaction
UDPglucose + N-acetyl-D-glucosaminyldiphosphoundecaprenol →
→ UDP + beta-D-glucosyl-1,4-N-acetyl-D-glucosaminyldiphosphoundeca-
prenol

Reaction type
Hexosyl group transfer

Natural substrates

Substrate spectrum
1 UDPglucose + N-acetylglucosaminylpyrophosphorylundecaprenol [1]

Product spectrum
1 UDP + beta-D-glucosyl-1,4-N-acetylglucosaminylpyrophosphorylundeca-
prenol [1]

Inhibitor(s)

Cofactor(s)/prosthetic group(s)/activating agents
Nonidet P-40 (0.1%: slight stimulatory effect) [1]; Glycerol (20%: stimulates
threefold) [1]

Metal compounds/salts
   $MgCl_2$ (40 mM: stimulation) [1]; KCl (0.6 M: stimulation) [1]; NaCl (stimulation) [1]; NH_4Cl (stimulation) [1]; $(NH_4)_2SO_4$ (stimulation) [1]

Turnover number (min^{-1})

Specific activity (U/mg)

K_m-value (mM)
   0.021 (UDPglucose) [1]

pH-optimum
   6.6–8.0 [1]

pH-range

Temperature optimum (°C)
   25 (assay at) [1]

Temperature range (°C)

---

3 ENZYME STRUCTURE

Molecular weight

Subunits

Glycoprotein/Lipoprotein
   –

---

4 ISOLATION/PREPARATION

Source organism
   Bacillus coagulans AHU 1366 [1]

Source tissue
   Cell membrane [1]

Localization in source
   Membrane-associated [1]

Purification
   Bacillus coagulans AHU 1366 (partial) [1]

Crystallization
   –

Cloned

–

Renatured

–

5 STABILITY

pH

Temperature (°C)

Oxidation

Organic solvent

General stability information
Stabilization of the solubilized enzyme by cations and glycerol [1]

Storage
–18°C, several months without significant loss of activity [1]

6 CROSSREFERENCES TO STRUCTURE DATABANKS

PIR/MIPS code

Brookhaven code

7 LITERATURE REFERENCES

[1] Kumita, K., Murazumi, N., Araki, Y., Ito, E.: J. Biochem.,104,985–988 (1988)

1 NOMENCLATURE

EC number
2.4.1.189

Systematic name
UDPglucuronate:luteolin 7-O-glucuronosyltransferase

Recommended name
Luteolin 7-O-glucuronosyltransferase

Synonyms
Glucuronosyltransferase, uridine diphosphoglucuronate-luteolin 7-O-LGT [1]

CAS Reg. No.
115490-49-8

---

2 REACTION AND SPECIFICITY

Catalysed reaction
UDPglucuronate + luteolin →
→ UDP + luteolin 7-O-beta-D-glucuronide

Reaction type
Hexosyl group transfer

Natural substrates
UDPglucuronic acid + luteolin (involved in the biosynthesis of luteolin tri-glucuronide in Secale cereale) [1]

Substrate spectrum
1 UDPglucuronic acid + luteolin (other UDP-sugars, like UDPgalactose, UDPglucose, UDPxylose are accepted with 10% of maximal activity) [1]
2 UDPglucuronic acid + apigenin (38% of activity compared to luteolin) [1]

Product spectrum
1 UDP + luteolin 7-O-glucuronide [1]
2 ?

Inhibitor(s)
UDP (strong) [1]; Luteolin (strong, at 0.14 mM: 60% inhibition) [1]

Cofactor(s)/prosthetic group(s)/activating agents

Metal compounds/salts
  Mg^{2+} (up to 1 mM: 20% stimulation) [1]; Ca^{2+} (0.25 mM: 25% stimulation) [1]

Turnover number (min^{-1})

Specific activity (U/mg)

K_m-value (mM)
  0.008 (luteolin) [1]; 0.012 (UDPglucuronic acid) [1]

pH-optimum
  6.5 [1]

pH-range
  5.6–9.0 (50% of maximal activity at pH 5.6 and pH 9.0) [1]

Temperature optimum (°C)
  50 [1]

Temperature range (°C)

3 ENZYME STRUCTURE

Molecular weight
  34500 (Secale cereale, gel filtration) [1]

Subunits

Glycoprotein/Lipoprotein
  –

4 ISOLATION/PREPARATION

Source organism
  Secale cereale [1]

Source tissue
  Leaf [1]

Localization in source
  Cytosol (probably) [1]

Purification
  Secale cereale (partial) [1]

Crystallization
  –

Cloned

–

Renatured

–

5 STABILITY

pH

Temperature (°C)

Oxidation

Organic solvent

General stability information
Freezing without glycerol: complete inactivation [1]

Storage
–20°C, 50% glycerol, 3 months: 30–50% loss of activity [1]

6 CROSSREFERENCES TO STRUCTURE DATABANKS

PIR/MIPS code

Brookhaven code

7 LITERATURE REFERENCES

[1] Schulz, M., Weissenböck, G.: Phytochemistry,27,1261–1267 (1988)

1 NOMENCLATURE

EC number
2.4.1.190

Systematic name
UDPglucuronate:luteolin-7-O-beta-D-glucuronide 7-O-glucuronosyltrans-
ferase

Recommended name
Luteolin-7-O-glucuronide 7-O-glucuronosyltransferase

Synonyms
Glucuronosyltransferase, uridine diphosphoglucuronate-luteolin 7-O-glucu-
ronide
LMT [1]
UDP-glucuronate:luteolin 7-O-glucuronide-glucuronosyltransferase [1]

CAS Reg. No.
115490-51-2

---

2 REACTION AND SPECIFICITY

Catalysed reaction
UDPglucuronate + luteolin 7-O-glucuronide →
→ UDP + luteolin 7-O-beta-D-diglucuronide

Reaction type
Hexosyl group transfer

Natural substrates
UDPglucuronic acid + luteolin 7-O-glucuronide (involved in the sequential
biosynthesis of luteolin triglucuronide) [1]

Substrate spectrum
1 UDPglucuronic acid + luteolin 7-O-glucuronide [1, 2]
2 UDPglucuronic acid + apigenin 7-O-glucuronide (88% of activity compared to luteolin 7-O-glucuronide) [1]
3 UDPglucuronic acid + chrysoeriol 7-O-glucuronide (65% of activity compared to luteolin 7-O-glucuronide) [1]

Product spectrum
1 UDP + luteolin diglucuronide [1, 2]
2 ?
3 ?

Inhibitor(s)
  Cu^{2+} (0.75 mM: 40% inhibition) [1]; Pb^{2+} (0.75 mM: 40% inhibition) [1]; Ag^+
  (0.75 mM: 40% inhibition) [1]; Luteolin (0.1 mM: 50% inhibition) [1]; UDP [1]

Cofactor(s)/prosthetic group(s)/activating agents

Metal compounds/salts

Turnover number (min^{-1})

Specific activity (U/mg)

K_m-value (mM)
  0.012 (luteolin 7-O-glucuronide) [1]; 0.04 (UDPglucuronic acid) [1]

pH-optimum
  6.5 [1]; 8.5 (in crude enzyme preparations the pH-optimum at pH 8.5 is absent) [1]

pH-range
  6–9 (50% of maximal activity at pH 6.0 and pH 9.0) [1]

Temperature optimum (°C)
  52 [1]

Temperature range (°C)

3 ENZYME STRUCTURE

Molecular weight
  37000 (Secale cereale, gel filtration) [1]

Subunits

Glycoprotein/Lipoprotein
  –

4 ISOLATION/PREPARATION

Source organism
  Secale cereale (rye) [1, 2]

Source tissue
  Leaf [1, 2]

Localization in source
  Vacuole [2]

Purification
  Secale cereale (partial) [1, 2]

Crystallization
–

Cloned
–

Renatured
–

---

5 STABILITY

pH

Temperature (°C)

Oxidation

Organic solvent

General stability information
Freezing without glycerol: complete inactivation [1]

Storage
–20°C, 50% glycerol, 3 months: 30–50% loss of activity [1]

---

6 CROSSREFERENCES TO STRUCTURE DATABANKS

PIR/MIPS code

Brookhaven code

---

7 LITERATURE REFERENCES

[1] Schulz, M., Weissenböck, G.: Phytochemistry,27,1261–1267 (1988)
[2] Anhalt, S., Weissenböck, G.: Planta,187,83–88 (1992)

---

1 NOMENCLATURE

EC number
2.4.1.191

Systematic name
UDPglucuronate:luteolin-7-O-beta-D-diglucuronide 4'-O-glucuronosyltrans-ferase

Recommended name
Luteolin-7-O-diglucuronide 4'-O-glucuronosyltransferase

Synonyms
Glucuronosyltransferase, uridine diphosphoglucuronate-luteolin 7-O-diglucu-ronide
UDP-glucuronate:luteolin 7-O-diglucuronide-glucuronosyltransferase [1]
UDPglucuronate:luteolin 7-O-diglucuronide-4'-O-glucuronosyl-transferase [2]
LDT [1, 2]

CAS Reg. No.
115490-50-1

---

2 REACTION AND SPECIFICITY

Catalysed reaction
UDPglucuronate + luteolin 7-O-beta-D-diglucuronide →
→ UDP + luteolin 7-O-[beta-D-glucuronosyl-(1→2)-beta-D-glucuronide]-4'-O-beta-D-glucuronide

Reaction type
Hexosyl group transfer

Natural substrates
UDPglucuronic acid + luteolin 7-O-diglucuronide (involved in the sequential biosynthesis of luteolin triglucuronide in Secale cereale) [1, 2]

Substrate spectrum
1 UDPglucuronic acid + luteolin 7-O-diglucuronide (other UDP-sugars are accepted to a lesser extent: about 10% of maximal activity for UDPglu-cose and UDPxylose [1]) [1, 2]

Product spectrum
1 UDP + luteolin 7-O-[beta-D-glucuronosyl-(1→2)-beta-D-glucuronide]-4'-O-beta-D-glucuronide

---

Inhibitor(s)
Luteolin (strong) [1]; UDP (strong) [1]

Cofactor(s)/prosthetic group(s)/activating agents

Metal compounds/salts
More (Ca^{2+} or Mg^{2+} have no influence) [1]

Turnover number (min^{-1})

Specific activity (U/mg)

K_m-value (mM)
0.009 (luteolin 7-O-diglucuronide) [1]; 0.09 (UDPglucuronic acid) [1]

pH-optimum
7 [1]

pH-range
6–9 (50% of maximal activity at pH 6 and pH 9) [1]

Temperature optimum (°C)
40 [1]

Temperature range (°C)

---

3 ENZYME STRUCTURE

Molecular weight
29000 (Secale cereale, gel filtration) [1]

Subunits

Glycoprotein/Lipoprotein
–

---

4 ISOLATION/PREPARATION

Source organism
Secale cereale (rye) [1, 2]

Source tissue
Leaf [1]; Mesophyll protoplasts [2]

Localization in source
Vacuole (probably) [2]

Purification
Secale cereale (partial) [1, 2]

Crystallization
–

Cloned
–

Renatured
–

5 STABILITY

pH

Temperature (°C)

Oxidation

Organic solvent

General stability information
Freezing without glycerol: complete loss of activity [1]

Storage
–20°C, 50% glycerol, 3 months: 30–50% loss of activity [1]

6 CROSSREFERENCES TO STRUCTURE DATABANKS

PIR/MIPS code

Brookhaven code

7 LITERATURE REFERENCES

[1] Schulz, M., Weissenböck, G.: Phytochemistry,27,1261–1267 (1988)
[2] Anhalt, S., Weissenböck, G.: Planta,187,83–88 (1992)

1 NOMENCLATURE

EC number
    2.4.1.192

Systematic name
    UDPglucose:(20S,22S,25S)-22,25-epoxyfurost-5-ene-3beta,26-diol 3-O-beta-
    D-glucosyltransferase

Recommended name
    Nuatigenin 3beta-glucosyltransferase

Synonyms
    Glucosyltransferase, uridine diphosphoglucose-nuatigenin
    More (not identical with EC 2.4.1.173 or EC 2.4.1.193)

CAS Reg. No.
    108891-57-2

---

2 REACTION AND SPECIFICITY

Catalysed reaction
    UDPglucose + (20S,22S,25S)-22,25-epoxyfurost-5-ene-3beta,26-diol →
    → UDP + (20S,22S,25S)-22,25-epoxyfurost-5-ene-3beta,26-diol
    3-O-beta-D-glucoside

Reaction type
    Hexosyl group transfer

Natural substrates
    UDPglucose + (20S,22S,25S)-22,25-epoxyfurost-5-ene-3beta,26-diol (partici-
    pates in the initiation of sugar chain formation during the biosynthesis of oat
    saponins: avenacosides A and B) [3]

Substrate spectrum
    1 UDPglucose + (20S,22S,25S)-22,25-epoxyfurost-5-ene-3beta,26-diol (i.e.
      nuatigenin, TDPglucose at 8%, ADPglucose at 11%, UDPgalactose at
      40% the rate of UDPglucose [3]) [1–3]
    2 20S,22S,25S-Spirost-5-ene-3beta,25-diol + UDPglucose (i.e. isonuatigenin,
      at 57% of the activity with nuatigenin [3]) [2, 3]
    3 25R-Spirost-5-en-3beta-ol + UDPglucose (i.e. diosgenin, at 17% of the
      activity with nuatigenin [3]) [2, 3]
    4 25R-5alpha-Spirostane-3beta,6alpha-diol + UDPglucose (i.e. chlorogenin,
      at 30% of the activity with nuatigenin) [3]
    5 25R-5alpha-Spirostan-3beta-ol + UDPglucose (i.e. tigogenin, at 12%
      of the activity with nuatigenin) [3]

6 22S,25S-5alpha-Tomatanin-3beta-ol + UDPglucose (i.e. tomatidine, at 57% of the activity with nuatigenin) [3]
7 22S,25S-Solanid-5-enin-3beta-ol + UDPglucose (i.e. solanidine, at 39% of the activity with nuatigenin) [3]
8 Pregn-5-en-3beta-ol-20-one + UDPglucose (i.e. pregnenolone, at 26% of the activity with nuatigenin) [3]
9 Androst-5-en-3beta-ol-17-one + UDPglucose (i.e. androsterone, at 15% of the activity with nuatigenin) [3]

Product spectrum
1 UDP + (20S,22S,25S)-22,25-epoxyfurost-5-ene-3beta,26-diol 3-O-beta-D-glucoside (i.e. nuatigenin 3beta-D-monoglucoside) [1–3]
2 ?
3 ?
4 ?
5 ?
6 ?
7 UDP + gamma-chaconine [3]
8 ?
9 ?

Inhibitor(s)
PCMB [3]; UDP [3]; UMP [3]; Triton X-100 [1–3]; Zn^{2+} [2, 3]; Hg^{2+} [2, 3]; More (not: EDTA, 2,2'-dipyridyl) [3]

Cofactor(s)/prosthetic group(s)/activating agents

Metal compounds/salts
More (no stimulation by divalent cations: Ca^{2+}, Mn^{2+} or Mg^{2+}) [1, 2]

Turnover number (min^{-1})

Specific activity (U/mg)

K_m-value (mM)

pH-optimum
6.5–8.5 [2, 3]; 7.5–8.5 [1]

pH-range
5.8–8.5 (5.8: about 50% of activity maximum, 6.5–8.5: activity maximum) [2]

Temperature optimum (°C)
30 [2, 3]

Temperature range (°C)
30–39 (30°C: activity maximum, 39°C: activity is 7times lower than at 30°C) [2]

3 ENZYME STRUCTURE

Molecular weight
58000–62000 (Avena sativa, gel filtration) [3]

Subunits

Glycoprotein/Lipoprotein
–

---

4 ISOLATION/PREPARATION

Source organism
Avena sativa (oat) [1–3]

Source tissue
Leaf [1–3]

Localization in source
Cytosol (70% of the activity [2]) [2, 3]

Purification
Avena sativa (partial) [3]

Crystallization
–

Cloned
–

Renatured
–

---

5 STABILITY

pH

Temperature (°C)
45 (unstable) [3]

Oxidation

Organic solvent

General stability information
Rather unstable when dissolved in buffer, 65% loss of activity after 24 h at 4°C [3]; 2-Mercaptoethanol, 10 mM, improves stability in solution, 35% loss of activity after 24 h at 4°C [3]

Storage

–20°C, in the form of dry acetone powder, stable for several weeks [3];
–20°C, cytosolic fraction stable for several months [2]

6 CROSSREFERENCES TO STRUCTURE DATABANKS

PIR/MIPS code

Brookhaven code

7 LITERATURE REFERENCES

[1] Kalinowska, M., Wojciechowski, Z.A.: Phytochemistry,25,2525–2529 (1986)
[2] Kalinowska, M., Wojciechowski, Z.A.: Phytochemistry,26,353–357 (1987)
[3] Kalinowska, M., Wojciechowski, Z.A.: Plant Sci.,55,239–245 (1988)

1 NOMENCLATURE

EC number
2.4.1.193

Systematic name
UDPglucose:(25S)-5beta-spirostan-3beta-ol 3-O-beta-D-glucosyltransferase

Recommended name
Sarsapogenin 3beta-glucosyltransferase

Synonyms
Glucosyltransferase, uridine diphosphoglucose-sarsapogenin
More (not identical with EC 2.4.1.173 and 2.4.1.192)

CAS Reg. No.
117698-14-3

2 REACTION AND SPECIFICITY

Catalysed reaction
UDPglucose + (25S)-5beta-spirostan-3beta-ol →
→ UDP + (25S)-5beta-spirostan-3beta-ol 3-O-beta-D-glucoside

Reaction type
Hexosyl group transfer

Natural substrates
UDPglucose + (25S)-5beta-spirostan-3beta-ol (involved in the biosynthesis
of plant saponins) [1]

Substrate spectrum
1 UDPglucose + (25S)-5beta-spirostan-3beta-ol (i.e. sarsapogenin) [1]
2 UDPglucose + (25R)-5beta-spirostan-3beta-ol (i.e. smilagenin) [1]
3 More (specific for 5beta-spirostanols, configuration of the C-25 methyl
group is rather unimportant) [1]

Product spectrum
1 UDP + (25S)-5beta-spirostan-3beta-ol 3-O-beta-D-glucoside [1]
2 ?
3 ?

Inhibitor(s)
Triton X-100 [1]; EDTA [1]; EGTA [1]; PCMB [1]; UDP [1]; UMP [1]

Cofactor(s)/prosthetic group(s)/activating agents

Metal compounds/salts
  Mn^{2+} (activates, highest stimulation at 1 mM) [1]

Turnover number (min^{-1})

Specific activity (U/mg)

K_m-value (mM)

pH-optimum
  9.6 [1]

pH-range
  7.2–9.9 (7.2: about 40% of activity maximum, 9.9: about 60% of activity maximum) [1]

Temperature optimum (°C)

Temperature range (°C)

3 ENZYME STRUCTURE

Molecular weight

Subunits

Glycoprotein/Lipoprotein
  –

4 ISOLATION/PREPARATION

Source organism
  Asparagus officinalis [1]

Source tissue
  Shoots (of 3-week-old plants) [1]

Localization in source
  Soluble [1]

Purification

Crystallization
  –

Cloned
  –

Renatured
  –

2

5 STABILITY

pH

Temperature (°C)

Oxidation

Organic solvent

General stability information

Storage

6 CROSSREFERENCES TO STRUCTURE DATABANKS

PIR/MIPS code

Brookhaven code

7 LITERATURE REFERENCES

[1] Paczkowski, C., Wojciechowski, Z.A.: Phytochemistry,27,2743–2747 (1988)

1 NOMENCLATURE

EC number
2.4.1.194

Systematic name
UDPglucose:4-hydroxybenzoate 4-O-beta-D-glucosyltransferase

Recommended name
4-Hydroxybenzoate 4-O-beta-D-glucosyltransferase

Synonyms
Glucosyltransferase, uridine diphosphoglucose-4-hydroxybenzoate
UDP-glucose:4-(beta-D-glucopyranosyloxy)benzoic acid glucosyltransferase
[2]
HBA glucosyltransferase [2]
p-Hydroxybenzoate glucosyltransferase [1]
PHB glucosyltransferase [1]
PHB-O-glucosyltransferase [1]

CAS Reg. No.
120860-68-6

---

2 REACTION AND SPECIFICITY

Catalysed reaction
UDPglucose + 4-hydroxybenzoate →
→ UDP + 4-(beta-D-glucosyloxy) benzoate

Reaction type
Hexosyl group transfer

Natural substrates
More (probably control of activity during pine pollen germination) [2]

Substrate spectrum
1 UDPglucose + 4-hydroxybenzoate [1, 2]

Product spectrum
1 UDP + 4-hydroxybenzoic acid O-beta-D-glucoside [1, 2]

Inhibitor(s)

4-Hydroxybenzoic acid [2]; Mg^{2+} (weak) [1]; Mn^{2+} (weak) [1]; Ca^{2+} (weak) [1]; Fe^{2+} (1 and 10 mM: strong) [1]; Co^{2+} (1 and 10 mM: strong) [1]; Ni^{2+} (1 and 10 mM: strong) [1]; Cu^{2+} (1 and 10 mM: strong) [1]; Zn^{2+} (1 and 10 mM:strong) [1]; UDP (product inhibition) [1]; 4-Hydroxybenzoic acid O-beta-D-glucoside (product inhibition) [1]; 3-Hydroxybenzoic acid [2]; Gallic acid [2]; Vanillic acid [2]; 4-Coumaric acid (weak) [2]; 2-Hydroxybenzoic acid (weak) [2]; EDTA (10 mM, negative influence on activity, but reversible) [1]

Cofactor(s)/prosthetic group(s)/activating agents

EDTA (0.1 mM: 60% increase of activity [2], 10 mM: negative influence on activity, but reversible [1]) [2]

Metal compounds/salts

More (no metal ions required [1], Mg^{2+} has no effect [2]) [1, 2]; Ca^{2+} (3 mM: 50% increase of activity) [2]

Turnover number (min⁻¹)

Specific activity (U/mg)

K_m-value (mM)

0.24 (UDPglucose) [2]; 0.264 (p-hydroxybenzoate) [1]; 0.268 (UDPglucose) [1]; 2.9 (p-hydroxybenzoate) [2]

pH-optimum

7.5 [2]; 7.8 [1]

pH-range

6.6–8.6 (half-maximal activity at pH 6.6 and 8.6) [1]

Temperature optimum (°C)

30 (assay at) [2]; 45 [1]

Temperature range (°C)

3 ENZYME STRUCTURE

Molecular weight

33000 (Pinus densiflora, gel filtration) [2]
47000 (Lithospermum erythrorhizon, gel filtration) [1]

Subunits

Glycoprotein/Lipoprotein

–

4 ISOLATION/PREPARATION

Source organism
Lithospermum erythrorhizon [1]; Pinus densiflora [2]

Source tissue
Callus (cell culture) [1]; Pollen [2]

Localization in source

Purification
Lithospermum erythrorhizon (partial) [1]; Pinus densiflora (pollen, partial) [2]

Crystallization
--

Cloned
--

Renatured
--

5 STABILITY

pH

Temperature (°C)

Oxidation

Organic solvent

General stability information
DTT, 1–10 mM, stabilizes [1]; PMSF, 0.02–1 mM, stabilizes [1]

Storage
4–8°C, 0.01 M Tris-HCl, pH 7.2, 15 days: 76% loss of activity [1]; –20°C, 0.05 M Tris-HCl, 1 mM DTT, 0.02 mM PMSF, pH 7.6, 3 months: 10% loss of activity [1]

6 CROSSREFERENCES TO STRUCTURE DATABANKS

PIR/MIPS code

Brookhaven code

7 LITERATURE REFERENCES

[1] Bechthold, A., Berger, U., Heide, L.: Arch. Biochem. Biophys.,288,39–47 (1991)
[2] Katsumata, T., Shige, H., Ejiri, S.: Phytochemistry,28,359–362 (1989)

1 NOMENCLATURE

EC number
2.4.1.195

Systematic name
UDPglucose:thiohydroximate S-beta-D-glucosyltransferase

Recommended name
Thiohydroximate beta-D-glucosyltransferase

Synonyms
Glucosyltransferase, desulfoglucosinolate-uridine diphosphate
Glucosyltransferase, uridine diphosphoglucose-thiohydroximate

CAS Reg. No.
9068-14-8

2 REACTION AND SPECIFICITY

Catalysed reaction
UDPglucose + phenylacetothiohydroximate →
→ UDP + desulfoglucotropeolin

Reaction type
Hexosyl group transfer

Natural substrates
UDPglucose + phenylacetothiohydroximate (integral step in biosynthesis of benzylglucosinolate [4], involved with EC 2.8.2.24 in final steps of thioglycoside biosynthesis in cruciferous plants [3]) [3, 4]

Substrate spectrum
1 UDPglucose + phenylacetothiohydroximate (best substrate, specific: no glucosyl donors are ADPglucose, CDPglucose and GDPglucose, galactose or xylose cannot replace glucose [4], phenylacetothiohydroximate can be substituted by propiothiohydroximate, butyrothiohydroximate, isobutyrothiohydroximate, benzothiohydroximate, 4-methylthiobutyrothiohydroximate with 50%, 70%, 41%, 10% and 77% effectiveness, respectively [4], no substrates are acetothiohydroximate, phenylacetohydroximic acid, ethanol, 2-mercaptoethanol, cysteine hydrochloride, Tropaelum majus [4]) [1–4]
2 TDPglucose + phenylacetothiohydroximate (glucosylation at 10% the rate of the reaction with UDPglucose, Tropaelum majus) [4]

Product spectrum
1 UDP + desulfoglucosinolate [1–4]
2 TDP + desulfoglucosinolate [4]

Inhibitor(s)
$HgCl_2$ (strong, Tropaelum majus) [4]; PCMB (strong, Tropaelum majus) [4];
N-Ethylmaleimide (Tropaelum majus) [4]; More (no inhibition by KCN or iodoacetic acid, Tropaelum majus) [4]

Cofactor(s)/prosthetic group(s)/activating agents
2-Mercaptoethanol (requirement, Tropaelum majus) [4]; DTT (activation,
1 mM, Tropaelum majus) [4]; Cysteine hydrochloride (activation, 1 mM,
Tropaelum majus) [4]; Glutathione (activation, 1 mM, Tropaelum majus) [4];
EDTA (activation, 0.01 mM, Tropaelum majus) [4]; 2,2'-Dipyridyl (activation,
1 mM, Tropaelum majus) [4]; o-Phenanthroline (activation, Tropaelum majus)
[4]

Metal compounds/salts
Sodium hydrosulfite (activation, 10 mM, Tropaelum majus) [4]; Ascorbic
acid (activation, 1 mM, Tropaelum majus) [4]

Turnover number (min^{-1})

Specific activity (U/mg)
More [1]; 0.045 (Tropaelum majus) [4]

K_m-value (mM)
0.65 (phenylacetothiohydroximate, Tropaelum majus) [4]; 1.05 (4-methyl-
thiobutyrothiohydroximate, Tropaelum majus) [4]; 1.47 (UDPglucose
(+ 4-methylthiobutyrothiohydroximate), Tropaelum majus) [4]; 1.54 (UDPglucose (+ phenylacetothiohydroximate), Tropaelum majus) [4]

pH-optimum
6.5 (phosphate buffer, Tropaelum majus) [4]

pH-range
5–9 (detectable activity in this range, Tris buffer, Tropaelum majus) [4]; 6–8
(about 80% of maximal activity at pH 6 and about 60% of maximal activity at
pH 8, phosphate buffer, Tropaelum majus) [4]

Temperature optimum (°C)
30 (assay at) [1–4]

Temperature range (°C)

3 ENZYME STRUCTURE

Molecular weight
  50000 (Tropaelum majus, gel filtration) [4]

Subunits

Glycoprotein/Lipoprotein
  –

4 ISOLATION/PREPARATION

Source organism
  Brassica juncea (cv. Cutlass [1–3], cv. Domo [1]) [1–3]; Brassica napus
  (rapeseed, cv. Westar) [1]; Brassica campestris cv. R-500 [1]; Brassica oleracea (savoy cabbage) [1]; Brassica nigra cv. 526 [1]; Tropaelum majus [4];
  Sinapis alba [4]; Nasturtium officinale R.Br. [4]; Armoracia lapathifolia [4]

Source tissue
  Seedlings (tissue distribution [2]) [1, 2]; Leaves [4]

Localization in source
  Cytoplasm (subcellular distribution [2]) [2, 3]

Purification
  Brassica napus (partial) [1]; Tropaelum majus (partial) [4]; More (persistent-
  ly co-purifies with EC 2.8.2.24) [3]

Crystallization
  –

Cloned
  –

Renatured
  –

5 STABILITY

pH
  4.5 ($t_{1/2}$: 1 h, 4°C) [3]; 7.5 (stable below) [3]; 9.5 ($t_{1/2}$: 1 h, 4°C) [3]

Temperature (°C)
  30 (up to, stable) [3]; 37 ($t_{1/2}$: 1 h, in 20 mM Tris-HCl buffer, pH 7.5, 14 mM
  2-mercaptoethanol) [3]

Oxidation

Organic solvent

General stability information

Storage
-15°C, freeze-dried enzyme, at least 3 months [4]

---

6 CROSSREFERENCES TO STRUCTURE DATABANKS

PIR/MIPS code

Brookhaven code

---

7 LITERATURE REFERENCES

[1] Jain, J.C., Reed, D.W., Groot Wassink, J.W.D., Underhill, E.W.: Anal. Biochem.,178, 137–140 (1989)
[2] Jain, J.C., Michayluk, M.R., Groot Wassink, J.W.D., Underhill, E.W.: Plant Sci.,64, 25–29 (1989)
[3] Jain, J.C., Groot Wassink, J.W.D., Reed, D.W., Underhill, E.W.: J. Plant Physiol.,136, 356–361 (1990)
[4] Matsuo, M., Underhill, E.W.: Phytochemistry,10,2279–2288 (1971)

1 NOMENCLATURE

EC number
2.4.1.196

Systematic name
UDPglucose:nicotinate N-glucosyltransferase

Recommended name
Nicotinate glucosyltransferase

Synonyms
Glucosyltransferase, uridine diphosphoglucose-nicotinate N-
UDP-glucose:nicotinic acid-N-glucosyltransferase [1]

CAS Reg. No.
120858-56-2

---

2 REACTION AND SPECIFICITY

Catalysed reaction
UDPglucose + nicotinate →
→ UDP + N-glucosylnicotinate

Reaction type
Hexosyl group transfer

Natural substrates
UDPglucose + nicotinate (the reversible glucosyltransferase reaction provides free nicotinate for the synthesis of NAD^+ and $NADP^+$ in pyridine nucleotide cycle) [1]

Substrate spectrum
1 UDPglucose + nicotinate (r, high specificity for UDPglucose as glucosyl
   donor [1]) [1, 2]
2 UDPglucose + nicotinic acid methyl ester [1]
3 UDPglucose + catechol [1]
4 UDPglucose + vanillin [1]
5 UDPglucose + vanillinic acid [1]

Product spectrum
1 UDP + N-glucosylnicotinate [1, 2]
2 ?
3 ?
4 ?
5 ?

Inhibitor(s)
    Zn^{2+} [1]; p-Chloromercuribenzoate [1]

Cofactor(s)/prosthetic group(s)/activating agents

Metal compounds/salts
    No requirement for divalent cations [1]

Turnover number (min^{-1})

Specific activity (U/mg)
    More [1]

K_m-value (mM)
    0.170 (nicotinic acid) [1]; 1.2 (UDPglucose) [1]

pH-optimum
    7.8–8.2 [1]

pH-range
    5.0–11 (5.0: nearly inactive, 11: about 50% of activity maximum) [1]

Temperature optimum (°C)
    30 [1]

Temperature range (°C)

3 ENZYME STRUCTURE

Molecular weight
    46000 (Petroselinum hortense, gel filtration) [1]

Subunits
    Monomer (1 × 46000, Petroselinum hortense, SDS-PAGE) [1]

Glycoprotein/Lipoprotein
    –

4 ISOLATION/PREPARATION

Source organism
    Petroselinum hortense (Hoffm., parsley) [1]; Solanum tuberosum (HH258 and F81) [2]

Source tissue
    Leaf [2]; Tuber [2]; Callus [2]; Heterotrophic cell suspension [1]; Suspension culture [2]

Localization in source
    Soluble [1]

Purification
  Petroselinum hortense (partial) [1]

Crystallization
  –

Cloned
  –

Renatured
  –

---

5 STABILITY

pH

Temperature (°C)

Oxidation

Organic solvent

General stability information

Storage

---

6 CROSSREFERENCES TO STRUCTURE DATABANKS

PIR/MIPS code

Brookhaven code

---

7 LITERATURE REFERENCES

[1] Upmeier, B., Thomzik, J.E., Barz, W.: Z. Naturforsch.,43c,835–842 (1988)
[2] Köster, S., Upmeier, B., Komossa, D., Barz, W.: Z. Naturforsch.,44c,623–628 (1989)

---

1 NOMENCLATURE

EC number
2.4.1.197

Systematic name
UDP-N-acetyl-D-glucosamine:high-mannose-oligosaccharide beta-1,4-N-ace-
tylglucosaminyltransferase

Recommended name
High-mannose-oligosaccharide beta-1,4-N-acetyl-glucosaminyltransferase

Synonyms
Acetylglucosaminyltransferase, uridine diphosphoacetylglucosamine-
oligosaccharide
Acetylglucosamine-oligosaccharide acetylglucosaminyltransferase
UDP-GlcNAc:oligosaccharide beta-N-acetylglucosaminyltransferase [1]

CAS Reg. No.
123425-54-7

2 REACTION AND SPECIFICITY

Catalysed reaction
Transfers an N-acetyl-D-glucosamine residue from UDP-N-acetyl-D-glucosamine to the 4-position of a mannose linked alpha-1,6 to the core mannose
of high-mannose oligosaccharides produced by Dictyostelium discoideum

Reaction type
Hexosyl group transfer

Natural substrates

Substrate spectrum
1 UDP-N-acetyl-D-glucosamine + Manalpha(1–2)Manalpha(1–2)Manalpha
(1–3)[Manalpha(1–2)Manalpha(1–3)][Manalpha(1–2)Manalpha(1–6)] Man-
alpha(1–6)]Manbeta(1–4)GlcNAc (i.e. Man$_9$GlcNAc oligosaccharide) [1]
2 UDP-N-acetyl-D-glucosamine + Manalpha(1–3)[Manalpha(1–3)[Man-
alpha(1–6)]Manalpha(1–6)]Manbeta (1–4)GlcNAc (i.e. Man$_5$GlcNAc
oligosaccharide, 20% of activity compared to MangGlcNAc) [1]
3 UDP-N-acetyl-D-glucosamine + Manalpha(1–3)[Manalpha(1–6)]Man-
beta(1–4)GlcNAc (i.e. Man$_3$GlcNAc) [1]
4 UDP-N-acetyl-D-glucosamine + Manalpha(1–3)[Manalpha(1–6)]Man-
alphamethyl (i.e. Man$_3$Me) [1]

5 UDP-N-acetyl-D-glucosamine + Manalpha(1–3)Manbeta(1–4)GlcNAc (i.e. $Man_2GlcNAc$) [1]

6 UDP-N-acetyl-D-glucosamine + Manalpha(1–3)Manalphamethyl (i.e. Man_2Me) [1]

7 UDP-N-acetyl-D-glucosamine + [GlcA-GlcNAc]$_n$-GlcA-aMan oligosaccharide (general structure: even-numbered, isolated from E. coli K5 capsular polysaccharide) [2]

8 More (enzyme does not discriminate between differently N-acetyled/N-sulfated/N-unsubstituted acceptor sequences, the incorporation of N-acetyl-D-glucosamine increases with increasing size of even-numbered oligosaccharide acceptors) [2]

Product spectrum

1 UDP + $Man_9GlcNAc$ oligosaccharide with an additional N-acetyl-D-glucosamine residue attached beta-1,4 to the mannose linked alpha-1,6 to the core mannose in an intersecting position [1]

2 UDP + $Man_5GlcNAc$ oligosaccharide with an additional N-acetyl-D-glucosamine residue in an intersected and/or bisected position, so that either the core beta-mannose or the branching alpha-mannose residue accepts the transferred N-acetyl-D-glucosamine residue [1]

3 ?

4 ?

5 ?

6 ?

7 ?

8 ?

Inhibitor(s)

EDTA (strong) [1]; Mg^{2+} (strong) [1]; DTT (with Mg^{2+} and Mn^{2+}: strong) [1]; EGTA (strong) [1]

Cofactor(s)/prosthetic group(s)/activating agents

Metal compounds/salts

Mn^{2+} (stimulation [1, 2], required for full enzyme activity [2]) [1, 2]; More (Ca^{2+} or Mg^{2+} cannot substitute for Mn^{2+}) [2]

Turnover number (min^{-1})

Specific activity (U/mg)

0.0000077 ($Man_5GlcNAc$, intersected) [1]; 0.0000129 ($Man_9GlcNAc$, intersected) [1]; 0.0000133 ($Man_5GlcNAc$, bisected) [1]

K_m-value (mM)

0.006 (hexadecasaccharide from E. coli K5 capsular polysaccharide) [2];
0.008 (decasaccharide acceptor)) [2]; 0.06 (hexasaccharide acceptor from
E. coli K5 capsular polysaccharide, UDP-N-acetyl-D-glucosamine (+ decasaccharide)) [2]; 0.3–0.43 ($Man_5GlcNAc$, depending wether N-acetyl-D-glucosamine residue is intersected or bisected) [1]; 0.65 ($Man_9GlcNAc$, intersected product) [1]; 1.2 (UDP-N-acetyl-D-glucosamine (+ $Man_9GlcNAc$)) [1];
1.42 (Man_3Me, intersected product) [1]; 2.54 ($Man_3GlcNAc$, bisected product) [1]; 18.52 (Man_2Me, intersected product) [1]; 19.48 ($Man_2GlcNAc$, bisected product) [1]

pH-optimum

6–6.9 [1]; 6.5 [2]

pH-range

5.2–8 (25% of maximal activity at pH 5.2, 35% of maximal activity at pH 8)
[2]; 5.5–7.2 (20% of maximal activity at pH 5.5, 70% of maximal activity at
pH 7.2) [1]

Temperature optimum (°C)

37 (assay at) [1, 2]

Temperature range (°C)

---

3 ENZYME STRUCTURE

Molecular weight

Subunits

Glycoprotein/Lipoprotein

–

---

4 ISOLATION/PREPARATION

Source organism

Dictyostelium discoideum [1]; Mouse [2]

Source tissue

Cell [1]; Mastocytoma [2]

Localization in source

Membrane associated [1]

Purification

Dictyostelium discoideum (strain M4, partial) [1]; Mouse (partial) [2]

Crystallization

–

Cloned

–

Renatured

–

5 STABILITY

pH

Temperature (°C)
   45 (2 min: 75% decrease of activity) [1]

Oxidation

Organic solvent

General stability information

Storage

6 CROSSREFERENCES TO STRUCTURE DATABANKS

PIR/MIPS code

Brookhaven code

7 LITERATURE REFERENCES

[1] Sharkey, D.J., Kornfeld, R.: J. Biol. Chem.,264,10411–10419 (1989)
[2] Lidholt, K., Lindahl, U.: Biochem. J.,287,21–29 (1992)

1 NOMENCLATURE

EC number
2.4.1.198

Systematic name
UDP-N-acetyl-D-glucosamine:phosphatidylinositol N-acetyl-D-glucosaminyl-transferase

Recommended name
Phosphatidylinositol N-acetylglucosaminyltransferase

Synonyms
Acetyl-D-glucosaminyltransferase, uridine diphosphoacetylglucosamine alpha1,6-

CAS Reg. No.
144388-35-2

---

2 REACTION AND SPECIFICITY

Catalysed reaction
UDP-N-acetyl-D-glucosamine + phosphatidylinositol →
→ UDP + N-acetyl-D-glucosaminylphosphatidylinositol

Reaction type
Hexosyl group transfer

Natural substrates
UDP-N-acetyl-D-glucosamine + phosphatidylinositol (first step in bio-synthesis of the glycosyl phosphatidylinositol membrane anchor of trypano-some variant surface glycoprotein) [1]

Substrate spectrum
1 UDP-N-acetyl-D-glucosamine + phosphatidylinositol [1–3]

Product spectrum
1 UDP + N-acetyl-D-glucosaminyl-phosphatidylinositol (in alpha1→6 linkage [2, 3]) [1–3]

Inhibitor(s)
SH-alkylating reagents [2]; N-Ethylmaleimide [2]; p-Chloromercuriphenylsul-fonic acid [2]; Iodoacetic acid [2]

Cofactor(s)/prosthetic group(s)/activating agents

Metal compounds/salts

Turnover number (min⁻¹)

Specific activity (U/mg)

K_m-value (mM)

pH-optimum

pH-range

Temperature optimum (°C)
    37 (assay at) [2]

Temperature range (°C)

---

3 ENZYME STRUCTURE

Molecular weight

Subunits

Glycoprotein/Lipoprotein
    –

---

4 ISOLATION/PREPARATION

Source organism
    Trypanosoma brucei (cloned variant ILTat 1.3, isolated from rat blood [1], variant MITat 1.4 [2]) [1, 2]; Human [3]

Source tissue
    Cells and cell-lysates [1–3]; Lymphoblastoid cell lines (with paroxysmal nocturnal hemoglobinuria phenotype obtained by Epstein-Barr immortalization of lymphocytes from PNH-patients) [3]

Localization in source
    Membrane-bound [1–3]; Microsomes [3]

Purification

Crystallization
    –

Cloned
    –

Renatured
    –

5 STABILITY

pH

Temperature (°C)

Oxidation

Organic solvent

General stability information

Storage

6 CROSSREFERENCES TO STRUCTURE DATABANKS

PIR/MIPS code

Brookhaven code

7 LITERATURE REFERENCES

[1] Doering, T.L., Masterson, W.J., Englund, P.T., Hart, G.W.: J. Biol. Chem.,264, 11168–11173 (1989)
[2] Milne, K.G., Ferguson, M.A.J., Masterson, W.J.: Eur. J. Biochem.,208,309–314 (1992)
[3] Hillmen, P., Bessler, M., Mason, P.J., Watkins, W.M., Luzzatto, L.: Proc. Natl. Acad. Sci. USA,90,5272–5276 (1993)

1 NOMENCLATURE

EC number
2.4.1.199

Systematic name
beta-D-Mannosylphosphodecaprenol:1,6-alpha-D-mannosyloligosaccharide
1,6-alpha-D-mannosyltransferase

Recommended name
beta-Mannosylphosphodecaprenol-mannooligosaccharide 6-mannosyl-
transferase

Synonyms
Mannosyltransferase, mannosylphospholipid-methylmannoside alpha-1,6-

CAS Reg. No.
125008-27-7

2 REACTION AND SPECIFICITY

Catalysed reaction
beta-D-Mannosylphosphodecaprenol + 1,6-alpha-D-mannosyloligosaccha-
ride →
→ decaprenol phosphate + 1,6-alpha-D-mannosyl-1,6-alpha-D-mannosyl-
oligosaccharide

Reaction type
Hexosyl group transfer

Natural substrates
Mannosylphosphodecaprenol + 1,6-alpha-D-mannooligosaccharides (involved in the formation of mannooligosaccharides in membrane of Mycobacterium smegmatis) [1]

Substrate spectrum
1 Mannosylphosphodecaprenol + 1,6-alpha-D-mannopentaose [1]
2 Mannosylphosphodecaprenol + 1,6-alpha-D-mannooligosaccharides
(poor substrates are mannose and 1,6-alpha-mannobiose, no substrates
are 1,2-, 1,3-alpha-mannooligosaccharides or myo-inositol-2,6-dimanno-
side) [1]
3 Mannosylphosphodecaprenol + 1,6-alpha-D-mannotriose [1]
4 Mannosylphosphodecaprenol + 1,6-alpha-D-mannotetraose [1]
5 Mannosylphosphodecaprenol + myo-inositol-trimannoside (with an
1,6-alpha-mannobiose unit as part of its structure) [1]
6 Mannosylphosphodecaprenol + methyl-alpha-D-mannoside [1]

Product spectrum

1 Decaprenol phosphate + 1,6-alpha-D-mannohexaose
2 Decaprenol phosphate + 1,6-alpha-D-mannosyl-1,6-alpha-D-manno-
  oligosaccharides [1]
3 Decaprenol phosphate + ?
4 Decaprenol phosphate + ?
5 Decaprenol phosphate + ?
6 Decaprenol phosphate + ?

Inhibitor(s)

Cofactor(s)/prosthetic group(s)/activating agents

Metal compounds/salts

Turnover number (min^{-1})

Specific activity (U/mg)

K_m-value (mM)

2 (1,6-alpha-D-mannotriose) [1]; 2.4 (1,6-alpha-D-mannotetraose) [1]; 2.6
(1,6-alpha-D-mannopentaose) [1]; 28.6 (methyl-alpha-D-mannoside) [1]; 35.7
(1,6-alpha-mannobiose) [1]; 86.9 (mannose) [1]

pH-optimum

7 (assay at) [1]

pH-range

Temperature optimum (°C)

37 (assay at) [1]

Temperature range (°C)

---

3 ENZYME STRUCTURE

Molecular weight

Subunits

Glycoprotein/Lipoprotein

–

---

4 ISOLATION/PREPARATION

Source organism

Mycobacterium smegmatis [1]

Source tissue
  Cell [1]

Localization in source
  Membrane [1]

Purification

Crystallization
  –

Cloned
  –

Renatured
  –

---

5 STABILITY

pH

Temperature (°C)

Oxidation

Organic solvent

General stability information

Storage

---

6 CROSSREFERENCES TO STRUCTURE DATABANKS

PIR/MIPS code

Brookhaven code

---

7 LITERATURE REFERENCES

[1] Yokoyama, K., Ballou, C.E.: J. Biol. Chem.,264,21621–21628 (1989)

---

1 NOMENCLATURE

EC number
2.4.1.200

Systematic name
Inulin D-fructosyl-D-fructosyltransferase (1,2':2',1-dianhydride-forming)

Recommended name
Inulin fructotransferase (depolymerizing, difructofuranose-1,2':2',1-dian-
hydride-forming)

Synonyms
Inulin fructotransferase (DFA-I-producing) [1]
More (cf. EC 2.4.1.93)

CAS Reg. No.
125008-19-7

---

2 REACTION AND SPECIFICITY

Catalysed reaction
Removes successive terminal D-fructosyl-D-fructofuranosyl groups from
inulin as the cyclic 1,2':2',1-dianhydride, leaving a residual tetra- or pentasaccharide

Reaction type

Natural substrates

Substrate spectrum
1 Inulin [1]

Product spectrum
1 Di-D-fructofuranose-1,2':2,1'-dianhydride (i.e. DFA I) [1]

Inhibitor(s)
Hg^{2+} (1 mM, strong) [1]; Fe^{3+} (1 mM, strong) [1]; Co^{2+} (weak) [1]; Cu^{2+}
(weak) [1]; Ni^{2+} (weak) [1]

Cofactor(s)/prosthetic group(s)/activating agents
EDTA (slight stimulation) [1]; More (SDS has no influence) [1]

Metal compounds/salts

Turnover number (min⁻¹)

Specific activity (U/mg)

K_m-value (mM)

pH-optimum
6.0 [1]

pH-range
3.5–8.0 (10% of maximal activity at pH 3.5, 25% of maximal activity at
pH 8.0) [1]

Temperature optimum (°C)
40 [1]

Temperature range (°C)
30–65 (80% of maximal activity at 30°C, 20% of maximal activity at 65°C) [1]

---

3 ENZYME STRUCTURE

Molecular weight
46000 (Arthrobacter globiformis, gel filtration) [1]

Subunits
Monomer (1 × 39000, Arthrobacter globiformis, SDS-PAGE) [1]

Glycoprotein/Lipoprotein
–

---

4 ISOLATION/PREPARATION

Source organism
Arthrobacter globiformis (S14–3) [1]

Source tissue
Supernatant [1]

Localization in source

Purification
Arthrobacter globiformis (S14–3, partial) [1]

Crystallization
–

Cloned
–

Renatured
–

5 STABILITY

pH
   3.0–9.0 (stable at 25°C) [1]

Temperature (°C)
   30–70 (stable, rapid inactivation above 80°C) [1]; 80 (rapid inactivation
   above) [1]

Oxidation

Organic solvent

General stability information

Storage

---

6 CROSSREFERENCES TO STRUCTURE DATABANKS

PIR/MIPS code

Brookhaven code

---

7 LITERATURE REFERENCES

[1] Seki, K., Haraguchi, K., Kishimoto, M., Koboyashi, S., Kainuma, K.: Agric. Biol. Chem.,
    53,2089–2094 (1989)

1 NOMENCLATURE

EC number
2.4.1.201

Systematic name
UDP-N-acetyl-D-glucosamine:glycoprotein (N-acetyl-D-glucosamine to D-mannose of N-acetyl-beta-D-glucosaminyl-1,6-beta-D-(N-acetyl-D-glucosaminyl-1,2-)-beta-D-mannosyl-R) beta-1,4-N-acetyl-D-glucosaminyl-transferase

Recommended name
Mannosyl-glycoprotein beta-1,4-N-acetylglucosaminyltransferase

Synonyms
N-Glycosyl-oligosaccharide-glycoprotein N-acetylglucosaminyltransferase VI
Acetylglucosaminyltransferase, uridine diphosphoacetylglucosamine-glyco-peptide beta-1→4-, VI
Acetylglucosaminyltransferase, uridine diphosphoacetylglucosamine-glyco-peptide beta-1.4-, VI
More (cf. EC 2.4.1.101, EC 2.4.1.143, EC 2.4.1.144, EC 2.4.1.145, EC 2.4.1.146)

CAS Reg. No.
119699-68-2

---

2 REACTION AND SPECIFICITY

Catalysed reaction
UDP-N-acetyl-D-glucosamine + N-acetyl-beta-D-glucosaminyl-1,6-beta-D-(N-acetyl-D-glucosaminyl-1,2)-beta-D-mannosyl-R →
→ UDP + N-acetyl-beta-D-glucosaminyl-1,6-beta-D-(N-acetyl-D-glucos-aminyl-1,2-beta)-(N-acetyl-D-glucosaminyl-1,4-beta)-D-mannosyl-R

Reaction type
Hexosyl group transfer

Natural substrates
UDP-N-acetyl-D-glucosamine + N-acetylglucosaminyl-1,6-beta-D-(N-ace-tylglucosaminyl-1,2)-beta-D-mannosyl-R (involved in N-glycan biosynthesis [2], reaction in biosynthesis of penta-antennary oligosaccharides of hen ovo-mucoid [1]) [1, 2]

Substrate spectrum
1 UDP-N-acetyl-D-glucosamine + N-acetylglucosaminyl-1,6-beta-D-(N-ace-tylglucosaminyl-1,2)-beta-D-mannosyl-R (R represents the remainder of the N-oligosaccharide core of asparagine-linked glycopeptides (+/- fucose), minimum structure requirement: N-acetylglucosaminyl-beta-1,6-(N-acetylglucosaminyl-beta-1,2)-alpha-mannosyl-R [2], acts on bisected and non-bisected substrates [1, 2], synthetic acceptors are N-acetylglucosaminyl-1,6-beta-D-(N-acetylglucosaminyl-1,2)-alpha-mannosyl-1,6-beta-D-mannosyl-R with-R: -(CH$_2$)$_8$-COOH or -CH$_3$ [1], substrate specificity [1], no acceptor substrates are compounds lacking beta-2-linked N-acetylglucosamine residues [2]) [1, 2]

Product spectrum
1 UDP + N-acetylglucosaminyl-1,6-beta-D-(N-acetylglucosaminyl-1,2-beta)-(N-acetyl-D-glucosaminyl-1,4)-beta-D-mannosyl-R [1, 2]

Inhibitor(s)
EDTA [1]; AMP (0.0125 mM, in the presence of 0.125 M N-acetylglucos-amine) [1]; ATP (0.0125 mM, in the presence of 0.125 M N-acetylglucos-amine) [1]; UDP (0.0125 mM, in the presence of 0.125 M N-acetylglucos-amine) [1]; UDPhexanolamine (0.0125 mM, in the presence of 0.125 M N-acetylglucosamine) [1]; N-Acetylglucosamine (weak, 0.125 M) [1]

Cofactor(s)/prosthetic group(s)/activating agents
Triton X-100 (activation, 0.125–1.25%) [1, 2]

Metal compounds/salts
Mn^{2+} (requirement, 0.05–0.1 M) [1, 2]; Ca^{2+} (activation, 12.5 mM, can substitute partially for Mn^{2+}) [1, 2]; Co^{2+} (activation, 12.5 mM, can substitute partially for Mn^{2+}) [1, 2]; Mg^{2+} (activation, 12.5 mM, can substitute partially for Mn^{2+}) [1, 2]; More (no activation with 12.5 mM Cd^{2+}, Ni^{2+} or Zn^{2+}) [1, 2]

Turnover number (min^{-1})

Specific activity (U/mg)

K$_m$-value (mM)
0.09 (N-acetylglucosaminyl-1,6-beta-D-(N-acetylglucosaminyl-1,2)-beta-D-mannosyl-R) [1]; 0.6 (UDP-N-acetyl-D-glucosamine) [1]

pH-optimum
7–8.4 [1, 2]

pH-range
5–8.4 (about half-maximal activity at pH 5 and about 80% of maximal activity at pH 8.4) [1]

Temperature optimum (°C)
37 (assay at) [1, 2]

Temperature range (°C)

3 ENZYME STRUCTURE

Molecular weight

Subunits

Glycoprotein/Lipoprotein

–

4 ISOLATION/PREPARATION

Source organism
Chicken (hen [2]) [1, 2]; Duck [1]; Turkey [1]; More (not in rat, pig or human colon, pig stomach, rat liver or bovine kidney) [1]

Source tissue
Oviduct [1, 2]; Liver (chicken) [1]; Colon (chicken, duck) [1]; Intestine (turkey) [1]; More (tissue distribution) [1]

Localization in source
Microsomes (hen) [1, 2]

Purification

Crystallization
–

Cloned
–

Renatured
–

5 STABILITY

pH

Temperature (°C)

Oxidation

Organic solvent

General stability information

Storage

---

6 CROSSREFERENCES TO STRUCTURE DATABANKS

PIR/MIPS code

Brookhaven code

---

7 LITERATURE REFERENCES

[1] Brockhausen, I., Hull, E., Hindsgaul, O., Schachter, H., Shah, R.N., Michnick, S.W., Carver, J.P.: J. Biol. Chem.,264,11211–11221 (1989)
[2] Schachter, H., Brockhausen, I., Hull, E.: Methods Enzymol.,179,351–397 (1989) (Review)

1 NOMENCLATURE

EC number
2.4.1.202

Systematic name
UDPglucose:2,4-dihydroxy-7-methoxy-2H-1,4-benzoxazin-3(4H)-one 2-D-glucosyltransferase

Recommended name
2,4-Dihydroxy-7-methoxy-2H-1,4-benzoxazin-3(4H)-one 2-D-glucosyltransferase

Synonyms
Glucosyltransferase, uridine diphosphoglucose-2,4-dihydroxy-7-methoxy-2H-1,4-benzoxazin-3(4H)-one 2-

CAS Reg. No.
122544-56-3

---

2 REACTION AND SPECIFICITY

Catalysed reaction
UDPglucose + 2,4-dihydroxy-7-methoxy-2H-1,4-benzoxazin-3(4H)-one →
→ UDP + 2,4-dihydroxy-7-methoxy-2H-1,4-benzoxazin-3(4H)-one 2-D-glucoside

Reaction type
Hexosyl group transfer

Natural substrates

Substrate spectrum
1 UDPglucose + 2,4-dihydroxy-7-methoxy-2H-1,4-benzoxazin-3(4H)-one [1]
2 UDPglucose + 2,4-dihydroxy-1,4-benzoxazin-3-one (peak 1 glucosyltransferase has 3.6% of activity compared to activity towards 2,4-dihydroxy-7-methoxy-2H-1,4-benzoxazin-3(4H)-one, peak 2 glucosyltransferase has 57% of activity compared to activity towards 2,4-dihydroxy-7-methoxy-2H-1,4-benzoxazin-3(4H)-one) [1]

Product spectrum
1 UDP + 2,4-dihydroxy-7-methoxy-2H-1,4-benzoxazin-3(4H)-one 2-D-glucoside [1]
2 ?

Inhibitor(s)
N-Ethylmaleimide (1 mM, 96% decrease of activity, addition of dithioerythritol: full activity) [1]; EDTA (16% decrease of activity) [1]; Fe^{2+} (5 mM, 5% remaining activity) [1]; Cu^{2+} (5 mM, 3% remaining activity) [1]

Cofactor(s)/prosthetic group(s)/activating agents
Ascorbate (5 mM, stimulation of 42%) [1]; Dithioerythritol (5 mM, stimulation of 96%) [1]; Glutathione (5 mM, stimulation of 92%) [1]; 2-Mercaptoethanol (5 mM, stimulation of 79%) [1]

Metal compounds/salts
$CaCl_2$ (5 mM, stimulation) [1]; $MgCl_2$ (5 mM, stimulation) [1]

Turnover number (min^{-1})

Specific activity (U/mg)
1.218 (enzyme in peak 1 + peak 2) [1]

K_m-value (mM)
0.174 (2,4-dihydroxy-7-methoxy-2H-1,4-benzoxazin-3(4H)-one, enzyme in peak 2) [1]; 0.217 (2,4-dihydroxy-7-methoxy-2H-1,4-benzoxazin-3(4H)-one, enzyme in peak 1) [1]; 0.28 (UDPglucose (+ 2,4-dihydroxy-1,4-benzoxazin-3-one), enzyme in peak 2) [1]; 0.286 (UDPglucose (+ 2,4-dihydroxy-7-methoxy-2H-1,4-benzoxazin-3(4H)-one), enzyme in peak 1) [1]; 0.638 (2,4-dihydroxy-1,4-benzoxazin-3-one, enzyme in peak 2) [1]

pH-optimum
8.5 [1]

pH-range
6.5–10.0 (10% of maximal activity at pH 6.5 and 30% of maximal activity at pH 10.0) [1]

Temperature optimum (°C)
37 (assay at) [1]; 45 [1]

Temperature range (°C)
25–55 (88% of maximal activity at 55°C, 60% of maximal activity at 25°C) [1]

3 ENZYME STRUCTURE

Molecular weight
50000 (Zea mays, gel filtration) [1]

Subunits

Glycoprotein/Lipoprotein
–

4 ISOLATION/PREPARATION

Source organism
  Zea mays [1]

Source tissue
  Seedling [1]

Localization in source

Purification
  Zea mays (partial, activity elutes from Q-Sepharose column in two peaks
  exhibiting different activities towards substrates) [1]

Crystallization
  –

Cloned
  –

Renatured
  –

---

5 STABILITY

pH

Temperature (°C)

Oxidation

Organic solvent

General stability information

Storage

---

6 CROSSREFERENCES TO STRUCTURE DATABANKS

PIR/MIPS code

Brookhaven code

---

7 LITERATURE REFERENCES

[1] Bailey, B.A., Larson, R.L.: Plant Physiol.,90,1071–1076 (1989)

---

1 NOMENCLATURE

EC number
2.4.1.203

Systematic name
UDPglucose:zeatin O-beta-D-glucosyltransferase

Recommended name
Zeatin O-beta-D-glucosyltransferase

Synonyms
Glucosyltransferase, uridine diphosphoglucose-zeatin O-
Zeatin O-glucosyltransferase
More (cf. EC 2.4.1.204)

CAS Reg. No.
123644-76-8

2 REACTION AND SPECIFICITY

Catalysed reaction
UDPglucose + zeatin →
→ UDP + O-beta-D-glucosylzeatin

Reaction type
Hexosyl group transfer

Natural substrates
UDPglucose + zeatin (important in regulating the level of active cytokinin (zeatin) in plant tissues) [1]

Substrate spectrum
1 UDPglucose + zeatin (trans-zeatin, not: cis-zeatin, ribosylzeatin, dihydrozeatin) [1]
2 UDPxylose + zeatin [1]
3 UDPgalactose + zeatin [1]

Product spectrum
1 UDP + O-beta-D-glucosylzeatin [1]
2 UDP + O-xylosylzeatin [1]
3 UDP + galactosyl derivative of zeatin [1]

Inhibitor(s)

Cofactor(s)/prosthetic group(s)/activating agents

Metal compounds/salts

Turnover number (min⁻¹)

Specific activity (U/mg)
    More [1]

K_m-value (mM)
    0.028 (trans-zeatin) [1]; 0.216 (UDPglucose) [1]; 2.7 (UDPxylose) [1]

pH-optimum
    8 [1]

pH-range

Temperature optimum (°C)
    27 (assay at) [1]

Temperature range (°C)

---

3 ENZYME STRUCTURE

Molecular weight
    44000 (Phaseolus lunatus, gel filtration) [1]

Subunits

Glycoprotein/Lipoprotein
    –

---

4 ISOLATION/PREPARATION

Source organism
    Phaseolus lunatus [1]

Source tissue
    Immature embryos [1]

Localization in source

Purification
    Phaseolus lunatus [1]

Crystallization
    –

Cloned

–

Renatured

–

5 STABILITY

pH

Temperature (°C)

Oxidation

Organic solvent

General stability information

Storage

6 CROSSREFERENCES TO STRUCTURE DATABANKS

PIR/MIPS code

Brookhaven code

7 LITERATURE REFERENCES

[1] Dixon, S.C., Martin, R.C., Mok, M.C., Shaw, G., Mok, D.W.S.: Plant Physiol.,90,
    1316–1321 (1989)

1 NOMENCLATURE

EC number
2.4.1.204

Systematic name
UDP-D-xylose:zeatin O-beta-D-xylosyltransferase

Recommended name
Zeatin O-beta-D-xylosyltransferase

Synonyms
Xylosyltransferase, uridine diphosphoxylose-zeatin
Zeatin O-xylosyltransferase
More (cf. EC 2.4.1.203)

CAS Reg. No.
110541-22-5

---

2 REACTION AND SPECIFICITY

Catalysed reaction
UDP-D-xylose + zeatin →
→ UDP + O-beta-D-xylosylzeatin

Reaction type
Pentosyl group transfer

Natural substrates
UDPxylose + trans-zeatin (may be involved in the nuclear-cytoplasmic transport of cytokinins and related molecules or possibly, with chromatin of rapidly dividing cells) [2]

Substrate spectrum
1 UDP-D-xylose + zeatin (trans-zeatin [1, 2], not: cis-zeatin, ribosylzeatin [1]) [1, 2]
2 UDP-D-xylose + dihydrozeatin [1]

Product spectrum
1 UDP + O-beta-D-xylosylzeatin [1]
2 ?

Inhibitor(s)

Cofactor(s)/prosthetic group(s)/activating agents

Metal compounds/salts

Turnover number (min^{-1})

Specific activity (U/mg)
  More [1]

K$_m$-value (mM)
  0.002 (trans-zeatin) [1]; 0.003 (UDPxylose) [1]; 0.010 (dihydrozeatin) [1]

pH-optimum
  8.0–8.5 [1]

pH-range

Temperature optimum (°C)
  25 (assay at) [1]

Temperature range (°C)

3 ENZYME STRUCTURE

Molecular weight
  50000 (Phaseolus vulgaris, gel filtration) [1]

Subunits

Glycoprotein/Lipoprotein
  –

4 ISOLATION/PREPARATION

Source organism
  Phaseolus vulgaris [1, 2]

Source tissue
  Embryos [1]; Endosperm (callus) [2]

Localization in source
  Soluble [1]; Nucleus [2]; Cytoplasm [2]

Purification

Crystallization
  –

Cloned

–

Renatured

–

5 STABILITY

pH

Temperature (°C)

Oxidation

Organic solvent

General stability information

Storage

6 CROSSREFERENCES TO STRUCTURE DATABANKS

PIR/MIPS code

Brookhaven code

7 LITERATURE REFERENCES

[1] Turner, J.E., Mok, D.W.S., Mok, M.C., Shaw, G.: Proc. Natl. Acad. Sci. USA,84, 3714–3717 (1987)
[2] Martin, R.C., Mok, M.C., Mok, D.W.S.: Proc. Natl. Acad. Sci. USA,90,953–957 (1993)

1 NOMENCLATURE

EC number
2.4.1.205

Systematic name
UDPgalactose:galactogen beta-1,6-D-galactosyltransferase

Recommended name
Galactogen 6beta-galactosyltransferase

Synonyms
Galactosyltransferase, uridine diphosphogalactose-galactogen
1,6-D-Galactosyltransferase [1]
beta-(1→6)-D-Galactosyltransferase [1]

CAS Reg. No.
88273-54-5

2 REACTION AND SPECIFICITY

Catalysed reaction
UDPgalactose + galactogen →
→ UDP + 1,6-beta-D-galactosylgalactogen

Reaction type
Hexosyl group transfer

Natural substrates
UDPgalactose + galactogen (Helix pomatia, galactogen biosynthesis) [1]

Substrate spectrum
1 UDP-D-galactose + galactogen (galactogen from Helix pomatia best
   acceptor, galactogen from Lymnaea stagnalis, 22% as effective as galactogen from Helix pomatia) [1]
2 UDP-D-galactose + galactan (galactan from bovine lung 64% as effective
   as galactogen from Helix pomatia) [1]
3 UDP-D-galactose + arabinogalactan (arabinogalactan from larch wood,
   22% as effective as galactogen from Helix pomatia) [1]

Product spectrum
1 UDP + 1,6-beta-D-galactosylgalactogen
2 ?
3 ?

Inhibitor(s)

Cofactor(s)/prosthetic group(s)/activating agents

Metal compounds/salts
  $MgCl_2$ (10 mM, 30% stimulation) [1]; More ($MnCl_2$ has no effect at 5 or
  10 mM) [1]

Turnover number (min^{-1})

Specific activity (U/mg)
  3.466 [1]

K_m-value (mM)
  0.36 (UDP-D-galactose) [1]

pH-optimum
  6.8–7.5 [1]

pH-range

Temperature optimum (°C)
  20 (assay at) [1]

Temperature range (°C)

---

3 ENZYME STRUCTURE

Molecular weight

Subunits
  ? (x × 70000, triplet band, Helix pomatia, SDS-PAGE) [1]

Glycoprotein/Lipoprotein
  –

---

4 ISOLATION/PREPARATION

Source organism
  Helix pomatia [1]

Source tissue
  Albumen gland [1]

Localization in source
  Endomembrane-associated (probably in Golgi complex) [1]

Purification
  Helix pomatia (partial) [1]

Crystallization
–

Cloned
–

Renatured
–

---

5 STABILITY

pH

Temperature (°C)

Oxidation

Organic solvent

General stability information
Glycerol, 20% v/v, stabilizes [1]

Storage
–80°C, crude homogenate, 6 months [1]; 4°C, purified enzyme, stable for
1 week [1]; 0°C, 24 h, 50% loss of activity [1]

---

6 CROSSREFERENCES TO STRUCTURE DATABANKS

PIR/MIPS code

Brookhaven code

---

7 LITERATURE REFERENCES

[1] Goudsmit, E.M., Ketchum, P.A., Grossens, M.K., Blake, D.A.: Biochim. Biophys. Acta,
992,289–297 (1989)

1 NOMENCLATURE

EC number
2.4.1.206

Systematic name
UDP-N-acetyl-D-glucosamine:D-galactosyl-1,4-beta-D-glucosylceramide beta-1,3-acetylglucosaminyltransferase

Recommended name
Lactosylceramide 1,3-N-acetyl-beta-D-glucosaminyltransferase

Synonyms
LA2 synthase [1]
beta1→3-N-Acetylglucosaminyltransferase [2]
Acetylglucosaminyltransferase, uridine diphosphoacetylglucosamine-lactosylceramide beta-
Lactosylceramide beta-acetylglucosaminyltransferase

CAS Reg. No.
83682-80-8

2 REACTION AND SPECIFICITY

Catalysed reaction
UDP-N-acetyl-D-glucosamine + D-galactosyl-1,4-beta-D-glucosylceramide →
→ UDP + N-acetyl-D-glucosaminyl-1,3-beta-D-galactosyl-1,4-beta-D-glucosylceramide

Reaction type
Hexosyl group transfer

Natural substrates
UDP-N-acetyl-D-glucosamine + lactosylceramide (ganglioside metabolism [1], may play a key role in regulating the level of various types of lactoseries tumor-associated antigens with the lacto type 1 or 2 chain [2]) [1, 2]

Substrate spectrum
1 UDP-N-acetyl-D-glucosamine + D-galactosyl-1,4-beta-D-glucosylceramide (i.e. lactosylceramide) [1, 2]

Product spectrum
1 UDP + N-acetyl-D-glucosaminyl-1,3-beta-D-galactosyl-1,4-beta-D-glucosylceramide [1, 2]

Inhibitor(s)

Cofactor(s)/prosthetic group(s)/activating agents

Metal compounds/salts

Turnover number (min^{-1})

Specific activity (U/mg)

K_m-value (mM)

pH-optimum
    7.3 (assay at) [1]

pH-range

Temperature optimum (°C)
    37 (assay at) [1]

Temperature range (°C)

---

3 ENZYME STRUCTURE

Molecular weight

Subunits

Glycoprotein/Lipoprotein
    –

---

4 ISOLATION/PREPARATION

Source organism
    Human [1, 2]

Source tissue
    Medulloblastoma cell line (Daoy and D341 Med, cultured both in vitro and
    as xenografts in nude mice) [1]; Colonic adenocarcinoma cell lines [2]

Localization in source

Purification

Crystallization
    –

Cloned
    –

Renatured
    –

5 STABILITY

pH

Temperature (°C)

Oxidation

Organic solvent

General stability information

Storage

6 CROSSREFERENCES TO STRUCTURE DATABANKS

PIR/MIPS code

Brookhaven code

7 LITERATURE REFERENCES

[1] Gottfries, J., Percy, A.K., Mansson, J.-E., Fredman, P., Wikstrand, C.J., Friedman, H.S., Bigner, D.D., Svennerholm, L.: Biochim. Biophys. Acta,1081,253–261 (1991)
[2] Holmes, E.H., Hakomori, S.-i., Ostrander, G.K.: J. Biol. Chem.,262,15649–15658 (1987)

1 NOMENCLATURE

EC number
2.4.2.1

Systematic name
Purine-nucleoside:orthophosphate ribosyltransferase

Recommended name
Purine-nucleoside phosphorylase

Synonyms
Inosine phosphorylase
PNPase [4, 21]
PUNPI [17, 22]
PUNPII [22]
Phosphorylase, purine nucleoside
Inosine-guanosine phosphorylase
Nucleotide phosphatase (2.4.2.1)
Purine deoxynucleoside phosphorylase
Purine deoxyribonucleoside phosphorylase
Purine nucleoside phosphorylase
Purine ribonucleoside phosphorylase

CAS Reg. No.
9030-21-1

2 REACTION AND SPECIFICITY

Catalysed reaction
Purine nucleoside + phosphate →
→ purine + alpha-D-ribose 1-phosphate (bimolecular nucleophilic substitution reaction involving a Walden inversion [4], sequential mechanism [2], rapid equilibrium random bi-bi reaction [7], Theorell-Chance mechanism [8, 32], ordered bi-bi mechanism with nucleoside being the first substrate to add and the base the last product to leave [26], rapid equilibrium random bi-bi abortive complexes [28], sequential bireactant mechanism [37])

Reaction type
Pentosyl group transfer

Natural substrates
Purine nucleoside + phosphate (mammalian enzyme concerned primarily with nucleoside breakdown, enzyme in fish skin plays a key role in the deposition of guanine and hypoxanthine crystals (reflectors in the scales, skin and eyes of fish), role in bacteria is unclear) [4]

Enzyme Handbook © Springer-Verlag Berlin Heidelberg 1996
Duplication, reproduction and storage in data banks are only
allowed with the prior permission of the publishers

Substrate spectrum

1   Purine nucleoside + phosphate (r [4, 12, 18, 28], nucleoside synthesis is favored, equilibrium is reached when 80–90% of the purine base is ribosylated [4], specificity overview [4], enzyme from mammalian tissues shows considerable specificity for hypoxanthine, guanine and xanthine and their respective nucleosides, but shows little or no activity with adenine or adenine nucleosides [4], an enzyme from rabbit bone marrow shows high specificity for guanine nucleosides, it reacts equally well with guanosine and deoxyguanosine but has negligible activity with inosine, xanthosine and pyrimidine nucleosides [4], an enzyme from rat ventral prostate catalyzes phosphate-dependent release of adenine from 5'-methylthioadenosine, human erythrocytic and bovine spleen enzymes do not react with this compound [4], overview: specificity of mammalian and E. coli enzyme [39], overview: 7-alkylguanosines with higher alkyl groups as substrates [33], structural requirements of substrates of mammalian and E. coli enzyme [39], 2 purine nucleoside phosphorylases: 1. inosine-guanosine phosphorylase, 2. adenosine phosphorylase specific for adenosine and deoxyadenosine [11], adenosine phosphorylase distinct from inosine-guanosine phosphorylase [41], inosine-guanosine phosphorylase different from deo-encoded purine nucleoside phosphorylase [43], specificity not completely determined, can also catalyze ribosyltransferase reactions of the type catalyzed by EC 2.4.2.5) [4, 11, 12, 18, 28, 33, 39, 41, 43]

2   Inosine + phosphate (inosine-guanosine phosphorylase, not adenine phosphorylase [11]) [1, 2, 4, 7, 10–13, 15, 17–26, 28, 29, 36–38, 42, 43]

3   Deoxyinosine + phosphate (2'- [43], 5'- [43], best substrate [10], inosine-guanosine phosphorylase, not adenine phosphorylase [11]) [1, 2, 10, 11, 17–24, 26, 28, 38, 43]

4   Guanosine + phosphate (inosine-guanosine phosphorylase, not adenine phosphorylase [11], at 47.4% of the activity with inosine [19]) [2, 8, 10–12, 15, 17, 19–26, 28, 29, 37, 38, 42, 43]

5   Deoxyguanosine + phosphate (inosine-guanosine phosphorylase, not adenine phosphorylase [11]) [2, 10, 11, 17, 20–22, 24, 25, 28, 37, 38, 42, 43]

6   Purine riboside + arsenate [4]

7   Adenosine + phosphate (not [23, 25, 28, 29, 37, 42, 43], adenosine phosphorylase, not inosine-guanosine phosphorylase [11], 2.6% of the activity with inosine [19]) [10, 11, 18–20, 22, 24, 38]

8   Deoxyadenosine + phosphate (adenosine phosphorylase, inosine-guanosine phosphorylase not [11]) [10, 11, 18, 20–22, 24, 38, 43]

9   Adenine arabinoside + phosphate [18]

10 Xanthosine + phosphate (not [21, 38, 43], poor substrate [25, 29, 37], at 43.5% of the reaction with inosine [19]) [19, 20, 23, 25, 28, 29, 37]

11 2',3'-Dideoxyinosine + phosphate [23, 43]
12 Ribavirin + phosphate [23]
13 6-Mercaptopurine riboside + phosphate [29]
14 N-1-Methylinosine + phosphate (not substrate for calf enzyme, substrate for E. coli enzyme) [33]
15 N^1-Methylguanosine + phosphate (not substrate for calf enzyme, substrate for E. coli enzyme) [33]
16 7-Methylguanosine + phosphate (ir) [35]
17 7-Methylinosine + phosphate (ir) [35]
18 3-Deazainosine + phosphate [39]
19 Purine riboside + phosphate (activity with E. coli enzyme no activity with calf enzyme) [39]
20 1,6-Dihydropurine riboside + phosphate [39]
21 N^6-Furfuryladenine + phosphate [41]
22 6-Mercaptopurine + phosphate [4]
23 6-Thioguanine + phosphate [4]
24 8-Azaguanine + phosphate (not [12]) [4]
25 More (not: allopurinol [12], adenine [4, 12], hypoxanthine arabinoside [43], pyrimidine nucleosides [4, 38], uridine [10], cytidine [10], deoxyuridine [10], deoxycytidine [10], deoxyribose 1-phosphate can replace ribose 1-phosphate in the reverse reaction [28]) [4, 10, 12, 28, 38, 43]

Product spectrum

1 Purine + alpha-D-ribose 1-phosphate [4, 12, 18, 28]
2 Hypoxanthine + alpha-D-ribose 1-phosphate [4, 7, 12, 26, 28, 37, 43]
3 ?
4 Guanine + alpha-D-ribose 1-phosphate [12, 28, 37, 43]
5 ?
6 Purine + riboside + arsenate (unstable intermediate ribose 1-arsenate hydrolyzes immediately upon formation to yield ribose and arsenate) [4]
7 Adenine + alpha-D-ribose 1-phosphate [4]
8 ?
9 ?
10 ?
11 ?
12 ?
13 ?
14 ?
15 ?
16 7-Methylguanine + alpha-D-ribose 1-phosphate [35]
17 7-Methylhypoxanthine + alpha-D-ribose 1-phosphate [35]
18 ?
19 ?
20 ?

21 ?
22 ?
23 ?
24 ?
25 ?

Inhibitor(s)

$HgCl_2$ [17, 21, 23]; Formycin A (inactive to mammalian enzyme, inhibition of E. coli enzyme [9, 39], not [12]) [9, 39]; Formycin B (competitive, inhibits phosphorolysis of both inosine and 6-mercaptopurine [4], moderate inhibition of mammalian enzyme effective inhibition of E. coli enzyme [9, 39]) [4, 9, 12, 29, 39]; Deoxyinosine [4]; 7-Deazainosine [4, 39]; Adenine [4]; $AgNO_3$ [21, 23]; $CuSO_4$ (slight) [21, 23]; Hypoxanthine (product inhibition [25]) [4, 25, 32, 37]; Guanine (product inhibition [25]) [4, 7, 25, 28, 32, 37]; 6-Mercaptopurine [4]; 6-Methylthiopurine [4]; 2-Amino-6-methylthiopurine [4]; 6-Thioxanthine [4]; Allopurinol [4]; Oxoallopurinol [4]; Diphosphate [4]; p-Substituted mercuribenzoate (no protection by addition of substrate) [4]; Guanosine [32]; p-Chloromercuriphenylsulfonic acid (dithiothreitol restores activity) [43]; Thiol reagents (reactivation by excess of beta-mercaptoethanol) [4]; Ribose 1-phosphate [7, 25, 28, 32, 37]; N(1)-Methylformycin A [9]; N(6)-Methylformycin A [9]; p-Chloromercuribenzoate (protection by: phosphate [11], adenosine [11], ribose 1-phosphate [11, 38], inosine [38], inactivation of adenosine phosphorylase [11], reversal by beta-mercaptoethanol [11, 25], no inhibition [22]) [11, 17, 21, 23, 25, 26, 32, 37, 38, 41, 43]; SH-reagents [23]; N-Ethylmaleimide [23]; 5,5'-Dithiobis(2-nitrobenzoic acid) (addition of a substrate protects [4]) [4, 26, 28]; Purine riboside (inhibition of E. coli enzyme, no inhibition of calf spleen enzyme) [39]; 1,6-Dihydropurine riboside (inhibition of E. coli enzyme, no inhibition of calf spleen enzyme) [39]; Folic acid (competitive) [40]; D,L-6-Methyl-5,6,7,8-tetrahydropterine (competitive) [40]; Aminopterin (noncompetitive) [40]; Erythro-9-(2-hydroxy-3-nonyl)adenine (competitive) [40]; Inosine (selectively inhibits the ribosylation of hypoxanthine and guanine catalyzed by Sarcoma 180 and rat liver extract [41]) [4, 41]; 2-Chloroadenosine (selectively inhibits ribosylation of adenine and N^6-furfuryladenine) [41]; 8-Amino-5'-deoxy-5'-chloroguanine [42]; 8-Amino-9-benzylguanine [42]; More (overview: structural requirement of inhibitors of E. coli and mammalian enzyme) [39]

Cofactor(s)/prosthetic group(s)/activating agents

Inosine (substrate activation) [1, 7]; Nucleosides (substrate activation at high concentration) [26]; More (no cofactors known) [26]

Metal compounds/salts

Co^{2+} (1 mM, stimulates) [10]; No metal requirement [26]

4

Turnover number (min^{-1})

7140 (inosine) [28]; 5100 (2'-deoxyinosine) [28]; 3540 (guanosine) [28]; 2700 (2'-deoxyguanosine) [28]; 240 (xanthosine) [28]; 6180 (hypoxanthine (+ ribose 1-phosphate)) [28]; 7620 (hypoxanthine (+ deoxyribose 1-phosphate), guanine (+ ribose 1-phosphate)) [28]; 9480 (guanine (+ deoxyribose 1-phosphate)) [28]

Specific activity (U/mg)

301 (Salmonella typhimurium) [38]; 77.5 [28]; 195 (E. coli) [38]; 160 (Salmonella typhimurium [24]) [10, 24]; 8.7 [12]; 95 [14]; 405 [17]; 195.1 [23]; 132 [29]; 50.47 [18]; 50.6 [22]; 55 [8]; 50 [40]; 143 (rat) [24]; 93 [26]; 78.3 [32]; 47.12 [37]; 60.0 [27, 30]; 0.014 [43]; 47.12 [25]; More [1, 4, 7, 11, 15, 19]

K_m-value (mM)

More (effect of pH [3, 35], temperature dependent [18], overview: K_m of 7-alkylguanosines with higher alkyl groups [33]) [1, 3, 4, 13, 15, 17–30, 33, 35–43]; 0.013 (phosphate (+ inosine)) [15]; 0.014 (guanine (+ ribose 1-phosphate)) [28]; 0.0147 (7-methylguanosine) [35]; 0.015 (phosphate (+ guanosine)) [15]; 0.017 (guanine (+ deoxyribose 1-phosphate)) [28]; 0.019 (hypoxanthine) [26]; 0.020 (hypoxanthine, guanine [26], hypoxanthine (+ ribose 1-phosphate) [28]) [26, 28]; 0.021 (hypoxanthine (+ deoxyribose 1-phosphate)) [28]; 0.022 (guanosine) [20]; 0.027 (deoxyinosine, deoxyguanosine) [20]; 0.028 (phosphate (+ inosine)) [28]; 0.029 (guanosine) [15]; 0.033 (inosine) [12]; 0.041 (ribose 1-phosphate (+ guanine)) [28]; 0.042 (ribose 1-phosphate (+ hypoxanthine)) [28]; 0.047 (deoxyinosine) [10]; 0.050 (inosine) [10]; 0.075 (adenosine) [22]; 0.082 (hypoxanthine) [12]; 0.095 (xanthosine) [20]; 0.10 (2'-deoxyguanosine) [17]; 0.129 (deoxyribose 1-phosphate (+ hypoxanthine)) [28]; 0.130 (adenosine) [20]; 0.133 (deoxyribose 1-phosphate (+ guanine)) [28]; 0.14 (guanosine) [17]; 0.145 (deoxyadenosine) [20]; 0.19 (2'-deoxyadenosine) [2]; 0.2 (inosine, inosine-guanosine phosphorylase [11], 2'-deoxyinosine [17], ribose 1-phosphate, phosphate [27]) [11, 17, 27]; 0.31 (inosine) [2]; 0.33 (deoxyguanosine) [2]; 0.37 (phosphate (+ inosine)) [10]; 0.45 (xanthosine) [28]; 0.55 (guanosine) [2]; 0.67 (deoxyinosine) [2]; 0.92 (2'-deoxyinosine) [22]; 1.9 (2'-deoxyguanosine) [22]; 2.08 (ribavirin) [23]; 3.9 (phosphate (+ inosine), inosine-guanosine phosphorylase) [11]

pH-optimum

5.2 (xanthosine) [20]; 5.5–8.0 [28]; 6.0–6.5 [12]; 6.0–8.0 [25, 37]; 6.5 [2]; 6.5–8.0 (in general PNPases show relatively broad pH optima between 6.5 and 8.0) [4]; 6.6 (inosine phosphorolysis) [43]; 6.8 (phosphate-stimulated ribosyl transfer from inosine to adenine [38], inosine synthesis [43]) [18, 38, 43]; 7.0 (inosine) [20]; 7.0–7.5 [15]; 7.0–11.0 [22]; 7.1 (deoxyinosine) [38]; 7.5 (Salmonella typhimurium [24], inosine [38]) [10, 24, 38]; 7.5–8.5 [19, 29]; 7.5–11 [17]; 8.2 (hypoxanthine) [38]; 8.5 [21]

pH-range

5–11.5 (5: about 50% of activity maximum, 11.5: about 80% of activity maximum) [22]

Temperature optimum (°C)

25 (assay at) [12]; 30 (assay at) [20]; 37 (assay at) [7, 11, 19, 27, 28]; 50 [19]; 60 [18]; 70 [21–23]; 80 [17]

Temperature range (°C)

30–80 (30°C: about 35% of activity maximum, 80°C: about 80% of activity maximum) [23]

---

3 ENZYME STRUCTURE

Molecular weight

31000 (Brevibacterium acetylicum, SDS-PAGE without 2-mercaptoethanol) [23]

46000 (rabbit, gel filtration) [37]

56000–58000 (Hyalomma dromedarii) [2]

58000 (Erwinia carotovora, SDS-PAGE without 2-mercaptoethanol) [21]

61000 (rabbit, gel filtration) [25]

67000 (human, gel filtration, equilibrium sedimentation, amino acid content) [8]

68000 (Chinese hamster, enzyme behaves as a mixture of dimers (MW 68000) and trimers (MW 89000) during gel filtration [27, 30], Bacillus stearothermophilus, gel filtration [17]) [17, 27, 30]

70000 (rabbit, gel filtration) [40]

75000–83000 (rabbit, gel filtration, sucrose density gradient centrifugation) [36]

78000–80000 (bovine, gel filtration) [32]

80000–92000 (human, gel filtration, sedimentation equilibrium analysis) [26]

81000 (human, gel filtration [1], calf spleen [4]) [1, 4]

87000 (Enterobacter cloacae, gel filtration) [19]

87000–92000 (human, gel filtration, ultracentrifugation, sucrose density gradient centrifugation) [31]

89000–90000 (chicken, gel filtration, determination of sedimentation coefficient, sedimentation equilibrium data) [16]

89000 (Chinese hamster, enzyme behaves as a mixture of dimers (MW 68000) and trimers (MW 89000) during gel filtration) [27, 30]

90000 (chicken, gel filtration [13], bovine, gel filtration [20]) [13, 20]

90400 (human, HPLC gel filtration) [14]

93800 (human, disc gel electrophoresis) [34]

94000 (Chinese hamster, sucrose gradient centrifugation) [27, 30]

95000 (Bacillus subtilis, gel filtration, inosine-guanosine phosphorylase) [11]

97000 (bovine, HPLC gel permeation) [28]

113000 (Bacillus stearothermophilus, PUNPII, gel filtration) [22]

120000 (Proteus vulgaris, gradient slab polyacrylamide gel electrophoresis) [15]

125300 (Plasmodium lophurae, gel filtration, nondenaturing PAGE) [12]
130000–141000 (Salmonella typhimurium, high speed sedimentation equilibrium studies, gel filtration) [10, 24]
138000 (E. coli, Salmonella typhimurium, gel filtration) [38]
141000 (Salmonella typhimurium) [24]
147630 (Plasmodium falciparum, sucrose gradient ultracentrifugation) [42]
150000–160000 (Bacillus subtilis, gel filtration, sucrose gradient centrifugation, adenosine phosphorylase) [11]
More [4, 6]

Subunits

Monomer (1 × 39000, rabbit, SDS-PAGE [37], 1 × 56000–58000, Hyalomma dromedarii [2], 1 × 31000, Brevibacterium acetylicum, SDS-PAGE in presence of 2-mercaptoethanol after heat treatment [23]) [2, 23, 37]
Dimer (2 × 34500, rabbit, SDS-PAGE [40], 2 × 30500, rabbit, SDS-PAGE [25], 2 × 34000, Bacillus stearothermophilus, SDS-PAGE [17], 2 × 32000, human, SDS-PAGE [8], 2 × or 3 × 30000, Chinese hamster, SDS-PAGE, enzyme behaves as a mixture of dimers and trimers during gel filtration (sucrose gradient centrifugation: trimer) [27], 2 × 44000, Camelus dromedarius [3], 2 × 24000, Bacillus cereus spore enzyme, in absence of phosphate, aggregates to a tetramer in presence of 10 mM phosphate, vegetative cell enzyme is a tetramer that does not undergo phosphate-induced association [4], 2 × 28000, Erwinia carotovora, SDS-PAGE after treatment at 100°C for 5 min in presence of 2-mercaptoethanol [21], 2 × 38000, bovine, SDS-PAGE [32]) [3, 4, 8, 17, 21, 25, 27, 32, 40]
Trimer (3 × 24000, chicken, SDS-PAGE [13], 3 × 31600, human, SDS-PAGE [14], 1 × 30000–32000 + 2 × 27000–28000, chicken, SDS-PAGE, 6 M guanidine hydrochloride gel filtration [16], 3 × 30000, human, SDS-PAGE [26, 31], 3 × 30000, bovine, SDS-PAGE [28], 3 × 29000, rat, SDS-PAGE [29], 3 × 29700, human, SDS-PAGE [34], rabbit hybridization studies [36]) [13, 14, 16, 26, 28, 29, 31, 34, 36]
Tetramer (4 × 24000, Bacillus cereus, vegetative cell enzyme, spore enzyme in presence of 10 mM phosphate, dimer in absence of phosphate [4], 4 × 28000, Bacillus subtilis, inosine-guanosine phosphorylase, SDS-PAGE [11], 4 × 28000, Bacillus stearothermophilus, enzyme PUNPII, SDS-PAGE [22]) [4, 11, 22]
Pentamer (5 × 23900, Plasmodium lophurae, SDS-PAGE) [12]
Hexamer (6 × 23500, Salmonella typhimurium [10, 24], SDS-PAGE [10], 6 × 25500, Bacillus subtilis, adenine phosphorylase, SDS-PAGE [11], 6 × 23700, E. coli, Salmonella typhimurium, SDS-PAGE [38]) [10, 11, 24, 38]

Glycoprotein/Lipoprotein

–

4 ISOLATION/PREPARATION

Source organism

Bacillus cereus [4]; Bacillus subtilis [11]; Camelus dromedarius [3]; Hyalomma dromedarii [2]; Human [1, 4, 6, 8, 14, 26, 31, 34, 39]; Bovine (calf [33, 35, 39]) [4, 20, 28, 32, 33, 35, 39]; E. coli (K-12 [43], inosine-guanosine phosphorylase different from deo-encoded purine nucleoside phosphorylase [43]) [5, 9, 33, 38, 39, 43]; Chicken [7, 13, 16]; Salmonella typhimurium [10, 24, 38]; Rat [24, 29, 41]; Enterobacter cloacae (KY 3074) [19]; Erwinia carotovora [21]; Plasmodium lophurae [12]; Brevibacterium acetylicum (ATCC 154) [23]; Rabbit [25, 36, 37, 40]; Chinese hamster [27, 30]; Plasmodium falciparum [42]; Proteus vulgaris [15]; Bacillus stearothermophilus (JTS 859) [17, 22]; Enterobacter aerogenes (AJ 11125) [18]; More (overview) [4]

Source tissue

Erythrocytes [1, 4, 8, 14, 26, 31, 34, 36]; Embryos [2]; Liver [3, 7, 13, 16, 24, 25, 29, 30, 37, 41]; Spores [4]; Vegetative cells [4, 11]; Brain [32, 40]; Cell [15]; Thyroid tissue [20]; V79 tissue culture cells [27, 30]; Ocular lens [28]; Novikoff hepatoma cells [29]; Kidney [30]; Spleen [4, 33, 35]; Sarcoma 180 cells [41]; More (overview) [4]

Localization in source

Purification

Rat (adenosine phosphorylase distinct from inosine guanosine phosphorylase [41]) [24, 29, 41]; Camelus dromedarius [3]; Hyalomma dromedarii [2]; Human [1, 8, 14, 26, 31, 34]; Chicken [7]; Salmonella typhimurium LT-2 [10, 24, 38]; Bacillus subtilis (2 purine nucleoside phosphorylases: 1. inosine-guanosine phosphorylase, 2. adenosine phosphorylase specific for adenosine and deoxyadenosine) [11]; E. coli (partial, inosine-guanosine phosphorylase different from deo-encoded purine nucleoside phosphorylase [43]) [38, 43]; Plasmodium lophurae [12]; Chicken [13]; Proteus vulgaris [15]; Bacillus stearothermophilus (2 forms: PUNPI [17], PUNPII [22]) [17, 22]; Enterobacter aerogenes (partial) [18]; Enterobacter cloacae (partial) [19]; Bovine [20, 28, 32, 35]; Erwinia carotovora (AJ 2922) [21]; Brevibacterium acetylicum [23]; Rabbit (partial [36]) [25, 36, 37, 40]; Chinese hamster [27, 30]; Plasmodium falciparum [42]

Crystallization

[1, 5–7, 26, 38]

Cloned

–

Renatured

–

5 STABILITY

pH
5.5–9.0 (40°C, 10 min, stable) [25]; 6–8 (45°C, $t_{1/2}$: 8–9 min) [43]; 6.2–10.0 (5 min, stable at room temperature) [1]; 7 (50% loss of activity after 4 weeks) [28]; 7.0–7.3 [19]

Temperature (°C)
30 (1 month, less than 20% loss of activity) [18]; 45 (pH 6–7, $t_{1/2}$: 8–9 min) [43]; 55 (rapid loss of activity) [32]; 57 (15 min, 50% inactivation) [1]; 60 (rapid loss of activity [29], half-life: around 1 week [18], 30 min, stable below [19]) [18, 19, 29]; 69 (half-life: in 2 M deoxyadenosine, pH 7.1, 59 min, in 0.2 M potassium phosphate, 19 min, in 0.2 M Tris-succinate, 11 min) [38]; 70 (30 h, no loss of activity [17], 30 min, almost complete loss of activity [19], enzyme PUNPII stable [22]) [17, 19, 22]; 80 (pH 7.0, 20 mM potassium phosphate, 1 mM inosine, half-life: 16 h [17], enzyme PUNPI stable, PUNPII unstable [22]) [17, 22]; More (inosine, 10 mM, or guanosine, 5 mM, protects against heat inactivation up to 65°C [29], adenine selectively protects adenosine phosphorylase activity of Sarcoma 180 and rat liver extract against heat inactivation, hypoxanthine selectively protects inosine-guanosine phosphorylase against heat inactivation [41], hypoxanthine, 1 mM, stabilizes against thermal inactivation, greatest stabilization in combination with 20 mM phosphate [43]) [29, 41, 43]

Oxidation
Susceptible to photooxidation in presence of methylene blue (pH-dependent [28, 32, 37, 40], maximal photooxidation near pH 8.5 [25]) [2, 25, 28, 32, 37, 40]

Organic solvent

General stability information
Freezing and thawing results in complete loss of activity [28]; Dilute solutions are irreversibly denatured by freezing [32]; Inosine, 10 mM, or guanosine, 5 mM, protects against heat inactivation up to 65°C [29]; Freezing of solubilized enzyme, especially in dilute solutions, causes total loss of activity [37]; Adenine selectively protects adenosine phosphorylase activity of Sarcoma 180 and rat liver extract against heat inactivation [41]; Hypoxanthine selectively protects inosine-guanosine phosphorylase against heat inactivation [41]; Crystals are stable to X-rays [5, 6]; Dithiothreitol, 0.1 mM, stabilizes in buffered system [20]; Mercaptoethanol, 10 mM, stabilizes [20]; Thioglycolate, 3 mM, stabilizes [20]; Absence of buffering salts results in loss of activity [20]; Enzyme PUNPII not stabilized by inosine [22]; Not stable when frozen, Salmonella typhimurium enzyme [24]; 2-Mercaptoethanol stabilizes during purification [24]; Freezing in presence of dithiothreitol results in greater loss of activity than in the absence of a sulfhydryl reagent [26]; Hypoxanthine, 1 mM, stabilizes against thermal inactivation, greatest stabilization in combination with 20 mM phosphate [43]

Storage

Stable as ammonium sulfate precipitate at 4°C or frozen for long periods of time [26]; 4°C, stable for at least 4 months (85% recovery in 10 mM sodium phosphate buffer, pH 7.0, a weekly dialysis with 2 mM DTT is necessary) [28]; 4°C, 0.2 M potassium phosphate, pH 6.0 [29]; 4°C, ammonium sulfate solution, Salmonella typhimurium enzyme stable [24]; –20°C, 20% glycerol, 10 mM inosine, rat enzyme stable [24]; 4°C, requires protection of essential cysteine SH by 2-mercaptoethanol or dithiothreitol, stable [32]; 4°C, 2-mercaptoethanol or dithiothreitol, stable for several weeks [37]; 4°C, precipitation with ammonium sulfate, near pH 7.0, stable for several months [37]

6 CROSSREFERENCES TO STRUCTURE DATABANKS

PIR/MIPS code

PIR2:A42708 (Bacillus subtilis (fragment)); PIR2:A27854 (Escherichia coli); PIR1:PHHUPN (human); PIR2:JC1115 (mouse); PIR3:A53312 (Mycoplasma pirum (strain BER) (SGC3) (fragment))

Brookhaven code

1ULA (Human (Homo Sapiens) erythrocytes); 1ULB (Human (Homo Sapiens) erythrocytes)

7 LITERATURE REFERENCES

[1] Agarwal, R.P., Parks, R.E.: J. Biol. Chem.,244,644–647 (1969)
[2] Kamel, M.Y., Fahmy, A.S., Ghazy, A.H., Mohamed, M.A.: Biochem. Cell Biol.,69, 223–231 (1991)
[3] Osman, A.M., Del Corso, A., Mohamed, A.S., Ipata, P.L., Mura, U.: Comp. Biochem. Physiol.,97B,177–182 (1990)
[4] Parks, R.E., Agarwal, R.P. in "Enzymes",3rd Ed. (Boyer, P.D., ed.) 7,483–514, Academic Press (1972) (Review)
[5] Cook, W.J., Ealick, S.E., Krenitsky, T.A., Stoeckler, J.D., Helliwell, J.R., Bugg, C.E.: J. Biol. Chem.,260,12968–12969 (1985)
[6] Cook, W.J., Ealick, S.E., Bugg, C.E., Stoeckler, J.D., Parks, R.E.: J. Biol. Chem.,256, 4079–4080 (1981)
[7] Murakami, K., Tsushima, K.: Biochim. Biophys. Acta,384,390–398 (1975)
[8] Lewis, A.S., Lowy, B.A.: J. Biol. Chem.,254,9927–9932 (1979)
[9] Bzowska, A., Kulikowska, E., Shugar, D.: Biochim. Biophys. Acta,1120,239–247 (1992)
[10] Robertson, B.C., Hoffee, P.A.: J. Biol. Chem.,248,2040–2043 (1973)
[11] Jensen, K.J.: Biochim. Biophys. Acta,525,346–356 (1978)
[12] Schimandle, C.M., Tanigoshi, L., Mole, L.A., Sherman, I.W.: J. Biol. Chem.,260, 4455–4460 (1985)
[13] Umemura, S., Nishino, T., Murakami, K., Tsushima, K.: J. Biol. Chem.,257, 13374–13378 (1982)

[14] Osborne, W.R.A.: J. Biol. Chem.,255,7089–7092 (1980)
[15] Surette, M., Gill, T., MacLean, S.: Appl. Environ. Microbiol.,56,1435–1439 (1990)
[16] Murakami, K., Tsushima, K.: Biochim. Biophys. Acta,435,205–210 (1976)
[17] Hori, N., Watanabe, M., Yamazaki, Y., Mikami, Y.: Agric. Biol. Chem.,53,2205–2210 (1989)
[18] Utagawa, T., Morisawa, H., Yamanaka, S., Yamazaki, A., Yoshinaga, F., Hirose, Y.: Agric. Biol. Chem.,49,3239–3246 (1985)
[19] Machida, Y., Nakanishi, T.: Agric. Biol. Chem.,45,1801–1807 (1981)
[20] Moyer, T.P., Fischer, A.G.: Arch. Biochem. Biophys.,174,622–629 (1976)
[21] Shirae, H., Yokozeki, K.: Agric. Biol. Chem.,55,1849–1857 (1991)
[22] Hori, N., Watanabe, M., Yamazaki, Y., Mikami, Y.: Agric. Biol. Chem.,53,3219–3224 (1989)
[23] Shirae, H., Yokozeki, K.: Agric. Biol. Chem.,55,493–499 (1991)
[24] Hoffee, P.A., May, R., Robertson, B.C.: Methods Enzymol.,51,517–524 (1978) (Review)
[25] Glantz, M.D., Lewis, A.S.: Methods Enzymol.,51,524–530 (1978) (Review)
[26] Stoeckler, J.D., Agarwal, R.P., Agarwal, K.C., Parks, R.E.: Methods Enzymol.,51, 530–538 (1978) (Review)
[27] Milman, G.: Methods Enzymol.,51,538–543 (1978) (Review)
[28] Barsacchi, D., Cappiello, M., Tozzi, M.G., Del Corso, A., Peccatori, M., Camici, M., Ipata, P.L., Mura, U.: Biochim. Biophys. Acta,1160,163–170 (1992)
[29] May, R.A., Hoffee, P.: Arch. Biochem. Biophys.,193,398–406 (1979)
[30] Milman, G., Anton, D.L., Weber, J.L.: Biochemistry,15,4967–4973 (1976)
[31] Stoeckler, J.D., Agarwal, R.P., Agarwal, K.C., Schmid, K., Parks, R.E.: Biochemistry, 17,278–283 (1978)
[32] Lewis, A.S., Glantz, M.D.: Biochemistry,15,4451–4457 (1976)
[33] Bzowska, A., Kulikowska, E., Darzynkiewicz, E., Shugar, D.: J. Biol. Chem.,263, 9212–9217 (1988)
[34] Zannis, V., Doyle, D., Martin, D.W.: J. Biol. Chem.,253,504–510 (1978)
[35] Kulikowska, E., Bzowska, A., Wierzchowski, J., Shugar, D.: Biochim. Biophys. Acta, 874,355–363 (1986)
[36] Savage, B., Spencer, N.: Biochem. J.,167,703–710 (1977)
[37] Lewis, A.S., Glantz, M.D.: J. Biol. Chem.,251,407–413 (1976)
[38] Jensen, K.F., Nygaard, P.: Eur. J. Biochem.,51,253–265 (1975)
[39] Bzowska, A., Kulikowska, E., Shugar, D.: Z. Naturforsch.,45c,59–70 (1990)
[40] Lewis, A.S.: Arch. Biochem. Biophys.,190,662–670 (1978)
[41] Divekar, A.Y.: Biochim. Biophys. Acta,422,15–28 (1976)
[42] Daddona, P.E., Wiesmann, W.P., Milhouse, W., Chern, J.-W., Townsend, L.B., Hershfield, M.S., Webster, H.K.: J. Biol. Chem.,261,11667–11673 (1986)
[43] Koszalka, G.W., Vanhooke, J., Short, S.A., Hall, W.W.: J. Bacteriol.,170,3493–3498 (1988)

1 NOMENCLATURE

EC number
2.4.2.2

Systematic name
Pyrimidine-nucleoside:orthophosphate alpha-D-ribosyltransferase

Recommended name
Pyrimidine-nucleoside phosphorylase

Synonyms
More (enzyme may be identical with EC 2.4.2.3 from some organisms, catalyzing both uridine and thymidine phosphorolysis)

CAS Reg. No.
9055-35-0

---

2 REACTION AND SPECIFICITY

Catalysed reaction
Pyrimidine nucleoside + phosphate →
→ pyrimidine + alpha-D-ribose 1-phosphate

Reaction type
Pentosyl group transfer

Natural substrates

Substrate spectrum
1 Uridine + phosphate (r, by measuring the conversion of uracil to uridine or deoxyuridine a slight preference for ribose 1-phosphate over deoxyribose 1-phosphate is observed) [1]
2 Deoxyuridine + phosphate (r, best substrate, by measuring the conversion of uracil to uridine or deoxyuridine a slight preference for ribose 1-phosphate over deoxyribose 1-phosphate is observed) [1]
3 Thymidine + phosphate [1]
4 5-Bromouridine + phosphate [1]
5 5-Bromodeoxyuridine + phosphate [1]
6 More (not: cytidine, deoxycytidine) [1]

Product spectrum
1 Uracil + alpha-D-ribose 1-phosphate [1]
2 ?
3 Thymine + 2-deoxy-D-ribose 1-phosphate
4 ?
5 ?
6 ?

Inhibitor(s)

Cofactor(s)/prosthetic group(s)/activating agents

Metal compounds/salts

Turnover number (min^{-1})

Specific activity (U/mg)

K_m-value (mM)
0.25 (uridine) [1]; 0.38 (thymidine) [1]

pH-optimum
7.2 (uridine, thymidine) [1]

pH-range

Temperature optimum (°C)
37 (assay at) [1]

Temperature range (°C)

3 ENZYME STRUCTURE

Molecular weight
78000 (Bacillus stearothermophilus, sedimentation velocity analysis) [1]

Subunits

Glycoprotein/Lipoprotein
–

4 ISOLATION/PREPARATION

Source organism
Bacillus stearothermophilus [1]

Source tissue
Cell [1]

Localization in source

Purification
    Bacillus stearothermophilus [1]

Crystallization
    –

Cloned
    –

Renatured
    –

---

5 STABILITY

pH

Temperature (°C)
    60 (60 min, stable) [1]

Oxidation

Organic solvent

General stability information

Storage

---

6 CROSSREFERENCES TO STRUCTURE DATABANKS

PIR/MIPS code

Brookhaven code

---

7 LITERATURE REFERENCES

[1] Saunders, P.P., Wilson, B.A., Saunders, G.F.: J. Biol. Chem.,244,3691–3697 (1969)

1 NOMENCLATURE

EC number
2.4.2.3

Systematic name
Uridine:orthophosphate alpha-D-ribosyltransferase

Recommended name
Uridine phosphorylase

Synonyms
Pyrimidine phosphorylase
Phosphorylase, uridine
UrdPase [6]
UPH [9]
UPase [23]
More (enzyme may be identical with EC 2.4.2.2)

CAS Reg. No.
9030-22-2

---

2 REACTION AND SPECIFICITY

Catalysed reaction
Uridine + phosphate →
→ uracil + alpha-D-ribose 1-phosphate (ordered bi bi mechanism [8, 19, 20], phosphate binds before uridine and ribose 1-phosphate is released after uracil [19, 20], rapid-equilibrium random mechanism [11], random mechanism [13], sequential rather than ping-pong mechanism, addition of substrate is random [16], sequential mechanism [24, 26]) [8, 11, 13, 16, 19, 20, 24, 26]

Reaction type
Pentosyl group transfer

Natural substrates
Uracil + alpha-D-ribose 1-phosphate (in conjunction with uridine kinase, the enzyme provides a route for the conversion of uracil to UMP via uridine [4], role in degradation of pyrimidine nucleosides as well as in the salvage pathway for nucleic acid synthesis [19], enzyme of pyrimidine salvage pathway [24]) [4, 19, 24]

Enzyme Handbook © Springer-Verlag Berlin Heidelberg 1996
Duplication, reproduction and storage in data banks are only
allowed with the prior permission of the publishers

Substrate spectrum

1 Uridine + phosphate (r [1–3, 11, 19, 24], equilibrium position favouring nucleoside synthesis [1]) [1–4, 6, 8, 11, 16, 19, 23, 24]

2 Deoxyuridine + phosphate (catalyzed by a different protein or by a different active center of the same enzyme [1], unlike cytosol enzyme, enzyme from plasma membranes shows little or no deoxyuridine-cleaving activity [21], activities of uridine, deoxyuridine and thymidine phosphorylase from Giardia lamblia remain associated throughout purification suggesting that a single enzyme is responsible for the 3 activities [7]) [1, 7, 21, 22]

3 5-Methyluridine + phoshate (27% of the activity with uridine [25]) [16, 25]

4 Uracil arabinoside + phosphate (10% of the activity with uridine) [23, 25]

5 5-Bromouridine + phosphate (69% [22], 40% of the activity with uridine [25], not [16]) [22, 25]

6 Thymine ribonucleoside + phosphate (19% of the activity with uridine) [22]

7 5-Bromo-2'-deoxyuridine + phosphate (27% [22], 75% of the activity with uridine [25]) [22, 25]

8 5-Fluoro-2'-deoxyuridine (14% of the activity with uridine) [22]

9 2'-Deoxyuridine + phosphate (6% [22], 18% [23], 12% of the activity with uridine [25]) [22, 23, 25]

10 Thymidine + phosphate (2% [22], 22% [23], 21% of the activity with uridine [25], the activities of uridine, deoxyuridine and thymidine phosphorylases from Giardia lamblia remain associated throughout purification, suggesting that a single enzyme is responsible for the 3 activities [7]) [7, 16, 22–25]

11 More (not: 3'-azido-2',3'-dideoxy-5-methyluridine [14], 3'-azido-2',3'-dideoxy-5-ethyluridine [14], 2',3'-dideoxy-5-ethyluridine [14], 3'-chloro-2',3'-dideoxy-5-ethyluridine [14], 3'-chloro-2',3'-dideoxy-5-methyluridine [14], 3'-bromo-2',3'-dideoxy-5-ethyluridine [14], 2'-deoxylyxofuranosyl-5-ethyluracil [14], arabinofuranosyl-5-ethyluracil [14], 3'-O-acetyl-2,2'-anhydro-5-ethyluridine [14], 2,3'-anhydro-2'-deoxy-5-ethyluridine [14], 2,5'-anhydro-2'-deoxy-5-ethyluridine [14], 5-substituted-2,2'-anhydrouridine [5], 2-deoxyuridine [16], adenosine [16], cytosine [24], deoxycytidine [24], orotidine [24], cytidine [16, 23–25]) [5, 14, 16, 23–25]

Product spectrum

1 Uracil + alpha-D-ribose 1-phosphate [1, 8, 25, 26]
2 Uracil + deoxyribose 1-phosphate [22]
3 ?
4 ?
5 ?
6 ?
7 ?
8 ?
9 ?
10 ?
11 ?

Inhibitor(s)

Pyrimidine acyclonucleosides (competitive) [3]; Acyclothymidine [3]; 5-Substituted 2,2'-anhydrouridine [5]; 2,2'-Anhydro-5-ethyluridine [5, 14, 24]; Benzylacyclouridines [6, 24]; 5-(3'-Benzyloxybenzyl)-1-[(1'-aminomethyl-2'-hydroxyethoxy)methyl]uracil [6]; Phosphate (product inhibition [11], substrate inhibition [24]) [8, 11, 24]; Uridine (product inhibition [11], substrate inhibition [24]) [11, 24]; Ribose 1-phosphate (product inhibition [11, 13]) [8, 11, 13]; Uracil (product inhibition [11, 13]) [11, 13, 19]; 3'-Azido-2',3'-dideoxy-5-methyluridine [14]; 3'-Azido-2',3'-dideoxy-5-ethyluridine [14]; 2',3'-Dideoxy-5-ethyluridine [14]; 3'-Chloro-2',3'-dideoxy-5-ethyluridine [14]; 3'-Chloro-2',3'-dideoxy-5-methyluridine [14]; 3'-Bromo-2',3'-dideoxy-5-ethyluridine [14]; 2'-Deoxylyxofuranosyl-5-ethyluracil [14]; Arabinofuranosyl-5-ethyluracil [14]; 3'-O-Acetyl-2,2'-anhydro-5-ethyluridine [14]; 2,3'-Anhydro-2'-deoxy-5-ethyluridine [14]; 2,5'-Anhydro-2'-deoxy-5-ethyluridine [14]; Acyclonucleoside analogues (consisting of 5- and 5,6-substituted uracils and different acyclic chains) [15]; 3-O-Methyl-alpha-D-glucopyranoside [16, 17]; p-Chloromercuribenzoate [16, 19]; Thymidine 5-monophosphate [17]; 5-Fluoro-2-deoxyuridine [17]; 6-Methyluracil [17]; Deoxyglucosylthymine (phosphorolysis of uridine and deoxyuridine, synthesis of uridine at concentrations of 0.10 mM, 0.018 mM and 0.14 mM, not: phosphorolysis of deoxyuridine or thymidine at 0.19 mM) [19]; p-Mercuriphenylsulfonate [19]; 5,5'-Dithiobis(2-nitrobenzoic acid) [19]; o-Iodosobenzoate [19]; N-Ethylmaleimide [19]; Iodoacetamide [19]; Iodoacetic acid [19]; Deoxythymidine [24]; 5-(Benzyloxybenzyloxybenzyl)acyluridine [24]

Cofactor(s)/prosthetic group(s)/activating agents

Triton X-100 (stimulates enzyme of isolated plasma membrane) [21]

Metal compounds/salts

Enzyme Handbook © Springer-Verlag Berlin Heidelberg 1996
Duplication, reproduction and storage in data banks are only
allowed with the prior permission of the publishers

Turnover number (min^{-1})

Specific activity (U/mg)
16.1 [13]; 133 [22]; 129 [26]; 182.8 [10]; More [1, 17, 23, 24]

K$_m$-value (mM)
0.016 (uridine) [4]; 0.06 (uracil) [8]; 0.076 (phosphate, mouse) [6]; 0.088 (ribose 1-phosphate) [4]; 0.130 (phosphate) [3]; 0.14 (ribose 1-phosphate) [8]; 0.143 (uridine, mouse) [6]; 0.24 (uridine) [8]; 0.242 (uridine, human) [6]; 0.279 (phosphate, human) [6]; 0.36 (uracil) [4]; 0.42 (phosphate) [8]; 0.76 (uridine) [1]; 3.9 (phosphate) [1]; 4 (5-methyluridine) [16]; More [13, 16, 17, 22–26]

pH-optimum
6.5 (deoxyuridine phosphorolysis) [19]; 6.9 (deoxynucleoside phosphorolysis) [25]; 7.0 [16, 17]; 7.3 (uridine phosphorolysis) [22]; 7.4 (assay at [1], phosphorolysis of ribonucleosides and uracil arabinoside [25]) [1, 25]; 8.2 (uridine phosphorolysis [19]) [1, 19]; 8.5 (uridine synthesis [19]) [19, 23]

pH-range
5.8–9.0 (5.8: about 50% of activity maximum, 9.0: about 40% of activity maximum) [22]; 6.5–10.5 (6.5: about 50% of activity maximum, 10.5: about 65% of activity maximum) [23]

Temperature optimum (°C)
37 (assay at) [1, 22]; 65 [23]

Temperature range (°C)
40–80 (40°C: about 40% of activity maximum, 80°C: about 10% of activity maximum) [23]

---

3 ENZYME STRUCTURE

Molecular weight
43000 (Giardia lamblia, gel filtration) [7]
45000 (rat, gel filtration) [4]
56000 (Schistosoma mansoni, gel filtration) [24]
65000 (Acholeplasma laidlawii, gel filtration) [26]
80000 (Lactobacillus casei, non-denaturing PAGE) [16]
102500 (rat, disc gel electrophoresis) [20]
110000 (rat, gel filtration) [19]
160000 (E. coli, gel filtration) [22]

Subunits
Monomer (1 × 38000, Giardia lamblia, SDS-PAGE) [7]
Tetramer (4 × 26000, rat, SDS-PAGE in presence of 4 M urea and 0.5% beta-mercaptoethanol [19], 4 × 26000, rat, SDS-PAGE in presence of 6 M urea, without 2-mercaptoethanol [20], 4 × 23000, E. coli, SDS-PAGE [22], 4 × 20000, Lactobacillus casei, SDS-PAGE in presence of 2-mercaptoethanol [16]) [16, 19, 20, 22]
Hexamer (6 × 27500, E. coli) [12]

Glycoprotein/Lipoprotein
–

---

4 ISOLATION/PREPARATION

Source organism
Dictyostelium discoideum [8]; E. coli (K-12 [9, 12], B [10, 11, 15], B-96, ATCC 13473 [13]) [9–13, 15, 18, 22]; Mouse [1, 5, 6]; Rat [2, 4, 14, 19–21]; Human [3, 5, 6]; Giardia lamblia [7]; Schistosoma mansoni [24]; Haemophilus influenzae [25]; Lactobacillus casei [16, 17]; Enterobacter aerogenes (AJ 11125) [23]; Acholeplasma laidlawii [26]

Source tissue
Novikoff hepatoma cells [4]; Ehrlich ascites tumor cells [1]; Liver [2, 6, 20, 21]; Tumors (human [5]) [3, 5]; Leukemic cells (mouse) [5]; Intestine (mucosa) [5, 14]

Localization in source
Soluble [2]; Cytosol [19, 20]; Plasma membrane [21]

Purification
E. coli [9, 10, 13, 22]; Mouse [1]; Rat (partial [2]) [2, 4, 19, 20]; Giardia lamblia [7]; Dictyostelium discoideum [8]; Lactobacillus casei [16, 17]; Enterobacter aerogenes (partial) [23]; Schistosoma mansoni [24]; Haemophilus influenzae [25]; Acholeplasma laidlawii [26]

Crystallization
[9, 12]

Cloned
–

Renatured
–

5 STABILITY

pH

Temperature (°C)
30 (1 month, 20% loss of activity) [23]; 55 (40 min, 95% loss of activity) [17]; 60 (20 min, complete loss of activity [17], half-life: around 1 week [23]) [17, 23]; More (aglycone substrates, nucleoside substrates, phosphate or pentose 1-phosphate ester substrates stabilize against heat inactivation) [18]

Oxidation

Organic solvent

General stability information
Aglycone substrates, nucleoside substrates, phosphate or pentose 1-phosphate ester substrates stabilize against heat inactivation [18]; Much less stable in Tris buffer than in phosphate buffer [20]; Use of dithiothreitol and glycerol is necessary for stabilization during purification [25]

Storage
5°C, 90% loss of activity after 3–4 days [16]; –40°C, 0.05 M potassium phosphate buffer, pH 7.0, 10 mM beta-mercaptoethanol, 1 mM EDTA, 10% loss of activity after 6 weeks [19]

---

6 CROSSREFERENCES TO STRUCTURE DATABANKS

PIR/MIPS code
PIR2:S05491 (Escherichia coli)

Brookhaven code

---

7 LITERATURE REFERENCES

[1] Pontis, H., Degerstedt, G., Reichard, P.: Biochim. Biophys. Acta,51,138–147 (1961)
[2] Canellakis, E.S.: J. Biol. Chem.,227,329–338 (1957)
[3] Niedzwicki, J.G., El Kouni, M.H., Chu, S.H., Cha, S.: Biochem. Pharmacol.,30, 2097–2101 (1981)
[4] McIvor, R.S., Wohlhueter, R.M., Plagemann, P.P.G.: J. Cell. Physiol.,122,397–404 (1985)
[5] Veres, Z., Szabolcs, A., Szinai, I., Denes, G., Kajtar-Peredy, M., Otvos, L.: Biochem. Pharmacol.,34,1737–1740 (1985)
[6] Naguib, F.N.M., El Kouni, M.H., Chu, S.H., Cha, S.: Biochem. Pharmacol.,36, 2195–2201 (1987)
[7] Lee, C.S., Jimenez, B.M., O'Sullivan, W.J.: Mol. Biochem. Parasitol.,30,271–287 (1988)
[8] Albe, K.R., Wright, B.E.: Exp. Mycol.,13,13–19 (1989)

[9] Mikhailov, A.M., Smirnova, E.A., Tsuprun, V.L., Tagunova, I.V., Vainshtein, B.K., Linkova, E.V., Komissarov, A.A., Siprashvili, Z.Z., Mironov, A.S.: Biochem. Int.,26, 607–615 (1992)
[10] Vita, A., Magni, G.: Anal. Biochem.,133,153–156 (1983)
[11] Vita, A., Huang, C.Y., Magni, G.: Arch. Biochem. Biophys.,226,687–692 (1983)
[12] Cook, W.J., Koszalka, G.W., Hall, W.W., Narayana, S.V.L., Ealick, S.E.: J. Biol. Chem., 262,2852–2853 (1987)
[13] Krenitsky, T.A.: Biochim. Biophys. Acta,429,352–358 (1976)
[14] Veres, Z., Neszmelyi, A., Szabolcs, A., Denes, G.: Eur. J. Biochem.,178,173–181 (1988)
[15] Drabikowska, A.K., Lissowska, L., Draminski, M., Zgit-Wroblewska, L., Shugar, D.: Z. Naturforsch.,42c,288–296 (1987)
[16] Avraham, Y., Grossowicz, N., Yashphe, J.: Biochim. Biophys. Acta,1040,287–293 (1990)
[17] Avraham, Y., Yashphe, J., Grossowicz, N.: FEMS Microbiol. Lett.,56,29–34 (1988)
[18] Krenitsky, T.A., Tuttle, J.V.: Biochim. Biophys. Acta,703,247–249 (1982)
[19] Yamada, E.W.: Methods Enzymol.,51,423–431 (1978) (Review)
[20] Bose, R., Yamada, E.W.: Biochemistry,13,2051–2056 (1974)
[21] Bose, R., Yamada, E.W.: Can. J. Biochem.,55,528–533 (1977)
[22] Leer, J.C., Hammer-Jespersen, K., Schwartz, M.: Eur. J. Biochem.,75,217–224 (1977)
[23] Utagawa, T., Morisawa, H., Yamanaka, S., Yamazaki, A., Yoshinaga, F., Hirose, Y.: Agric. Biol. Chem.,49,3239–3246 (1985)
[24] El Kouni, M.H., Fardos, N.M., Naguib, F.N.M., Niedzwicki, J.G., Iltzsch, M.H., Cha, S.: J. Biol. Chem.,263,6081–6086 (1988)
[25] Scocca, J.J.: J. Biol. Chem.,246,6606–6610 (1971)
[26] McIvor, R.S., Wohlhueter, R.M., Plagemann, P.G.W.: J. Bacteriol.,156,198–204 (1983)

1 NOMENCLATURE

EC number
2.4.2.4

Systematic name
Thymidine:orthophosphate deoxy-D-ribosyltransferase

Recommended name
Thymidine phosphorylase

Synonyms
Pyrimidine phosphorylase
Thymidine-orthophosphate deoxyribosyltransferase [13]
Phosphorylase, thymidine
Animal growth regulators, blood platelet-derived endothelial cell growth
factors
Blood platelet-derived endothelial cell growth factor
Deoxythymidine phosphorylase
Gliostatins
Pyrimidine deoxynucleoside phosphorylase

CAS Reg. No.
9030-23-3

---

2 REACTION AND SPECIFICITY

Catalysed reaction
Thymidine + phosphate →
→ thymine + 2-deoxy-D-ribose 1-phosphate (ordered bi bi mechanism with
the nucleoside the first substrate to add, and the pyrimidine base the last
product to leave [12], mechanism [16], ordered sequential reaction mechanism, phosphate is the first substrate to bind to and deoxyribose the last
product to dissociate from the enzyme [6], rapid equilibrium random bi-bi
mechanism with an enzyme-phosphate-thymine dead-end complex [10])

Reaction type
Pentosyl group transfer

Natural substrates
Thymine + 2-deoxy-D-ribose 1-phosphate (in E. coli and Salmonella typhimurium, thymidine phosphorylase plays an important role in metabolism of
thymine auxotrophs and is necessary for the conversion of exogenous thymine to thymidine) [5]
Thymidine + phosphate (intact platelets degrade thymidine but are not able
to synthesize thymidine from thymine) [11]

Substrate spectrum

1 Thymidine + phosphate (ir [11], in the reverse reaction only alpha-D-2-deoxyribose 1-phosphate is accepted as substrate [2]) [1, 2, 5, 11–13, 16]

2 Uridine + phosphate (the activities of uridine, deoxyuridine and thymidine phosphorylases from Giardia lamblia remain associated throughout purification, suggesting that a single enzyme is responsible for the 3 activities [3], not [5, 11–13]) [3]

3 Deoxyuridine + phosphate [5, 11–13, 16]

4 Thymine arabinoside + phosphate [16]

5 5-Deoxythymidine + phosphate [16]

6 5-Fluorouridine + phosphate [16]

7 Bromodeoxyuridine + phosphate (at 40% the rate of thymidine [5, 12]) [5, 11, 12, 16]

8 Iododeoxyuridine + phosphate [5, 11, 12]

9 Deoxythymidine + phosphate [13]

10 More (the enzyme in some tissues also catalyzes deoxyribonucleosyltransferase reaction of the type catalyzed by EC 2.4.2.6: 2-deoxy-D-ribosyl-base[1] + base[2] = 2-deoxy-D-ribosyl-base[2] + base[1] [1, 7], arsenate can replace phosphate [5, 6, 11, 12], specificity towards the deoxyribosyl moiety of the substrate [16], nonsubstituted pyrimidine moiety or one which is substituted in position 5 required [16], purine deoxyribonucleosides and ribonucleosides are not cleaved [6], specific for deoxyribonucleosides [6, 12], the phosphorolytic acitvities towards thymidine, 5'-deoxy-5-fluorouridine and 1-(tetrahydro-2-furanyl)-5-fluorouracil remain closely parallel during purification [4], not: deoxycytidine [5, 6, 12, 13], deoxyadenosine [5, 12], deoxyguanosine [5, 12]) [1, 4–7, 11–13, 16]

Product spectrum

1 Thymine + 2-deoxy-D-ribose 1-phosphate [1, 2, 5, 12, 13]

2 ?

3 ?

4 ?

5 ?

6 ?

7 ?

8 ?

9 ?

10 ?

Inhibitor(s)
    Thymine (substrate inhibition [10], product inhibition [10, 12]) [10, 12, 16];
    Bromouracil [12]; Uracil derivatives [12]; Uridine (competitive) [12]; Deoxy-
    adenosine [12]; Ribose 1-phosphate (competitive [6]) [6, 12]; Deoxyribose
    1-phosphate (product inhibition [12], competitive [16]) [12, 16]; 5-Bromo-2-
    deoxyuridine [17]; 5-Fluorodeoxyuridine [17]; [6]; p-Chloromercuribenzoate
    [7]; SO_4^{2-} (inhibition reversed by phosphate) [7]; Urea (inhibition at high
    concentration, 4 M, stimulation at low concentration) [7]; 6-Methyluracil [17];
    More (photoinactivation in presence of thymine, thymidine and some halo-
    genated analogs) [15]

Cofactor(s)/prosthetic group(s)/activating agents
    Urea (stimulates at low concentration, inhibits at high concentration, 4 M) [7]

Metal compounds/salts

Turnover number (min^{-1})

Specific activity (U/mg)
    10.0 [11]; 122 [6]; 465 [5, 12]; 1100 [16]; 5.3 [13]; More [7]

K$_m$-value (mM)
    0.168 (thymidine) [4]; 0.38 (thymidine) [6]; 0.89 (phosphate) [6]; 1.3 (arse-
    nate) [5, 12]; 2.1 (thymidine) [5, 12]; 2.3 (phosphate) [5, 12]; 8.0 (deoxyuri-
    dine) [5, 12]; More [16, 17]

pH-optimum
    5.7 [11]; 6.0 (around [1]) [1, 16, 17]; 6.3 [6]; 7.1 (assay at) [6]; 7.4 (assay at)
    [1, 5]; 7.5–8.0 (rapid decrease of activity above and below this range [5])
    [5, 12]

pH-range
    7.5–8.0 (rapid decrease of activity above and below this range) [5]

Temperature optimum (°C)
    37 (assay at) [5, 6, 11]

Temperature range (°C)

---

3 ENZYME STRUCTURE

Molecular weight
    80000 (Lactobacillus casei, gel filtration) [16]
    90000 (E. coli, gel filtration) [6]
    98000 (E. coli) [8]
    100000 (Salmonella typhimurium, gel filtration) [12]
    110000 (human, gel filtration) [11]
    120000 (human, gel filtration) [13]

Subunits

Dimer (2 × 47000, Salmonella typhimurium, SDS disc gel electrophoresis [12], 2 × 46000, E. coli [8], E. coli, three-dimensional structure [9], 2 × 60000, human, SDS-PAGE [11], 2 × 28000, Lactobacillus casei, SDS-PAGE [16], 2 × 58000, human, SDS-PAGE, enzyme is capable of being converted to a less active form larger in MW and possibly trimeric or tetrameric in structure [13]) [8, 9, 11–13, 16]

Glycoprotein/Lipoprotein

–

---

4 ISOLATION/PREPARATION

Source organism

Human [1, 4, 7, 11, 13]; Rat [1, 7]; Guinea pig [1]; Hamster [1]; E. coli (strain B [7], K12 [8], strain W, ATCC 9637 [15]) [2, 6–9, 14, 15]; Giardia lamblia [3]; Salmonella typhimurium (LT-2 [12]) [5, 12]; Mouse [10]; Lactobacillus casei [16, 17]

Source tissue

Spleen [1, 7]; Liver [1, 7]; Kidney [1]; Tumors (overview) [1]; Gastric cancer (poorly differentiated adenocarcinoma) [4]; Cell [5, 6, 16, 17]; Blood platelets [11]; Amniochorion [13]

Localization in source

Cytoplasm [11]

Purification

Human (partial [11]) [4, 7, 11, 13]; Rat [7]; Giardia lamblia [3]; Salmonella typhimurium [5, 12]; E. coli [6, 7]; Lactobacillus casei [16, 17]

Crystallization

[8, 9]

Cloned

[2]

Renatured

–

---

5 STABILITY

pH

6 (maximal stability, rat) [7]

Temperature (°C)
    55 (40 min, 10% loss of activity) [17]; 60 (20 min, complete loss of activity)
    [17]; 65 (30 min, complete inactivation) [11]; More (phosphate or pentose
    1-phosphate ester substrates stabilize against heat inactivation) [14]

Oxidation

Organic solvent

General stability information
    Crystals are stable to X-rays at room temperature for at least 5 days [8];
    Quite labile during latter states of purification [5]; Phosphate or pentose
    1-phosphate ester substrates stabilize against heat inactivation [14];
    2-Mercaptoethanol and sucrose stabilize during preincubation procedure
    with ammonium sulfate [5]; Stability depends on protein concentration [6];
    Human enzyme is stable to repeated freezing and thawing [7]

Storage
    5°C, 90% loss of activity after 3–4 days [16]; 4°C, 10 mM potassium phos-
    phate, pH 7.5, 10 mM 2-mercaptoethanol, 20% sucrose, stable for 3 months
    [5]; –20°C, 10 mM Tris-HCl, pH 7.3, stable for several months [6]; 4°C, grad-
    ual loss of activity [6]; –20°C, phosphate buffer containing 2% mannitol, sta-
    ble for over 1 year, human enzyme [7]

---

6 CROSSREFERENCES TO STRUCTURE DATABANKS

PIR/MIPS code
    PIR2:A37131 (Escherichia coli); PIR2:JX0275 (human (fragment));
    PIR3:S42196 (Mycoplasma hominis (SGC3)); PIR3:C53312 (Mycoplasma pi-
    rum (strain BER) (SGC3))

Brookhaven code
    1TPT ((Escherichia Coli) k12 strain)

---

7 LITERATURE REFERENCES

[1] Zimmerman, M., Seidenberg, J.: J. Biol. Chem.,239,2618–2621 (1964)
[2] Barbas, C.F.III, Wong, C.H.: Bioorg. Chem.,19,261–269 (1991)
[3] Lee, C.S., Jimenez, B.M., O'Sullivan, W.J.: Mol. Biochem. Parasitol.,30,271–287
    (1988)
[4] Sugata, S., Kono, A., Hara, Y., Karube, Y., Matsushima, Y.: Chem. Pharm. Bull.,34,
    1219–1222 (1986)
[5] Hoffee, P.A., Blank, J.: Methods Enzymol.,51,437–442 (1978) (Review)
[6] Schwartz, M.: Methods Enzymol.,51,442–445 (1978) (Review)
[7] Zimmerman, M.: J. Biol. Chem.,239,2622–2627 (1964)
[8] Cook, W.J., Koszalka, G.W., Hall, W.W., Burns, C.L., Ealick, S.E.: J. Biol. Chem.,
    262,3788–3789 (1987)

[9] Walter, M.R., Cook, W.J., Cole, L.B., Short, S.A., Koszalka, G.W., Krenitsky, T.A., Ealick, S.E.: J. Biol. Chem.,265,14016–14022 (1990)
[10] Iltzsch, M.H., El Kouni, M.H., Cha, S.: Biochemistry,24,6799–6807 (1985)
[11] Desgranges, C., Razaka, G., Rabaud, M., Bricaud, H.: Biochim. Biophys. Acta, 654,211–218 (1981)
[12] Blank, J.G., Hoffee, P.A.: Arch. Biochem. Biophys.,168,259–265 (1975)
[13] Kubilus, J., Lee, L.D., Baden, H.P.: Biochim. Biophys. Acta,527,221–228 (1978)
[14] Krenitsky, T.A., Tuttle, J.V.: Biochim. Biophys. Acta,703,247–249 (1982)
[15] Voytek, P.: J. Biol. Chem.,250,3660–3665 (1975)
[16] Avraham, Y., Grossowics, N., Yashphe, J.: Biochim. Biophys. Acta,1040,287–293 (1990)
[17] Avraham, Y., Yashphe, J., Grossowicz, N.: FEMS Microbiol. Lett.,56,29–34 (1988)

1 NOMENCLATURE

EC number
2.4.2.5

Systematic name
Nucleoside:purine(pyrimidine) D-ribosyltransferase

Recommended name
Nucleoside ribosyltransferase

Synonyms
Ribosyltransferase, nucleoside
Nucleoside N-ribosyltransferase

CAS Reg. No.
9030-31-3

2 REACTION AND SPECIFICITY

Catalysed reaction
D-Ribosyl-base[1] + base[2] →
→ D-ribosyl-base[2] + base[1]

Reaction type
Pentosyl group transfer

Natural substrates

Substrate spectrum
1 Inosine + adenine [1]
2 Inosine + xanthine [1]
3 Inosine + thymine [1]
4 Inosine + 4,5-diaminouracil [1]
5 Inosine + 5-bromouracil [1]
6 Inosine + 4,6-aminouracil [1]
7 Guanine + inosine [1, 2]
8 Guanine + uridine [2]

Product spectrum
1 Hypoxanthine + adenosine
2 Hypoxanthine + xanthosine
3 Hypoxanthine + ?
4 Hypoxanthine + ?
5 Hypoxanthine + ?
6 Hypoxanthine + ?
7 Guanosine + hypoxanthine [2]
8 Guanosine + uracil [2]

Inhibitor(s)

Cofactor(s)/prosthetic group(s)/activating agents

Metal compounds/salts

Turnover number (min⁻¹)

Turnover number (min^{-1})

Specific activity (U/mg)

K_m-value (mM)

pH-optimum
9.0 (in absence of inorganic phosphate) [2]; 10.5 (in presence of inorganic phosphate) [2]

pH-range

Temperature optimum (°C)
50–55 (guanosine formation) [2]

Temperature range (°C)
40–60 (40°C: about 60% of activity maximum, 60°C: about 80% of activity maximum, guanosine formation) [2]

3 ENZYME STRUCTURE

Molecular weight

Subunits

Glycoprotein/Lipoprotein
—

4 ISOLATION/PREPARATION

Source organism
  E. coli [1]; Pseudomonas trifolii (IAM-1555) [2]

Source tissue

Localization in source

Purification

Crystallization
  –

Cloned
  –

Renatured
  –

5 STABILITY

pH

Temperature (°C)

Oxidation

Organic solvent

General stability information

Storage

6 CROSSREFERENCES TO STRUCTURE DATABANKS

PIR/MIPS code

Brookhaven code

7 LITERATURE REFERENCES

[1] Koch, A.L.: J. Biol. Chem.,223,535–549 (1956)
[2] Kamimura, A., Mitsugi, K., Okumura, S.: Agric. Biol. Chem.,37,2063–2072 (1973)

1 NOMENCLATURE

EC number
2.4.2.6

Systematic name
Nucleoside:purine(pyrimidine) deoxy-D-ribosyltransferase

Recommended name
Nucleoside deoxyribosyltransferase

Synonyms
Purine(pyrimidine) nucleoside:purine(pyrimidine) deoxyribosyl transferase [6]
Deoxyribosyltransferase, nucleoside
Deoxyribose transferase
Nucleoside trans-N-deoxyribosylase
trans-Deoxyribosylase
trans-N-Deoxyribosylase
trans-N-Glycosidase
Transdeoxyribosylase
Nucleoside deoxyribosyltransferase I (purine nucleoside:purine deoxyribosyl-transferase: strictly specific for transfer between purine bases) [3]
Nucleoside deoxyribosyltransferase II (purine(pyrimidine) nucleo-side:purine(pyrimidine) deoxyribosyltransferase: catalyzes the transfer of the deoxyribosyl moiety between purines or pyrimidines as well as from a purine to a pyrimidine) [3]

CAS Reg. No.
9026-86-2

2 REACTION AND SPECIFICITY

Catalysed reaction
2-Deoxy-D-ribosyl-base^1 + base^2 →
→ 2-deoxy-D-ribosyl-base^2 + base^1 (ping pong bi-bi mechanism [3, 5, 6])

Reaction type
Pentosyl group transfer

Natural substrates

Substrate spectrum

1 2-Deoxy-D-ribosyl-base[1] + base[2] (r [3], deoxyribosyl donor: deoxyinosine [1, 3, 5–7], deoxyguanosine [1, 3, 6, 7], deoxyadenosine [1, 3, 6], deoxycytidine [1, 3, 5–7], deoxyuridine [7], deoxythymidine [7], deoxyribosyl acceptor: adenine [1, 3, 5–7, 9], guanine [1, 7], uric acid [1], azaguanine [1], cytosine [1, 3, 6, 7], thymine [1], uracil [1, 7], xanthine (products are 9-deoxyribosylxanthine (20%) and 7-deoxyribosylxanthine (80%)) [3], hypoxanthine [7], at low reaction rate: 8-methyladenine [9], 8-bromoadenine [9], 8-chloroadenine [9], 8-trifluoromethyladenine [9], specificity overview [3, 4, 11], substrate specificity studies [4, 11]: purine bases as acceptors [4], pyrimidine bases as acceptors [11], 2 different enzymes: 1. nucleoside deoxyribosyltransferase I (purine nucleoside:purine deoxyribosyltransferase): strictly specific for transfer between purine bases, 2. nucleoside deoxyribosyltransferase II (purine(pyrimidine) nucleoside:purine(pyrimidine) deoxyribosyltransferase): catalyzes the transfer of the deoxyribosyl moiety between purines or pyrimidines as well as from a purine to a pyrimidine [3], formation of 3-(2'-deoxyribofuranosyl) and 9-(2'-deoxyribofuranosyl)nucleosides of 8-substituted purines [9]) [1, 3, 5–7, 9, 11]

2 Deoxyinosine + adenine (r) [3, 5]

3 Deoxycytidine + adenine (r) [3, 6]

Product spectrum

1 2-Deoxy-D-ribosyl-base[2] + base[1]

2 Hypoxanthine + deoxyadenosine [3, 5]

3 Cytosine + deoxyadenosine [3, 6]

Inhibitor(s)

Deoxyadenosine (linear competitive inhibitor of deoxyinosine [5], linear non-competitive inhibitor of adenine [5], linear competitive inhibitor of deoxycytidine [6]) [5, 6]; Imidazole (inhibits deoxycytidine→ adenosine transfer) [4]; Hydroxypyrrolo-pyrimidine (inhibits deoxycytidine→ adenosine transfer) [4]; Amino-pyrrolo-pyrimidine (inhibits deoxycytidine→ adenosine transfer) [4]; Tris(hydroxymethyl)aminomethane (completely inhibits transfer involving pyrimidines and partly inhibits transfer where only purines are involved as substrates, phosphate partly reverses inhibition of pyrimidine transfer reactions [1]) [1, 3]

Cofactor(s)/prosthetic group(s)/activating agents

Metal compounds/salts

More (no metal ion requirement) [7]

Turnover number (min^{-1})

Specific activity (U/mg)

More [3, 5–7]

K_m-value (mM)
  0.0073 (adenine (+ deoxyinosine)) [6]; 0.008 (adenine (+ deoxyguanosine)) [6]; 0.019 (adenine (+ deoxycytidine)) [3, 6]; 0.021 (adenine (+ deoxyadenosine)) [6]; 0.041 (adenine + (deoxyinosine)) [3, 5]; 0.073 (cytosine (+ deoxyinosine)) [6]; 0.077 (cytidine (+ deoxyadenosine)) [6]; 0.086 (hypoxanthine (+ deoxyadenosine)) [3, 5]; 0.090 (deoxycytidine (+ adenine)) [3, 6]; 0.092 (deoxyadenosine (+ adenine)) [6]; 0.095 (deoxycytidine (+ cytosine)) [6]; 0.12 (deoxyadenosine (+ cytosine)) [3, 6]; 0.17 (cytosine (+ deoxycytidine)) [6]; 0.22 (cytosine (+ deoxyadenosine)) [3, 6]; 0.29 (deoxycytidine (+ hypoxanthine)) [7]; 0.346 (deoxyguanosine (+ adenine)) [6]; 0.35 (deoxyinosine (+ adenine)) [3, 5]; 0.37 (deoxyguanosine (+ cytosine)) [6]; 0.45 (deoxyadenosine (+ hypoxanthine)) [3, 5]; 3.4 (deoxyinosine (+ adenine)) [6]; 3.5 (deoxyinosine (+ cytosine)) [6]

pH-optimum
  5.8 [1, 3]

pH-range

Temperature optimum (°C)
  40 (assay at) [5, 6]

Temperature range (°C)

---

3 ENZYME STRUCTURE

Molecular weight
  82000 (Lactobacillus helveticus, gel filtration) [7]
  86000 (Lactobacillus helveticus, deoxyribosyltransferase I) [3]
  110000 (Lactobacillus leichmanii, gel filtration) [8]

Subunits
  Hexamer (6 × 18000, Lactobacillus leichmanii, calculated from ntd open reading frame and confirmed by SDS-PAGE) [10]

Glycoprotein/Lipoprotein
  –

---

4 ISOLATION/PREPARATION

Source organism
  Lactobacillus delbrueckii [1]; Thermobacterium acidophilus [1]; Lactobacillus leichmanii [8–10]; E. coli [1]; Lactobacillus helveticus (ATCC 8018 and ATCC 10386 [1], CNRZ 66 (NCDO30) or NDC 30 [4, 11], CNRZ 303 [3, 5, 6]) [1, 3–7, 11]; More (Crithidia luculia purine-2'-deoxyribonucleosidase also has trans-N-deoxyribosylase activity) [2]

Source tissue
    Cell [1]

Localization in source

Purification
    Lactobacillus helveticus [1, 3, 5, 7]

Crystallization
    [7, 8]

Cloned
    (Lactobacillus leichmanii gene cloned and expressed in E. coli) [8]

Renatured
    –

---

5 STABILITY

pH
    6.5 (highest stability) [1]

Temperature (°C)
    55 (pH 6.3, 10 min, stable below) [1]; 90 (pH 6.3, 10 min, about 75% loss of
    activity) [1]

Oxidation

Organic solvent

General stability information
    Crystals stable to X-rays at room temperature [8]

Storage
    –50°C, either in lyophilized form or dissolved in freshly distilled water, stable
    for several months [7]

---

6 CROSSREFERENCES TO STRUCTURE DATABANKS

PIR/MIPS code

Brookhaven code

7 LITERATURE REFERENCES

[1] Roush, A.H., Betz, R.F.: J. Biol. Chem.,233,261–266 (1958)
[2] Steenkamp, D.J.: Eur. J. Biochem.,197,431–439 (1991)
[3] Cardinaud, R.: Methods Enzymol.,51,446–455 (1978) (Review)
[4] Holguin, J., Cardinaud, R.: Eur. J. Biochem.,54,515–520 (1975)
[5] Danzin, C., Cardinaud, R.: Eur. J. Biochem.,48,255–262 (1974)
[6] Danzin, C., Cardinaud, R.: Eur. J. Biochem.,62,365–372 (1976)
[7] Uerkvitz, W.: Eur. J. Biochem.,23,387–395 (1971)
[8] Cook, W.J., Short, S.A., Ealick, S.E.: J. Biol. Chem.,265,2682–2683 (1990)
[9] Huang, M.-C., Montgomery, J.A., Thorpe, M.C., Stewart, E.L., Secrist III, J.A., Blakley, R.L.: Arch. Biochem. Biophys.,222,133–144 (1983)
[10] Singer, J.T., Barbier, C.S., Short, S.A.: J. Bacteriol.,163,1095–1100 (1985)
[11] Cardinaud, R., Holguin, J.: Biochim. Biophys. Acta,568,339–347 (1979)

Enzyme Handbook © Springer-Verlag Berlin Heidelberg 1996
Duplication, reproduction and storage in data banks are only
allowed with the prior permission of the publishers

1 NOMENCLATURE

EC number
2.4.2.7

Systematic name
AMP:pyrophosphate phospho-D-ribosyltransferase

Recommended name
Adenine phosphoribosyltransferase

Synonyms
Transphosphoribosidase
AMP pyrophosphorylase
APRT [2]
AMP-pyrophosphate phosphoribosyltransferase [9]
Phosphoribosyltransferase, adenine
Adenine phosphoribosylpyrophosphate transferase
Adenosine phosphoribosyltransferase
Adenylate pyrophosphorylase
Adenylic pyrophosphorylase

CAS Reg. No.
9027-80-9

---

2 REACTION AND SPECIFICITY

Catalysed reaction
Adenine + 5-phospho-alpha-D-ribose 1-diphosphate →
→ AMP + diphosphate (ping-pong mechanism [3])

Reaction type
Pentosyl group transfer

Natural substrates
Adenine + 5-phospho-alpha-D-ribose 1-diphosphate (enzyme mediates the
translocation of adenine into the cell as AMP [3], adenine salvage enzyme
[8], may play a role in maintaining the supply of adequate levels of active
cytokinin [9], necessary for appropriate regulation of purine biosynthesis
[3]) [3, 8, 9]

Substrate spectrum

1 Adenine + 5-phospho-alpha-D-ribose 1-diphosphate (equilibrium lies far in the direction of nucleotide synthesis [1], specific for adenine or 2,6-diamino-purine [3]) [1–19]
2 5-Amino-4-imidazolecarboxamide + 5-phospho-alpha-D-ribose 1-diphosphate (not [3]) [1, 2, 10]
3 4-Amino-5-imidazolecarboxamide + 5-phospho-alpha-D-ribose 1-diphosphate [4, 7]
4 4-Aminopyrazolo-(3,4-d)-pyrimidine + 5-phospho-alpha-D-ribose 1-diphosphate [10]
5 6-Amino-2-hydroxypurine + 5-phospho-alpha-D-ribose 1-diphosphate [10]
6 6-Methylpurine + 5-phospho-alpha-D-ribose 1-diphosphate [10]
7 Isopentenyladenine + 5-phospho-alpha-D-ribose 1-diphosphate (N^6-(delta-isopentenyl)adenine [9]) [2, 9]
8 Benzyladenine + 5-phospho-alpha-D-ribose 1-diphosphate (N^6-benzyl-adenine [9]) [2, 9]
9 6-Mercaptopurine + 5-phospho-alpha-D-ribose 1-diphosphate [4]
10 2,6-Diaminopurine + 5-phospho-alpha-D-ribose 1-diphosphate [4, 7, 10]
11 N^6-Furfuryladenine + 5-phospho-alpha-D-ribose 1-diphosphate [9]
12 8-Azaadenine + 5-phospho-alpha-D-ribose 1-diphosphate [10]
13 More (not: hypoxanthine [4, 9], guanine [4, 9], adenosine [4], D-ribose 5-phosphate [7], ribose 1-phosphate [7]) [4, 7, 9]

Product spectrum

1 AMP + diphosphate (5'-AMP [3]) [1–5, 11]
2 5-Amino-4-imidazolecarboxamide ribotide + diphosphate [1]
3 ?
4 ?
5 ?
6 ?
7 ?
8 ?
9 ?
10 ?
11 ?
12 ?
13 ?

Inhibitor(s)

Na$^+$ [4]; SO$_4^{2-}$ [4]; Succinate [4]; Citrate [4]; Hg^{2+} (in presence of MnCl$_2$ [2]) [2, 4, 9, 14]; HgCl$_2$ (reversed by 2-mercaptoethanol) [12]; Iodoacetate [4]; p-Hydroxymercuribenzoate (not enzyme from monkey liver [4], reversed by: 2-mercaptoethanol [12], DTT [15], no effect [13]) [4, 12, 15]; ADP [8, 17]; GMP [8]; UMP [8]; Adenine (at high concentrations, at low 5-phospho-alpha-D-ribose 1-diphosphate concentration) [11]; N-Ethylmaleimide [15]; ATP [17]; dAMP [17]; dATP [17]; 6-Mercaptopurine [14]; 2,6-Diaminopurine [14]; Co^{2+} [14]; Sulfhydryl reagents [15]; Mg^{2+} (inhibition above 2 mM, activation below [8], inhibition in presence of MnCl$_2$, activation in absence [2]) [2, 8]; Ca^{2+} (in presence of MnCl$_2$ [2], activation [9]) [2]; Ba^{2+} (in presence of MnCl$_2$ [2]) [2, 3]; Cd^{2+} (in presence of MnCl$_2$ [2]) [2, 14]; 5-Phosphoribose 1-diphosphate (substrate inhibition) [2]; Zn^{2+} (weak [14]) [3, 14]; AMP [3, 8, 11, 13, 17, 18]; Diphosphate [3]; Nucleotides (nucleotide mono-, di- and triphosphates of adenine, guanine and hypoxanthine [4], effect is strongly influenced by pH, inhibition at pH 7.1: ATP, GMP, activation at pH 7.1: GTP, UMP, UTP, CMP, CTP, IMP, no effect at pH 7.1: AMP, inhibition at pH 8.0: AMP, ATP, GMP, GTP, UTP, CTP, IMP, no effect at pH 8.0: UMP, CMP [11], higher concentrations of all 5'-nucleotides are most inhibitory, 6-OH purine nucleotides are moderately inhibitory, pyrimidine nucleotides are least inhibitory [3]) [3, 4, 11]; More (no substrate inhibition by adenine) [2]

Cofactor(s)/prosthetic group(s)/activating agents

Nucleotides (effect is strongly influenced by pH, inhibition at pH 7.1: ATP, GMP, activation at pH 7.1: GTP, UMP, UTP, CMP, CTP, IMP, no effect at pH 7.1: AMP, inhibition at pH 8.0: AMP, ATP, GMP, GTP, UTP, CTP, IMP, no effect at pH 8.0: UMP, CMP) [11]

Metal compounds/salts

Mg^{2+} (required [1, 3, 8, 13, 15, 17, 18], activity depends on presence of divalent cations [2–4], Mn^{2+} or Mg^{2+} [2, 3], Mg^{2+} most effective [4], competitive to 5-phospho-alpha-D-ribose 1-diphosphate [3], order of effectiveness: Mg^{2+} > Mn^{2+} > Ca^{2+} > Co^{2+} > Ni^{2+} > Zn^{2+} [4], K$_m$: 0.4 mM [13], 0.15–0.18 mM [18], optimal concentration: 2 mM [8], 5 mM [9], 1–2 mM [17], 3 mM at 60°C, at 37°C activity varies little in the range 3–50 mM MgCl$_2$ [15], inhibition above 2 mM [8], maximal activity if Mg^{2+} is twice the concentration of 5-phospho-alpha-D-ribose 1-diphosphate [3, 4], at 0°C in the absence of Mg^{2+} but in presence of substrates enzyme catalyzes a rapid and limited synthesis of AMP [7]) [1–4, 7–9, 13, 15, 17, 18]; Mn^{2+} (activity depends on presence of divalent cations, Mn^{2+} or Mg^{2+} activates [9, 18], K$_m$: 0.04–0.05 mM [18]) [9, 18]; Ca^{2+} (activates [9], inhibition [3, 18], inhibition in presence of MnCl$_2$ [2]) [9]; Zn^{2+} (activates, K$_m$: 0.045–0.05 mM) [18]

Turnover number (min^{-1})
    560 (adenine + 5-phospho-alpha-D-ribose 1-diphosphate) [3]

Specific activity (U/mg)
    2.8 [10]; 14.0 [3]; 9.58 [4, 7]; 0.156 [8]; 0.0413 [9]; 1.1 [12]; 9.15 [15]; 18.0
    [17]; More [5, 6, 13]

K$_m$-value (mM)
    0.0045 (adenine) [2]; 0.006 (5-phospho-alpha-D-ribose 1-diphosphate) [6];
    0.020 (adenine) [3]; 0.036 (4-aminopyrazolo-(3,4-d)pyrimidine) [10]; 0.074
    (adenine) [9]; 0.110 (N^6-furfuryladenine) [9]; 0.125 (5-phospho-alpha-D-ri-
    bose 1-diphosphate, in presence of a 2-fold excess of Mg^{2+}) [3]; 0.130
    (N^6-(DELTA^2-isopentenyl)adenine) [9]; 0.154 (N^6-benzyladenine) [9]; 0.24
    (8-azaadenine) [10]; 0.29 (5-phospho-alpha-D-ribose 1-diphosphate) [2];
    0.73 (benzyladenine) [2]; 0.89 (2,6-diaminopurine) [10]; 1.0 (5-aminoimida-
    zole-4-carboxamide) [10]; 1.4 (6-amino-2-hydroxypurine) [10]; 3.7 (6-methyl-
    purine) [10]; More [5, 8, 10–18]

pH-optimum
    7.4–9.5 [4]; 7.5 (two zones of pH-optima at pH 7.5 and pH 8.5) [17]; 7.6–8.0
    [8]; 7.8 [3]; 8 (around, broad, E. coli) [11]; 8.5 (two zones of pH-optima at
    pH 7.5 and 8.5) [17]; 9.0 [14]; 9.2 [15]; 10.0 [12]; More (Mycoplasma myco-
    ides: pH-optimum depends on 5-phospho-alpha-D-ribose 1-diphosphate
    concentration [11], temperature dependent [2]) [2, 11]

pH-range
    5.5–10 (active over a broad range increasing progressively in activity from
    pH 5.5–10) [5]; 7.0–9.0 [13]; 7.4–9.5 [7]

Temperature optimum (°C)
    30 (assay at) [5, 8, 13]; 37 (assay at) [3, 4, 9, 11, 17]; 60 [15]; 65 [2]

Temperature range (°C)

---

3 ENZYME STRUCTURE

Molecular weight
    18000 (Plasmodium falciparum, gel filtration) [14]
    22000 (rat, sucrose density gradient sedimentation) [5, 12]
    23000 (Triticum aestivum, gel filtration) [9]
    25000 (Leishmania donovani, gel filtration) [10]
    28000 (Artemia sp., gel filtration) [18]
    34000 (human, gel filtration, sucrose density gradient ultracentrifugation [7],
    Plasmodium chabaudi [16]) [7, 16]
    38200 (human, sedimentation equilibrium centrifugation) [6]
    40000 (E. coli, gel filtration) [3]

44000 (mouse, gel filtration [13], Schizosaccharomyces pombe, gel filtration, 2 forms: 44000 and 50000, in presence of 5 mM $MgCl_2$ dissociation into a single form of apparent MW 30000 [17]) [13, 17]

50000 (Schizosaccharomyces pombe, gel filtration, 2 forms: 44000 and 50000, in presence of 5 mM $MgCl_2$ dissociation into a single form of apparent MW 30000) [17]

54000 (Arabidopsis thaliana [2], Brassica juncea, gel filtration [15]) [2, 15]

Subunits

Dimer (Arabidopsis thaliana [2], 2 × 17000–18000, human, SDS-PAGE, gel filtration in guanidine hydrochloride, peptide mapping data suggest that the subunits are quite similar if not identical [6], 2 × 15000, Artemia sp., SDS-PAGE [18], 2 × 23000, mouse, SDS-PAGE [13], 2 × 27000, Brassica juncea [15]) [2, 6, 13, 15, 18]

Trimer (3 × 11100, human, SDS-PAGE) [7]

? (x × 17500, rat, SDS-PAGE [5, 12], x × 19481, human, determination of amino acid sequence [19]) [5, 12, 19]

More (Mycoplasma mycoides: association of subunits in presence of 5-phospho-alpha-D-ribose 1-diphosphate) [11]

Glycoprotein/Lipoprotein
–

4 ISOLATION/PREPARATION

Source organism
Catharanthus roseus [8]; Triticum aestivum [9]; Brassica juncea [15]; Bovine [1]; Arabidopsis thaliana [2]; E. coli [3, 11]; Human [4, 6, 7, 19]; Rat [5, 12]; Leishmania donovani [10]; Mycoplasma mycoides [11]; Mouse [13]; Plasmodium falciparum [14]; Plasmodium chabaudi [16]; Schizosaccharomyces pombe [17]; Artemia sp. [18]

Source tissue
Cell [3, 11]; Promastigotes [10]; Liver [1, 5, 12]; Leaf [2, 15]; Erythrocytes [4, 6, 7, 19]; Cultured cells [8]; Germ [9]; Mammary tumor FM3A cell culture [13]; Late trophozoites [14]; Cysts [18]; Nauplii [18]; More (human enzyme found in all tissues with the highest specific activity in nucleated cells) [4]

Localization in source
Cell membrane [3]; Cytoplasm (exclusively) [4]; Cytosol [9]

Purification
Catharanthus roseus [8]; Mouse [13]; Bovine [1]; Brassica juncea [15]; Arabidopsis thaliana [2]; Human [4, 6, 7, 19]; Rat [5, 12]; Triticum aestivum (partial) [9]; Leishmania donovani [10]; Plasmodium chabaudi [16]; Schizosaccharomyces pombe [17]; Artemia sp. [18]

Crystallization
−

Cloned
−

Renatured
−

5 STABILITY

pH

Temperature (°C)
50 (40 min, 5 mM MgCl$_2$, more than 90% loss of activity, + 0.4 mM 5-phos-pho-alpha-D-ribose 1-diphosphate, 30% loss of activity) [11]; 60 (5 min, stable [3], 10 min, stable in presence of 10 mM AMP [17]) [3, 17]

Oxidation

Organic solvent

General stability information
Ammonium sulfate, 1 M, partially protects against inactivation during storage [7]; Dithiothreitol, 10 mM, or dimethylsulfoxide, 5%, or 5-phospho-alpha-D-ribose 1-diphosphate, 1 mM, protects against inactivation during storage [7]; Inactivation by isoelectric focusing [10]; Freezing and thawing have little effect on E. coli enzyme, but inactivate Mycoplasma mycoides enzyme [11]; Bovine serum albumin or adenine stabilizes Mycoplasma mycoides enzyme slightly [11]; 5-Phospho-alpha-D-ribose 1-diphosphate stabilizes Mycoplasma mycoides enzyme markedly [11]; Dialysis, against 20 mM Tris-HCl, pH 7.5, 20 mM (NH$_4$)$_2$SO$_4$ + 5 mM 2-mercaptoethanol inactivates [12]; Dilution inactivates [12]

Storage
−15°C, stable for months, E. coli enzyme [11]; 2–4°C, average half-life of 2–4 weeks [12]; −70°C, 50 mM Tris-HCl, pH 7.4, 5 mM MgCl$_2$, 10 mM KCl, 10% glycerol, stable [15]; −20°C, 6 mM 5-phospho-alpha-D-ribose 1-diphosphate, 2 mM MgCl$_2$, stable for at least a month [17]; −70°C, 5 mM MgCl$_2$, 0.1 mM 5-phospho-alpha-D-ribose 1-diphosphate, stable for 2 months [7]; −80°C, stable for at least 3 weeks [8]; −20°C, 2–3 weeks, considerable loss of activity [3]; −79°C, 0.1 mg/ml protein, 0.1 M potassium phosphate buffer, pH 7, stable for several months [3]; Unstable at 4°C or at −20°C [8]

6 CROSSREFERENCES TO STRUCTURE DATABANKS

PIR/MIPS code

PIR2:S20867 (Arabidopsis thaliana); PIR2:S36334 (Chinese hamster); PIR1:RTECA (Escherichia coli); PIR2:A29596 (fruit fly (Drosophila melanogaster)); PIR3:S34831 (fruit fly (Drosophila melanogaster)); PIR1:RTHUA (human); PIR1:RTMSA (mouse); PIR2:B49927 (Pseudomonas stutzeri); PIR3:JC4213 (1 yeast (Saccharomyces cerevisiae)); PIR2:S44328 (APT2 yeast (Saccharomyces cerevisiae))

Brookhaven code

7 LITERATURE REFERENCES

[1] Flaks, J.G., Erwin, M.J., Buchanan, J.M.: J. Biol. Chem.,228,201–213 (1957)
[2] Lee, D., Moffatt, B.A.: Physiol. Plant.,87,483–492 (1993)
[3] Hochstadt, J.: Methods Enzymol.,51,558–567 (1978) (Review)
[4] Arnold, W.J., Kelley, W.N.: Methods Enzymol.,51,568–574 (1978) (Review)
[5] Groth, D.P., Young, L.G., Kenimer, J.G.: Methods Enzymol.,51,574–580 (1978) (Review)
[6] Holden, J.A., Meredith, G.S., Kelley, W.N.: J. Biol. Chem.,254,6951–6955 (1979)
[7] Thomas, C.B., Arnold, W.J., Kelley, W.N.: J. Biol. Chem.,248,2529–2535 (1973)
[8] Hirose, F., Ashihara, H.: Z. Pflanzenphysiol.,110,135–145 (1983)
[9] Chen, C.-M., Melitz, D.K., Clough, F.W.: Arch. Biochem. Biophys.,214,634–641 (1982)
[10] Tuttle, J.V., Krenitsky, T.A.: J. Biol. Chem.,256,909–916 (1980)
[11] Sin, I.L., Finch, L.R.: J. Bacteriol.,112,439–444 (1972)
[12] Kenimer, J.G., Young, L.G., Groth, D.P.: Biochim. Biophys. Acta,384,87–101 (1975)
[13] Okada, G., Kaneko, I., Koyama, H.: Biochim. Biophys. Acta,884,304–310 (1986)
[14] Queen, S.A., Vander Jagt, D.L., Reyes, P.: Biochim. Biophys. Acta,996,160–165 (1989)
[15] Moffatt, B.A., Somerville, C.R.: Arch. Biochem. Biophys.,283,484–490 (1990)
[16] Walter, R.D., Koenigk, E.: Tropenmed. Parasitol.,25,227–235 (1974)
[17] Nagy, M., Ribet, A.-M.: Eur. J. Biochem.,77,77–85 (1977)
[18] Montero, C., LLorente, P.: Biochem. J.,275,327–334 (1991)
[19] Wilson, J.M., O'Toole, T.E., Argos, P., Shewach, D.S., Daddona, P.E., Kelley, W.N.: J. Biol. Chem.,261,13677–13683 (1986)

1 NOMENCLATURE

EC number
2.4.2.8

Systematic name
IMP:pyrophosphate phospho-D-ribosyltransferase

Recommended name
Hypoxanthine phosphoribosyltransferase

Synonyms
IMP pyrophosphorylase
Transphosphoribosidase
GPRT (phosphoribosyltransferases for hypoxanthine and guanine are separate enzymes) [10]
HPRT (phosphoribosyltransferases for hypoxanthine and guanine are separate enzymes) [10]
Hypoxanthine-guanine phosphoribosyltransferase
Guanine phosphoribosyltransferase
Guanosine 5'-phosphate pyrophosphorylase [1]
IMP-GMP pyrophosphorylase [1]
HGPRTase [1]
Phosphoribosyltransferase, hypoxanthine
6-Hydroxypurine phosphoribosyltransferase
6-Mercaptopurine phosphoribosyltransferase
GMP pyrophosphorylase
Guanine-hypoxanthine phosphoribosyltransferase
Guanosine 5'-phosphate pyrophosphorylase
Guanosine phosphoribosyltransferase
Guanylate pyrophosphorylase
Guanylic pyrophosphorylase
Inosinate pyrophosphorylase
Inosine 5'-phosphate pyrophosphorylase
Inosinic acid pyrophosphorylase
Inosinic pyrophosphorylase
Phosphoribosyltransferase, 6-mercaptopurine
Purine-6-thiol phosphoribosyltransferase

CAS Reg. No.
9016-12-0

2 REACTION AND SPECIFICITY

Catalysed reaction
Hypoxanthine + 5-phospho-alpha-D-ribose 1-diphosphate →
→ IMP + diphosphate

Reaction type
Pentosyl group transfer

Natural substrates

Substrate spectrum
1 Hypoxanthine + 5-phospho-alpha-D-ribose 1-diphosphate (r [1], equilibrium lies far in direction of IMP formation [1], binds hypoxanthine 67-times less effectively than guanine and 4-times less effectively than xanthine [19], phosphoribosyltransferases for hypoxanthine and guanine are separate enzymes [10]) [1, 2, 4–7, 9–13, 15, 16, 19–22]
2 Guanine + 5-phospho-alpha-D-ribose 1-diphosphate (binds hypoxanthine 67-times less effectively than guanine and 4-times less effectively than xanthine [19], phosphoribosyltransferases for hypoxanthine and guanine are separate enzymes [10]) [1–4, 7, 9–13, 15, 16, 19–22]
3 6-Mercaptopurine + 5-phospho-alpha-D-ribose 1-diphosphate [1, 2, 7, 19]
4 6-Thioguanine + 5-phospho-alpha-D-ribose 1-diphosphate [7]
5 8-Azahypoxanthine + 5-phospho-alpha-D-ribose 1-diphosphate [7]
6 Xanthine + 5-phospho-alpha-D-ribose 1-diphosphate (poor substrate [12], not [1, 20]) [12, 19, 22]
7 2-Amino-6-mercaptopurine + 5-phospho-alpha-D-ribose 1-diphosphate [19]
8 2-Hydroxy-6-mercaptopurine + 5-phospho-alpha-D-ribose 1-diphosphate [19]
9 More (not: adenine [1, 12, 20], 5-formamido-4-imidazolecarboxamide [1], uric acid [1], 8-azaguanine [1], 2,6-diaminopurine [1], orotic acid [1]) [1, 12, 20]

Product spectrum
1 IMP + diphosphate [1]
2 GMP + diphosphate [1]
3 6-Mercaptopurine ribotide + diphosphate [2]
4 ?
5 ?
6 ?
7 ?
8 ?
9 ?

Inhibitor(s)

AMP [3]; GMP [3, 4, 6, 12]; GDP [3, 12]; GTP [3, 6, 12]; IMP [3, 12]; IDP [3, 12]; ITP (weak) [3]; UDP (weak) [3]; XMP [3]; CDP [3]; Diphosphate [3, 6]; Ca^{2+} [4, 10]; Ba^{2+} [10]; Zn^{2+} [10]; 6-Mercaptopurine (competitive to hypoxanthine) [5]; 6-Thioguanine (competitive to hypoxanthine) [5]; 2-Amino-6-mercaptopurine (competitive to hypoxanthine) [5]; 6-Thioinosine [7]; 9-beta-Arabinofuranosylhypoxanthine [7]; 6-Chloropurine [7]; Xanthine [7]; Azaguanine [7]; 6-Hydroxypurine nucleotides [10]; 6-Aminopurine nucleotides [10]; ATP [12]; p-Chloromercuribenzoate (reversed by dithiothreitol or 2-mercaptoethanol) [19]; EDTA [12]; More (overview: inhibition constant of purines and purine analogs) [19]

Cofactor(s)/prosthetic group(s)/activating agents

Metal compounds/salts

Mg^{2+} (required [1, 3, 4, 13], absolute requirement for Mg^{2+} or Mn^{2+} [10, 11], K_m: 0.04–0.05 mM [4], activating effect of Mn^{2+} is higher than that of Mg^{2+} [12], maximal activity at: 0.5–1 mM $MgCl_2$ [3], 1–10 mM [13], 1 mM [4]) [1, 3, 4, 10–13]; Mn^{2+} (absolute requirement for Mg^{2+} or Mn^{2+} [10, 11], can replace Mg^{2+} [4], K_m: 0.015–0.020 mM [4], activating effect of Mn^{2+} is higher than that of Mg^{2+} [12]) [4, 10–12]; Zn^{2+} (activates, can replace Mg^{2+}, K_m: 0.066–0.075 mM) [4]

Turnover number (min^{-1})

Specific activity (U/mg)

66.5 [11]; 698.1 [11]; 5.746 [3]; 17.5 (human erythrocytes) [9, 18]; 9 (Chinese hamster brain) [9]; 13.25 (guanine) [21]; 9.25 (hypoxanthine) [21]; More [1, 2, 4, 9, 13, 22, 24]

K_m-value (mM)

0.00052 (hypoxanthine) [9]; 0.001 (hypoxanthine, 6-mercaptopurine) [1]; 0.0011 (guanine) [9]; 0.0020 (guanine) [5]; 0.0024 (guanine) [7]; 0.0025 (hypoxanthine) [5]; 0.0038 (hypoxanthine) [7]; 0.0053 (5-phospho-alpha-D-ribose 1-diphosphate) [9]; 0.0062 (6-mercaptopurine) [7]; 0.0076 (6-thioguanine) [7]; 0.015 (5-phospho-alpha-D-ribose 1-diphosphate) [4]; 0.028 (guanine) [3]; 0.1 (5-phospho-alpha-D-ribose 1-diphosphate) [3]; 0.360 (8-azahypoxanthine) [7]; 0.5 (guanine) [1]

pH-optimum

6.0–10.0 (hypoxanthine) [7]; 7.0 (2 buffer-independent pH-optima: 7.0 and 9.5) [4]; 7.4 (assay at) [1]; 7.5–9.5 (guanine) [7]; 7.6–8.0 (2 zones of pH-optima: 7.6–8.0 and 9.2–9.5) [3]; 7.9 (guanine phosphoribosyltransferase) [10]; 8.4 (hypoxanthine phosphoribosyltransferase) [10]; 8.5 [12, 13]; 9.2–9.5 (2 zones of pH-optima: 7.6–8.0 and 9.2–9.5) [3]; 9.5 (2 buffer-independent pH-optima: 7.0 and 9.5) [4]; 10 (Chinese hamster [9]) [9, 10, 16]

pH-range
5.5–11 (Chinese hamster) [9, 16]; 6.3–9.7 (6.3: about 50% of activity maximum, 9.7: about 75% of activity maximum) [12]; 7.0–8.5 (7.0: about 80% of activity maximum, 8.5: about 70% of activity maximum) [19]

Temperature optimum (°C)
37 (assay at) [4, 9, 11, 16–18, 21, 24]; 38 (assay at) [1]

Temperature range (°C)

---

3 ENZYME STRUCTURE

Molecular weight
42000 (Schizosaccharomyces pombe, gel filtration in presence of $MgCl_2$) [3]
48000 (Schizosaccharomyces pombe, gel filtration in absence of $MgCl_2$) [3]
51000 (Saccharomyces cerevisiae, gel filtration) [13]
54500–54800 (Saccharomyces cerevisiae, equilibrium sedimentation analysis, gel filtration) [20]
58000–63000 (Giardia lamblia, gel filtration, isokinetic sucrose gradient) [22]
66000 (Artemia sp., gel filtration) [4]
68000 (human, gel filtration) [15]
71000 (Plasmodium chabaudi) [5]
79000 (Plasmodium lophurae) [7]
80000 (mouse, gel filtration) [17]
80000–85000 (Chinese hamster, gel filtration, acrylamide gel electrophoresis) [9]
81000–83000 (human, sedimentation equilibrium centrifugation) [9, 18]
85000 (chicken, gel filtration [11], human, sedimentation equilibrium method [24]) [11, 24]
100000 (human, gradient gel electrophoresis) [8]
105000 (Schistosoma mansoni, pore gradient electrophoresis) [21]
150000 (Streptomyces cyanogenus, gel filtration) [12]

Subunits
Monomer (1 × 51000, Saccharomyces cerevisiae, SDS-PAGE) [13]
Dimer (2 × 34000–34700, human, SDS-PAGE [15], 2 × 29000, Giardia lamblia, SDS-PAGE [22], 2 × 29500, Saccharomyces cerevisiae, SDS-PAGE [20], 2 × 64000, Schistosoma mansoni, SDS-PAGE [21]) [15, 20–22]
Trimer (3 × 27000, mouse, SDS-PAGE [17], 3 × 26000, human, SDS-PAGE [18]) [17, 18]
Tetramer (4 × 19000, Artemia sp. [4], SDS-PAGE, 4 × 24000, human, SDS-PAGE [8], 4 × 26000, chicken, SDS-PAGE [11]) [4, 8, 11]
Octamer (8 × 18000, Streptomyces cyanogenus, SDS-PAGE) [12]

? (x × 25000–27000, human erythrocytes, rat brain, SDS-PAGE [14],
x × 24000, Leishmania donovani, SDS-PAGE [6], x × 25000, Chinese hamster, SDS-PAGE [9], 2 × 26000, human, SDS-PAGE [9], x × 41000–45000,
human, sedimentation equilibrium in presence of guanidine-HCl, x × 26000,
human, SDS-PAGE [24]) [6, 9, 10, 14, 24]

Glycoprotein/Lipoprotein
More (no carbohydrate [15], no glucosamine, sialic acid and hexose [24])
[15, 24]

---

4 ISOLATION/PREPARATION

Source organism
Schistosoma mansoni [21, 23]; Mouse [17]; Bovine [1, 2]; Rat [14]; Plasmodium lophurae [7]; Human [8, 9, 14, 15, 18, 24]; Schizosaccharomyces
pombe [3]; Artemia sp. [4]; Plasmodium chabaudi [5]; Leishmania donovani
[6]; Giardia lamblia [22]; Streptomyces cyanogenus [12]; Saccharomyces
cerevisiae [13, 20]; Chinese hamster [9, 16]; E. coli (K12 [10], phosphoribosyltransferases for hypoxanthine and guanine are separate enzymes [10],
unlike the hypoxanthine-guanine phosphoribosyltransferase from other
sources this enzyme binds hypoxanthine 67-times less effectively than guanine and 4-times less effectively than xanthine [19]) [10, 19]; Salmonella typhimurium LT-2 (phosphoribosyltransferases for hypoxanthine and guanine
are separate enzymes) [10]; Chicken [11]

Source tissue
V79 tissue cells [16]; Schistosomules [21]; Erythrocytes [9, 14, 15, 18, 24];
Liver [1, 2, 16, 17]; Cysts [4]; Nauplii [4]; Promastigotes [6]; Brain [8, 9, 11,
14, 16]

Localization in source

Purification
Schistosoma mansoni [21, 23]; Streptomyces cyanogenus [12]; Saccharomyces cerevisiae [13, 20]; Giardia lamblia [22]; Bovine [1]; Schizosaccharomyces pombe [3]; Plasmodium chabaudi [5]; Leishmania donovani [6];
Plasmodium lophurae [7]; Rat [14]; Human (3 isoenzymes [15]) [8, 9, 14,
15, 18, 24]; Chinese hamster [9]; E. coli [10]; Mouse [17]; Salmonella typhimurium [10]; Chicken [11]

Crystallization
–

Cloned
  (Schistosoma mansoni enzyme expressed in E. coli [23], mouse neuroblas-
  toma hypoxanthine-guanine phosphoribosyltransferase cDNA clone [21])
  [21, 23]

Renatured
  –

---

5 STABILITY

pH
  4.2 (10 min, 50% loss of activity) [19]; 7.0–9.3 (10 min, stable) [19]; 11 (10
  min, 50% loss of activity) [19]

Temperature (°C)
  30 (pH 7.4, stable) [12]; 40 (pH 7.4, gradual decrease of activity) [12]; 60
  (1.5 mM GMP, 10 min, complete loss of activity) [3]; 60–65 (stable) [1]; 85
  (stable, if first incubated in 1 mM 5-phospho-alpha-D-ribose 1-diphosphate,
  remarkably stable [9, 16], half-life: 3 min [11]) [9, 11, 16]; More [19]

Oxidation

Organic solvent

General stability information
  5-Phospho-alpha-D-ribose 1-diphosphate stabilizes [10];
  5-Phospho-alpha-D-ribose 1-diphosphate stabilizes against heat inactivation
  [12]; Glycerol or sucrose or dimethylsulfoxide stabilizes the purified enzyme
  at –70°C [15]

Storage
  –20°C, several months [13]; –80°C [22]

---

6 CROSSREFERENCES TO STRUCTURE DATABANKS

PIR/MIPS code
  PIR1:RTHYG (Chinese hamster); PIR3:S14402 (Chinese hamster);
  PIR2:S09614 (fluke (Schistosoma mansoni)); PIR2:S04278 (fluke (Schistoso-
  ma mansoni)); PIR2:A37114 (fluke (Schistosoma mansoni) (fragment));
  PIR1:RTHUG (human); PIR2:A32728 (human); PIR2:S30100 (Lactococcus
  lactis); PIR3:S21474 (long tailed hamster); PIR1:RTMSG (mouse);
  PIR2:JN0085 (Plasmodium falciparum); PIR2:S06315 (Plasmodium falcipar-
  um); PIR2:S06601 (Plasmodium falciparum); PIR2:S18140 (rat); PIR3:S18888
  (Salmonella typhimurium); PIR3:S41631 (Trypanosoma brucei); PIR2:S10993
  (Vibrio harveyi)

Brookhaven code
1HMP (Human (Homo Sapiens) recombinant form expressed in (Escherichia coli))

---

7 LITERATURE REFERENCES

[1] Flaks, J.G.: Methods Enzymol.,6,136–158 (1963) (Review)
[2] Lukens, L.N., Herrington, K.A.: Biochim. Biophys. Acta,24,432–433 (1957)
[3] Nagy, M., Ribet, A.-M.: Eur. J. Biochem.,77,77–85 (1977)
[4] Montero, C., LLorente, P.: Biochem. J.,275,327–334 (1991)
[5] Walter, R.D., Koenigk, E.: Tropenmed. Parasitol.,25,227–235 (1974)
[6] Allen, T., Henschel, E.V., Coons, T., Cross, L., Conley, J., Ullman, B.: Mol. Biochem. Parasitol.,33,273–281 (1989)
[7] Schimandle, C.M., Mole, L.A., Sherman, I.W.: Mol. Biochem. Parasitol.,23,39–45 (1987)
[8] Smithers, G.W., O'Sullivan, W.J.: Biochem. Med.,32,106–121 (1984)
[9] Olsen, A.S., Milman, G.: Methods Enzymol.,51,543–549 (1978) (Review)
[10] Hochstadt, J.: Methods Enzymol.,51,549–558 (1978) (Review)
[11] Veres, G., Monostori, E., Rasko, I.: FEBS Lett.,184,299–303 (1985)
[12] Ohe, T., Watanabe, Y.: Agric. Biol. Chem.,44,1999–2006 (1980)
[13] Schmidt, R., Wiegand, H., Reichert, U.: Eur. J. Biochem.,93,355–361 (1979)
[14] Gutensohn, W., Huber, M., Jahn, H.: Hoppe-Seyler's Z. Physiol. Chem.,357, 1379–1385 (1976)
[15] Arnold, W.J., Kelley, W.N.: J. Biol. Chem.,246,7398–7404 (1971)
[16] Olsen, A.S., Milman, G.: J. Biol. Chem.,249,4030–4037 (1974)
[17] Hughes, S.H., Wahl, G.M., Capecchi, M.R.: J. Biol. Chem.,250,120–126 (1975)
[18] Olsen, A.S., Milman, G.: Biochemistry,16,2501–2505 (1977)
[19] Miller, R.L., Ramsey, G.A., Krenitsky, T.A., Elion, G.B.: Biochemistry,11,4723–4731 (1972)
[20] Nussbaum, R.L., Caskey, C.T.: Biochemistry,20,4584–4590 (1981)
[21] Dovey, H.F., McKerrow, J.H., Aldritt, S.M., Wang, C.C.: J. Biol. Chem.,261,944–948 (1986)
[22] Aldritt, S.M., Wang, C.C.: J. Biol. Chem.,261,8528–8533 (1986)
[23] Yuan, L., Craig, S.P., McKerrow, J.H., Wang, C.C.: J. Biol. Chem.,265,13528–13532 (1990)
[24] Muensch, H., Yoshida, A.: Eur. J. Biochem.,76,107–112 (1977)

1 NOMENCLATURE

EC number
2.4.2.9

Systematic name
UMP:pyrophosphate phospho-alpha-D-ribosyltransferase

Recommended name
Uracil phosphoribosyltransferase

Synonyms
UMP pyrophosphorylase
UPRTase [2]
UMP:pyrophosphate phosphoribosyltransferase [8]
Phosphoribosyltransferase, uracil
Uridine 5'-phosphate pyrophosphorylase
Uridine monophosphate pyrophosphorylase
Uridylate pyrophosphorylase
Uridylic pyrophosphorylase

CAS Reg. No.
9030-24-4

2 REACTION AND SPECIFICITY

Catalysed reaction
Uracil + 5-phospho-alpha-D-ribose 1-diphosphate →
→ UMP + diphosphate

Reaction type
Pentosyl group transfer

Natural substrates
Uracil + 5-phospho-alpha-D-ribose 1-diphosphate (pyrimidine salvage enzyme) [4, 5]

Substrate spectrum

1 Uracil + 5-phospho-alpha-D-ribose 1-diphosphate (r [1], equilibrium lies far in the direction of UMP formation, [1], highly specific for uracil [2, 3, 5], and some uracil analogues [5]) [1–8]

2 6-Azauracil + 5-phospho-alpha-D-ribose 1-diphosphate (not [3], 97% of the activity with uracil [5]) [5]

3 5-Fluorouracil + 5-phospho-alpha-D-ribose 1-diphosphate (216% of the activity with uracil) [5]

4 2-Thiouracil + 5-phospho-alpha-D-ribose 1-diphosphate [7]

5 More (not: cytosine [3, 5], orotic acid [3, 5], thymine [5], hypoxanthine [5]) [3, 5]

Product spectrum

1 UMP + diphosphate (5'-UMP [5]) [5, 6]

2 ?

3 ?

4 2-Thio-5'-UMP + diphosphate [7]

5 ?

Inhibitor(s)

Phosphate (0.005 M, Lactobacillus bulgaricus) [1]; GTP (slight [2], lowers K_m value for 5-phospho-alpha-D-ribose 1-diphosphate and increases V_{max} 2-fold, no effect on K_m for uracil, effect is pH-dependent [5]) [2]; UTP [3]; dUTP [4]; UMP [8]; dUMP [8]; dCMP [8]; UDP [8]; TTP [8]; dCTP [8]; More (not: cytosine [3], orotic acid [3], 6-azauracil [3], diphosphate [5]) [3, 5]

Cofactor(s)/prosthetic group(s)/activating agents

GTP (lowers K_m value for 5-phospho-alpha-D-ribose 1-diphosphate and increases V_{max} 2-fold, no effect on K_m for uracil, effect is pH-dependent [5], slight inhibition [2]) [5]

Metal compounds/salts

Mg^{2+} (required [1, 6], only 5% of the activity remaining in absence [1], optimal concentration: 5 mM [6]) [1, 6]; More (broad specificity for activating divalent metal ions) [2]

Turnover number (min^{-1})

Specific activity (U/mg)

0.25 [1]; 6.602 [5]; 0.22 [6]

K_m-value (mM)

0.0004 (uracil) [3]; 0.0042 (uracil) [6]; 0.0069 (5-phospho-alpha-D-ribose 1-diphosphate) [3]; 0.007 (uracil) [5]; 0.0077 (uracil) [7]; 0.044 (5-phospho-alpha-D-ribose 1-diphosphate) [1]; 0.066 (5-phospho-alpha-D-ribose 1-diphosphate) [6]; 0.30 (5-phospho-alpha-D-ribose 1-diphosphate) [5]; 0.67 (thiouracil) [7]

pH-optimum
6.2 (uracil) [7]; 7.4 (assay at) [1]; 7.5 [3, 6]; 7.5–8.5 [5]; 7.6–7.8 [8]; 7.8
(2-thiouracil) [7]

pH-range
7.2–8 (7.2: about 50% of activity maximum, 8: about 70% of activity maximum) [6]; 6.5–10 [8]

Temperature optimum (°C)
37 (assay at) [1, 5–7]; 40 [3]

Temperature range (°C)

---

3 ENZYME STRUCTURE

Molecular weight
36000 (Tetrahymena pyriformis) [3]
75000 (E. coli K12, gel filtration) [5]
80000 (Crithidia luciliae [2], Acholeplasma laidlawii, gel filtration [6],
Saccharomyces cerevisiae, gel filtration [8]) [2, 6, 8]

Subunits
Dimer (Crithidia luciliae) [2]
Trimer (3 × 23500, E. coli K12, SDS-PAGE) [5]
? (x × 22500, E. coli K12, calculation from nucleotide sequence) [4]

Glycoprotein/Lipoprotein
–

---

4 ISOLATION/PREPARATION

Source organism
Lactobacillus leichmanii (ATCC 4797) [7]; Crithidia luciliae [2]; Tetrahymena
pyriformis (GL-7) [3]; Acholeplasma laidlawii [6]; Lactobacillus bifidus
(ATCC 4963) [1]; E. coli (ATCC 9637 [1], K12 [4, 5], not: E. coli B [1]) [1, 4,
5]; Lactobacillus bulgaricus (strain 09X) [1]; Saccharomyces cerevisiae [8]

Source tissue
Cells [1, 3, 5, 6]

Localization in source
Cytosol [6]

Purification
E. coli (K12) [5]; Saccharomyces cerevisiae [8]; Lactobacillus bifidus
(ATCC 4963) [1]; Crithidia luciliae [2]; Tetrahymena pyriformis (GL-7) [3];
Acholeplasma laidlawii [6]

Crystallization
–

Cloned
[4]

Renatured
–

---

5 STABILITY

pH

Temperature (°C)
40 (irreversible inactivation above) [3]; 60 (2 min, complete loss of activity) [8]

Oxidation

Organic solvent

General stability information
beta-Mercaptoethanol, not essential for stability in long-term storage [5]; GTP, 5 mM with 10 mM Tris-HCl, pH 7.5, 10 mM MgCl$_2$, 2 mM beta-mercaptoethanol, labilizes with only 15% activity remaining after 13 days, labilizing effect abolished in presence of 5-phospho-alpha-D-ribose 1-diphosphate [5]; Ethylene glycol, significantly increases stability with virtually no loss of activity after storage at 4°C for 13 days or after 1 year at –70°C [5]; Enzyme is highly unstable, dimethyl sulfoxide stabilizes during purification [8]; Dialysis against buffers of low ionic strength invariably results in complete inactivation, whereas after 3 h dialysis against 0.5 M KCl only 40% of the activity is lost [1]

Storage
4°C, 10 mM Tris-HCl, pH 7.5, 10 mM MgCl$_2$, 2 mM 2-mercaptoethanol, 44% loss of activity after 8 d [5]; –15°C, stable for 2 weeks [1]; 4°C for 13 days or 1 year at –70°C, ethylene glycol, significantly increases stability with virtually no loss of activity [5]

---

6 CROSSREFERENCES TO STRUCTURE DATABANKS

PIR/MIPS code
PIR2:S23412 (Escherichia coli); PIR3:S42198 (Mycoplasma hominis (SGC3)); PIR2:JH0147 (chain FUR1 yeast (Saccharomyces cerevisiae))

Brookhaven code

7 LITERATURE REFERENCES

[1] Flaks, J.G.: Methods Enzymol.,6,136–158 (1963) (Review)
[2] Asai, T., Lee, C.S., Chandler, A., O'Sullivan, W.J.: Comp. Biochem. Physiol., B, Comp. Biochem.,95B,159–163 (1990)
[3] Plunkett, W., Moner, J.G.: Arch. Biochem. Biophys.,187,264–271 (1978)
[4] Andersen, P.S., Smith, J.M., Mygind, B.: Eur. J. Biochem.,204,51–56 (1992)
[5] Rasmussen, U.B., Mygind, B., Nygaard, P.: Biochim. Biophys. Acta,881,268–275 (1986)
[6] McIvor, R.S., Wohlhueter, R.M., Plagemann, P.G.W.: J. Bacteriol.,156,192–197 (1983)
[7] Lindsay, R.H., Tillery, C.R., Yu, M.-Y.W.: Arch. Biochem. Biophys.,148,466–474 (1972)
[8] Natalini, P., Ruggieri, S., Santarelli, I., Vita, A., Magni, G.: J. Biol. Chem.,254, 1558–1563 (1979)

1 NOMENCLATURE

EC number
2.4.2.10

Systematic name
Orotidine-5'-phosphate:pyrophosphate phospho-alpha-D-ribosyltransferase

Recommended name
Orotate phosphoribosyltransferase

Synonyms
Orotidylic acid phosphorylase
Orotidine-5'-phosphate pyrophosphorylase
Phosphoribosyltransferase, orotate
OPRTase
Orotate phosphoribosyl pyrophosphate transferase
Orotic acid phosphoribosyltransferase
Orotidine 5'-monophosphate pyrophosphorylase
Orotidine 5'-phosphate pyrophosphorylase
Orotidine monophosphate pyrophosphorylase
Orotidine phosphoribosyltransferase
Orotidylate phosphoribosyltransferase
Orotidylate pyrophosphorylase
Orotidylic acid pyrophosphorylase
Orotidylic phosphorylase
Orotidylic pyrophosphorylase
More (constitutes together with orotidylate decarboxylase (EC 4.1.1.23)
UMP-synthase, former complex U or multienzyme pyr-5,6 [1], enzyme from
mammals is a bifunctional polypeptide, it also catalyzes the reaction listed
as EC 4.1.1.23 [2, 10, 11, 14]) [1, 2, 10, 11, 14]

CAS Reg. No.
9030-25-5

---

2 REACTION AND SPECIFICITY

Catalysed reaction
Orotidine 5'-phosphate + diphosphate →
→ orotate + 5-phospho-alpha-D-ribose 1-diphosphate

Reaction type
Pentosyl group transfer

Enzyme Handbook © Springer-Verlag Berlin Heidelberg 1996
Duplication, reproduction and storage in data banks are only
allowed with the prior permission of the publishers

Natural substrates

Orotate + 5-phospho-alpha-D-ribose 1-diphosphate (involved in de novo synthesis of pyrimidine nucleotides [8, 12], final steps in biosynthesis of UMP [10], nucleotide-forming step in pyrimidine biosynthesis [13]) [8, 10, 12, 13]

Substrate spectrum

1 Orotate + 5-phospho-alpha-D-ribose 1-diphosphate (r [4, 5, 7, 8, 12], high specificity for orotate [7], predominant species of phosphoribose diphosphate: metal ion complex [7], catalyzes stereospecific formation of beta-glycosidic bond between orotate and ribose 5'-phosphate portion of phospho-ribose diphosphate [7]. Reverse reaction: tri-, tetrapoly- or trimetaphosphate can replace diphosphate with 29%, 70% or 78% efficiency [8]) [1–8]
2 Orotate methylester + phosphoribose diphosphate [3]
3 Uracil + phosphoribose diphosphate [3]
4 5-Fluoroorotate + phosphoribose diphosphate [7]
5 More (enzyme from mammals is a bifunctional polypeptide, it also catalyzes the reaction listed as EC 4.1.1.23) [2, 10, 11, 14]

Product spectrum

1 Orotidine 5'-phosphate + diphosphate (i.e. orotidylate or OMP) [1, 4, 5, 7]
2 ?
3 ?
4 5-Fluoroorotidine 5'-phosphate + diphosphate [7]
5 ?

Inhibitor(s)

Phosphoribose diphosphate (phosphorolysis, product inhibition, kinetics) [7, 12]; Orotate (phosphorolysis, product inhibition, kinetics) [7, 12]; Orotidylate (product inhibition, kinetics [7, 12]) [4, 7, 12]; Orotidine (10 mM) [1]; Diphosphate (product inhibition, kinetics [7, 12]) [1, 7, 12]; Phosphate (at higher concentrations [7]) [1, 7]; Dihydroorotate (10 mM) [1]; 5-Fluoroorotate (0.05 M) [1]; Barbituric acid [1]; Allopurinol [1]; Oxopurinol [1]; 5-Chlorouracil [1]; Adenine (not [8]) [1]; Uracil [1]; Azauracil [1]; 5-Bromouracil [1]; Cytosine [1]; Adenosine (not [8]) [1]; Uridine [1]; Azauridine [1]; UMP (1 mM, not 0.1 mM) [1]; dUMP [1]; IMP [1]; AMP (not [8]) [1]; GMP [1]; UDP [1]; UTP (not [8]) [1]; ADP (not [8]) [1]; ATP (not [8]) [1]; PCMB (reversible by DTT [8], not [10]) [8]; Co^{2+} (inhibits Mg^{2+}-activation) [5]; Zn^{2+} [8]; $HgCl_2$ (slight, mutant, not wild-type) [8]; EDTA [14]; IAA (slight, mutant, not wild-type) [8]; NEM (slight, mutant, not wild-type) [8]; Ribose 5-phosphate (at higher concentrations [7], not [1]) [7]; Erythrose 4-phosphate (at higher concentrations) [7]; Arabinose 5-phosphate (at higher concentrations) [7]; Fructose 1-phosphate (at higher concentrations) [7]; Fructose 6-phosphate (at higher concentrations) [7]; Nicotinate (weak) [7]; Anthranilate (weak) [7]; More (no inhibition by SH-group reagents [10], CMP, TMP, 6-aza-UMP or 5-bromo-UMP [1]) [1, 10]

Cofactor(s)/prosthetic group(s)/activating agents

Metal compounds/salts

Mg^{2+} (requirement [1, 3, 5, 7, 8, 14], 2 mM [3], K_m-value: 3 mM [1], mechanism [3, 5], no orotate-Mg-complex formation, weak Mg-enzyme-complex [3]) [1, 3, 5, 7, 8, 14]; Mn^{2+} (activation [3, 5, 8], can replace Mg^{2+} [3, 8], mechanism [3, 5]) [3, 5, 8]; Ca^{2+} (activation, 34% as effective as Mn^{2+} or Mg^{2+} [8], not [5]) [8]; Co^{2+} (slight activation [8], not [5]) [8]; Ba^{2+} (slight activation) [8]

Turnover number (min^{-1})

Specific activity (U/mg)

More [8, 14]; 0.115 [1]; 0.41 (mutant) [8]; 0.69 [2]; 15.2 (wild-type) [8]; 17 [6]; 27 [9]; 40 [5]; 45 (in the absence of added OMP-decarboxylase) [7]; 65 (in the presence of added OMP-decarboxylase) [7]; 81.6 [4]; 95.7 [12]

K_m-value (mM)

More (kinetic study [7]) [7, 12]; 0.0016 (phosphoribose diphosphate) [14]; 0.002 (orotate) [1]; 0.0031–0.0036 (orotidine 5'-phosphate, wild-type [8]) [8, 12]; 0.0045 (orotate) [14]; 0.008–0.0083 (orotidine 5'-phosphate, phosphorolysis) [4, 7]; 0.013 (diphosphate, wild-type) [8]; 0.016 (phosphoribose diphosphate) [1]; 0.027–0.0275 (orotate [3, 12], 5-fluoroorotate [7]) [3, 7]; 0.03–0.035 (orotate (wild-type [8]) [4, 7, 8], orotidine 5'-monophosphate [8]) [4, 7, 8]; 0.038–0.04 (phosphoribose diphosphate, wild-type [8]) [7, 8]; 0.044 (phosphoribose diphosphate) [3, 12]; 0.062 (phosphoribose diphosphate) [4]; 0.096 (diphosphate, phosphorolysis) [7]; 0.19 (orotate methylester) [3]; 0.22–0.25 (diphosphate, phosphorolysis) [4, 8]; 0.36 (phosphoribose diphosphate, mutant) [8]; 0.44 (orotate, mutant) [8]; 2.63 (uracil) [3]

pH-optimum

More (2 isozymes: pI: 5.65 (minor form), pI: 5.85 (major form)) [2]; 7–7.5 (preferred buffer: Tris-HCl, when substrates are non-saturating, phosphate buffer, when substrate are saturating, not: maleate, imidazole, N-tris[hydroxymethyl] methyl-2-aminoethane-sulfonic acid (i.e. TES) or 3-[N-morpholino]-2-hydroxypropanesulfonic acid (i.e. MOPS) buffer) [1]; 8 [14]; 8.5 (mutant) [8]; 8.5–9 [4]; 9.5 (wild-type) [8]

pH-range

7.2–9.2 (about half-maximal activity at pH 7.2 and about 70% of maximal activity at pH 9.2) [14]; 7.2–9.5 (about half-maximal activity at pH 7.2 and 9.5, mutant) [8]; 8.5–10.5 (about half-maximal activity at pH 8.5 and about 70% of maximal activity at pH 10.5, wild-type) [8]

Temperature optimum (°C)

20 (assay at) [9]; 25 (assay at) [4, 7]; 30 (assay at) [12, 14]; 37 (assay at) [1, 2, 6]

Temperature range (°C)

---

3 ENZYME STRUCTURE

Molecular weight
40000 (Saccharomyces cerevisiae, gel filtration) [7]
47000 (E. coli K-12, gel filtration) [8]
50000 (mouse, Ehrlich ascites carcinoma cells, sucrose density gradient centrifugation) [2]

Subunits
Monomer (1 × 51500, mouse, Ehrlich ascites carcinoma cells, SDS-PAGE) [2]
Dimer (2 × 20000, Saccharomyces cerevisiae, SDS-PAGE [7], 2 × 22000, E. coli K-12, SDS-PAGE [9], 2 × 23000, Salmonella typhimurium, SDS-PAGE [12], 2 × 23500, E. coli K-12, SDS-PAGE [8]) [7–9, 12]

Glycoprotein/Lipoprotein
–

---

4 ISOLATION/PREPARATION

Source organism
Bacteria [3]; E. coli K-12 (overproducing strains C-600(pNE24) (wild-type) and its purine-sensitive mutant PS100(pNE31), both harbouring hybrid multi-copy plasmids [8]) [8, 9]; Salmonella typhimurium (structural gene pyrE cloned and overexpressed via multi-copy plasmid in E. coli strain MB13 [12]) [12, 13]; Plasmodium falciparum (FCB strain) [10]; Saccharomyces cerevisiae (Budweiser brand [5, 7]) [4–7]; Phaseolus mungo (black gram) [14]; Wheat [15]; Human [10]; Mouse [1, 2, 11]

Source tissue
Cell [1–10, 12, 13]; Ehrlich ascites carcinoma cells [1, 2, 11]; Embryo [15]; Erythrocytes [10]; Seedlings [14]

Localization in source
Soluble [6, 9, 10, 14]; Cytosol [14]

Purification
E. coli K-12 (OMP-agarose affinity chromatography) [8, 9]; Salmonella typhimurium [12]; Saccharomyces cerevisiae (partial, affinity chromatography on Blue-Dextran- or Cibacron Blue F3GA-Sepharose [6]) [4, 6]; Mouse (Ehrlich ascites carcinoma cells, partial, rapid tandem affinity column method) [1, 2]; Phaseolus mungo (partial, not separable from EC 4.1.1.23) [14]

Crystallization
(Salmonella typhimurium) [13]

Cloned

(Salmonella typhimurium, structural gene pyrE, subcloned from a genomic library in pBR328(pAV002), transduced into pyr-auxotroph Salmonella typhimurium strain SA2434 with phage P22, finally cloned to and expressed in overproducing strain E. coli MB13) [12]

Renatured

–

---

5 STABILITY

pH

4–9 (stable) [1]; 7.5–9.5 (at least 6 months stable, –20°C) [4]

Temperature (°C)

Oxidation

Organic solvent

General stability information

5 mM Mg^{2+}, 1 mM phosphoribose diphosphate and 2 mM DTT stabilize dilute enzyme solutions [1]; Dialysis against phosphate buffer, pH 7 or 7.5, unstable to, 2 mM DTT protect [1]; Dilution inactivates [1]; Phosphate, 0.3 M and above, stabilizes [1]; Proteins, e.g. albumin, do not stabilize dilute enzyme solutions [1]; 0.05 mM UMP and 1% polyethyleneglycol stabilize during final stage of purification [2]

Storage

Frozen, partially purified enzyme in phosphate buffer, pH 7 or 7.5, 2 months [1]; –76°C, at least 6 months [5]; –60°C to –20°C, crude or ammonium sulfate preparation, 2 years [1]; –25°C, in 50% v/v glycerol, 0.1 mM DTT, about 35% loss of activity within 40 days [9]; –20°C, concentrated enzyme solution, at least 4 months [2]; –20°C, pH 7.5–9.5, at least 6 months [4]; 4°C, gradual loss of activity at pH 7.5–9.5 [4]; 4°C, Plasmodium falciparum, $t_{1/2}$: 1.5 days [10]

6 CROSSREFERENCES TO STRUCTURE DATABANKS

PIR/MIPS code

PIR3:S44156 (yeast (Yarrowia lipolytica)); PIR2:A30148 (/orotidine-5' phosphate decarboxylase (EC 4.1.1.23) human); PIR2:S30118 (anthracnose fungus (Colletotrichum graminicola)); PIR3:S34325 (Bacillus caldolyticus); PIR2:A30492 (Bacillus subtilis); PIR1:XJEC (Escherichia coli); PIR2:A36459 (fungus (Filobasidium floriforme)); PIR3:S13091 (imperfect fungus (Trichoderma reesei)); PIR2:A60993 (Lactobacillus plantarum); PIR2:A29459 (Podospora anserina); PIR3:S31508 (Salmonella typhimurium); PIR3:S32801 (Salmonella typhimurium); PIR2:B21911 (Salmonella typhimurium (fragment)); PIR2:S03826 (slime mold (Dictyostelium discoideum)); PIR2:JS0175 (Sordaria macrospora); PIR2:S46440 (/orotidine-5' phosphate decarboxylase (EC 4.1.1.23) Arabidopsis thaliana); PIR2:A30148 (/orotidine-5' phosphate decarboxylase (EC 4.1.1.23) human); PIR1:XJBY10 (URA10 yeast (Saccharomyces cerevisiae)); PIR1:XJBY5 (URA5 yeast (Saccharomyces cerevisiae))

Brookhaven code

1STO ((Salmonella Typhimurium) recombinant form expressed in (Escherichia coli))

---

7 LITERATURE REFERENCES

[1] Jones, M.E., Kavipurapu, P.R., Traut, T.W.: Methods Enzymol.,51,155–167 (1978) (Review)

[2] McClard, R.W., Black, M.J., Livingstone, L.R., Jones, M.E.: Biochemistry,19, 4699–4706 (1980)

[3] Bhatia, M.B., Grubmeyer, C.: Arch. Biochem. Biophys.,303,321–325 (1993)

[4] Yoshimoto, A., Amaya, T., Kobayashi, K., Tomita, K.: Methods Enzymol.,51,69–74 (1978) (Review)

[5] Victor, J., Leo-Mensah, A., Sloan, D.L.: Biochemistry,18,3597–3604 (1979)

[6] Reyes, P., Sandquist, R.B.: Anal. Biochem.,88,522–531 (1978)

[7] Victor, J., Greenberg, L.B., Sloan, D.L.: J. Biol. Chem.,254,2647–2655 (1979)

[8] Shimosaka, M., Fukuda, Y., Murata, K., Kimura, A.: J. Biochem.,98,1689–1697 (1985)

[9] Dodin, G.: FEBS Lett.,134,20–24 (1981)

[10] Rathod, P.K., Reyes, P.: J. Biol. Chem.,258,2852–2855 (1983)

[11] Floyd, E.E., Jones, M.E.: J. Biol. Chem.,260,9443–9451 (1985)

[12] Bhatia, M.B., Vinitsky, A., Grubmeyer, C.: Biochemistry,29,10480–10487 (1990)

[13] Scapin, G., Sacchettini, J.C., Dessen, A., Bhatia, M., Grubmeyer, C.: J. Mol. Biol., 230,1304–1308 (1993)

[14] Ashihara, H.: Z. Pflanzenphysiol.,87,225–241 (1978)

[15] Kapoor, M., Waygood, E.R.: Can. J. Biochem.,43,143–151 (1965)

1 NOMENCLATURE

EC number
2.4.2.11

Systematic name
Nicotinate-nucleotide:pyrophosphate phospho-alpha-D-ribosyltransferase

Recommended name
Nicotinate phosphoribosyltransferase

Synonyms
Phosphoribosyltransferase, nicotinate
Niacin ribonucleotidase
Nicotinic acid mononucleotide glycohydrolase
Nicotinic acid mononucleotide pyrophosphorylase
Nicotinic acid phosphoribosyltransferase

CAS Reg. No.
9030-26-6

2 REACTION AND SPECIFICITY

Catalysed reaction
Nicotinate D-ribonucleotide + diphosphate →
→ nicotinate + 5-phospho-alpha-D-ribose 1-diphosphate (mechanism [4, 6, 9])

Reaction type
Pentosyl group transfer

Natural substrates
Nicotinate + 5-phospho-alpha-D-ribose 1-diphosphate (first step in NAD+-
(salvage) biosynthesis) [5, 6]

Substrate spectrum
1 Nicotinate + 5-phospho-alpha-D-ribose 1-diphosphate (r [2, 10], favored
  reaction [10], the pyridine N and carboxyl groups are important for inter-
  action with the enzyme [4], no phosphoribosylation of nicotinamide, qui-
  nolic acid, adenine or hypoxanthine [7]) [1–10]
2 More (ATPase activity in the presence of either product and in the absence of phosphoribose diphosphate [6], one mol ATP is cleaved per mol
  product formed [4, 5, 10], no ATPase activity in the absence of substrates
  [4, 5], ATP and phosphoribose diphosphate compete for ATP-binding site
  [6]) [4–6, 10]

Product spectrum

    1 Nicotinate D-ribonucleotide + diphosphate [1–5, 9, 10]

    2 ?

Inhibitor(s)

Phosphoribose diphosphate (at high concentrations, when Mg^{2+} below 0.03 mM) [9]; UTP (weak) [2]; ATP (in the presence of Mn^{2+} [7], at high concentrations with Mg^{2+} below 0.03 mM [9]) [7, 9]; Mg-ADP (non-competitive to Mg-ATP or Mg-diphosphate) [9]; Diphosphate [2]; Tripolyphosphate [2]; Nucleoside diphosphates (e.g. ADP or GDP) [10]; Mg^{2+} (at higher concentrations than ATP, when free Mg^{2+}-ions occur) [2]; Nicotinate mononucleotide (competitive to both substrates [2] or Mg-ATP [9], non-competitive to nicotinate [9], kinetics [6, 9], not [10]) [2, 6, 7, 9]; Nicotinate analogues (not [7]) [4]; Pyrazine 2-carboxylic acid [7]; Pyridazine-4,5-dicarboxylic acid (weak) [7]; 3-Pyridylcarboxaldehyde [10]; 3-Acetylpyridine [10]; 3-Fluoro-nicotinic acid [10]; Methyl 3-pyridylcarbinol (weak) [10]; 3-Pyridylcarbinol (weak) [10]; NAD^+ (weak [7], not [10]) [7]; NADH (weak) [7]; $NADP^+$ (weak [7], not [10]) [7]; NADPH (weak) [7]; Urea [10]; KCN [10]; N-Ethylmaleimide (0.1 mM and above) [7]; PCMB (0.01 mM and above) [7]; Monoiodoacetic acid (1 mM and above) [7]; More (no inhibition by nicotinate (up to 10 mM) [2], nicotinamide [2, 10], nicotinamide mononucleotide, deamino-NAD [7], 4-hydroxy-nicotinate [4], quinolinic acid, AMP, GMP [10] or antibodies against pig kidney quinolinate phosphoribosyltransferase [7]) [2, 4, 7, 10]

Cofactor(s)/prosthetic group(s)/activating agents

ATP (activation [1, 2, 5, 7], requirement (in the absence of phosphate [8], not [5]) [3, 8, 10], together with Mg^{2+} [1–4, 7] or Co^{2+} [7], 1 mol ATP is hydrolyzed to ADP plus phosphate per mol product formed [4, 5, 10]) [1–5, 7, 8, 10]; ATP-Mg-complex (maximal activation at equimolar levels of Mg^{2+} and ATP, presumably enzyme structure stabilizing) [2]; GTP (activation, equally effective as ATP [10], not [7]) [2, 3, 10]; ITP (activation, about 90% as effective as ATP [10], not [7]) [2, 3, 10]; CTP (activation, about 70% as effective as ATP [10]) [2, 3, 10]; TTP (activation, about 30% as effective as ATP [10], not [3]) [10]; UTP (activation, about 50% as effective as ATP [10]) [1, 10]; Purine or purine nucleoside triphosphates (activation, can replace ATP with 30 to 90% efficiency, not TTP) [3]; GSH (activation) [1]; Tripolyphosphate (activation, can replace ATP) [10]; More (no activation by ADP [2, 3, 7], AMP, 2'-AMP [7], 3'-AMP, 5'-AMP [3, 7], GDP, GMP, IDP, IMP [7] or 3,5-cycloadenosine [2]) [2, 3, 7]

Metal compounds/salts

Mn^{2+} (requirement, 7.5 mM, no synergism with ATP, more effective than Mg^{2+} with or without ATP) [7]; Co^{2+} (requirement, slight synergism with ATP, more effective than Mg^{2+} with or without ATP) [7]; Mg^{2+} (requirement [1–4, 7, 9], 1–10 mM [3], synergism with ATP [1–4, 7], K_m-value: 0.4 mM [1]) [1–4, 7, 9]; Mg-ATP-complex (maximal activation at equimolar levels of Mg^{2+} and ATP, presumably enzyme structure stabilizing) [2]; NaF (activation) [1]; Phosphate (activation, 0.01–0.1 M, in the presence of Mg^{2+} [3]) [2, 3]; Fe^{2+} (slight stimulation, synergism with ATP) [7]; Zn^{2+} (slight stimulation) [7]; More (no activation by borate, sulfate [2], Ba^{2+}, Ca^{2+}, Cu^{2+}, Cd^{2+}, Al^{3+}, Fe^{3+} [7]) [2, 7]

Turnover number (min^{-1})

Specific activity (U/mg)

0.0012 [2]; 0.052 [5]; 2.3 [3]

K_m-value (mM)

0.0005 (nicotinate, in the presence of ATP) [5]; 0.0008–0.0015 (nicotinate) [2, 10]; 0.002 (nicotinate) [1]; 0.005 (phospho-alpha-D-ribose 1-diphosphate, in the presence of ATP) [5]; 0.024 (nicotinate) [4, 5]; 0.03 (phosphoribose diphosphate) [1]; 0.05 (phosphoribose diphosphate) [2]; 0.06–0.1 (phosphoribose diphosphate) [5, 10]

pH-optimum

More (pI: 4.8 [7], pI: 6.9 [6]) [6, 7]; 6.5–8 (broad) [5]; 7.2 (broad, phosphorolysis) [2]; 7.3–7.4 [7]; 7.5–8.5 (broad) [3]

pH-range

5.5–8 (about half-maximal activity at pH 5.5 and about 85% of maximal activity at pH 8) [2]; 5.5–10 (about half-maximal activity at pH 5.5 and 10) [3]

Temperature optimum (°C)

37 (assay at) [1–5, 7–10]

Temperature range (°C)

3 ENZYME STRUCTURE

Molecular weight

43000 (Saccharomyces cerevisiae, gel filtration) [3]
86000 (human, gel filtration) [5]

Subunits

Glycoprotein/Lipoprotein

–

4 ISOLATION/PREPARATION

Source organism
Bacillus subtilis SB19 [10]; E. coli K-12 [1]; Saccharomyces cerevisiae (Bud-weiser brand [6]) [3, 6, 9]; Bovine [2]; Human [4, 5]; Pig [7, 8]; Rat (male Wistar) [7]

Source tissue
Blood clot (rat) [7]; Blood platelets [4]; Cell [1, 3, 6, 9, 10]; Erythrocytes [5]; Heart (rat) [7]; Small intestine (rat) [7]; Kidney (rat) [7]; Liver [2, 7, 8]; Pancreas (rat) [7]; Spleen (rat) [7]; More (distribution in rat tissues: trace activities in brain, lung or stomach, not in thigh muscle, testis or serum) [7]

Localization in source
Soluble [2]

Purification
Bacillus subtilis (partial) [10]; E. coli (partial) [1]; Bovine (partial) [2]; Saccharomyces cerevisiae [3]; Human [5]

Crystallization
–

Cloned
–

Renatured
–

---

5 STABILITY

pH
5–10 (stable, 0.2 M potassium phosphate buffer) [5]; 6.5 (inactivation below, phosphorolysis) [2]

Temperature (°C)
50 (5 min: inactivation, ATP-Mg, Mg-phosphoribose diphosphate, ATP, GTP or nicotinic acid protects, with 27%, 37%, 52%, 54% or 61% loss of activity, respectively. 10 min: inactivation, ATP-Mg, Mg-phosphoribose diphosphate, ATP, GTP or nicotinic acid protects, with 43%, 52%, 60%, 63% or 69% loss of activity, respectively) [3]

Oxidation

Organic solvent

General stability information
Heat, unstable to, substrates protect [3]; Salt concentrations below 50 mM, inactivation, 20–35% glycerol protects [5]; SDS-PAGE, unstable to [5]; Glycerol, 20–35%, stabilizes during dialysis against low buffer concentrations [5]

Storage
-70°C, $t_{1/2}$: 3 months [5]; -25°C, at least 1 month [3]; -12°C, partially purified enzyme preparation, $t_{1/2}$: at least 3 months [2]

6 CROSSREFERENCES TO STRUCTURE DATABANKS

PIR/MIPS code
PIR2:JQ0756 (Escherichia coli); PIR2:A39130 (Salmonella typhimurium);
PIR2:S51845 (yeast (Saccharomyces cerevisiae))

Brookhaven code

7 LITERATURE REFERENCES

[1] Imsande, J.: J. Biol. Chem.,236,1494–1497 (1961)
[2] Imsande, J., Handler, P.: J. Biol. Chem.,236,525–530 (1961)
[3] Kosaka, A., Spivey, H.O., Gholson, R.K.: J. Biol. Chem.,246,3277–3283 (1971)
[4] Gaut, Z.N., Solomon, H.M.: Biochem. Pharmacol.,20,2903–2906 (1971)
[5] Niedel, J., Dietrich, L.S.: J. Biol. Chem.,248,3500–3505 (1973)
[6] Hanna, L.S., Hess, S.L., Sloan, D.L.: J. Biol. Chem.,258,9745–9754 (1983)
[7] Hayakawa, T., Shibata, K., Iwai, K.: Agric. Biol. Chem.,48,445–453 (1984)
[8] Hayakawa, T., Shibata, K., Iwai, K.: Agric. Biol. Chem.,48,455–460 (1984)
[9] Kosaka, A., Spivey, H.O., Gholson, R.K.: Arch. Biochem. Biophys.,179,334–341
    (1977)
[10] Imsande, J.: Biochim. Biophys. Acta,85,255–264 (1964)

1 NOMENCLATURE

EC number
2.4.2.12

Systematic name
Nicotinamide-nucleotide:pyrophosphate phospho-alpha-D-ribosyltransferase

Recommended name
Nicotinamide phosphoribosyltransferase

Synonyms
NMN pyrophosphorylase
Phosphoribosyltransferase, nicotinamide
Nicotinamide mononucleotide pyrophosphorylase
Nicotinamide mononucleotide synthetase
NMN synthetase

CAS Reg. No.
9030-27-7

---

2 REACTION AND SPECIFICITY

Catalysed reaction
Nicotinamide D-ribonucleotide + diphosphate →
→ nicotinamide + 5-phospho-alpha-D-ribose 1-diphosphate (mechanism [3])

Reaction type
Pentosyl group transfer

Natural substrates

Substrate spectrum
1 Nicotinamide + 5-phospho-alpha-D-ribose 1-diphosphate (beta-nicotin-
   amide [3], specific for nicotinamide, no substrates: thymine, 5-bromoura-
   cil, nicotinic acid adenine dinucleotide [4]) [1–6]

Product spectrum
1 Nicotinamide D-ribonucleotide + diphosphate [1–6]

Inhibitor(s)

Hypoxanthine (phosphorolysis) [1]; NAD$^+$ (strong (thigh muscle) [5], 1 mM (lung and liver, not at 2 mM) [5], kinetics [3]) [3–5]; Nicotinamide mononucleotide (kinetics) [3]; NADP$^+$ [3]; NADH (weak) [3]; NADPH (weak) [3]; alpha-NAD$^+$ (weak) [3]; Nicotinate adenine dinucleotide (weak [3], not [4]) [3]; Nicotinamide hypoxanthine dinucleotide [3]; Thionicotinamide adenine dinucleotide [3]; 3-Acetylpyridine adenine dinucleotide (kinetics) [3]; Nicotinamide mononucleotide-H$_2$ (strong, kinetics) [3]; Nicotinamide riboside (kinetics) [3]; Thionicotinamide (weak, kinetics) [3]; 5-Fluoro-nicotinamide (weak, kinetics) [3]; 3-Acetylpyridine (weak, kinetics) [3]; 6-Aminonicotinamide (weak, kinetics) [3]; More (no inhibition by NaF [1], nicotinic acid, thymine, 5-bromouracil [4]) [1, 4]

Cofactor(s)/prosthetic group(s)/activating agents

ATP (requirement [2–5], activation, up to 0.4 mM [6]) [2–6]; EDTA (activation, 1 mM) [2]

Metal compounds/salts

Mg^{2+} (requirement [2–5], activation, up to 10 mM [6], K_m-value: 0.6 mM [1]) [1–6]

Turnover number (min^{-1})

Specific activity (U/mg)

More (activity per g hemoglobin) [6]; 0.000225 [2]; 0.003 [3]

K_m-value (mM)

0.000067 (nicotinamide) [2]; 0.001 (nicotinamide, in the presence of ATP) [3]; 0.00127 (nicotinamide) [6]; 0.0016 (nicotinamide) [4]; 0.0038 (phosphoribose diphosphate) [2]; 0.00537 (phosphoribose diphosphate) [6]; 100 (nicotinamide, in the absence of ATP [3]) [1, 3]

pH-optimum

6.8–7.6 [1]; 8.5–9 [2]

pH-range

6.1–8.5 (about half-maximal activity at pH 6.1 and 8.5) [1]; 7.8–9.2 (about half-maximal activity at pH 7.8 and about 80% of maximal activity at pH 9.2) [2]

Temperature optimum (°C)

35 (assay at) [1]; 37 (assay at) [2–6]

Temperature range (°C)

3 ENZYME STRUCTURE

Molecular weight

Subunits

Glycoprotein/Lipoprotein

–

---

4 ISOLATION/PREPARATION

Source organism
Human [1, 4, 6]; Rat (male Sprague-Dawley [5]) [2, 3, 5]

Source tissue
Erythrocytes [1, 2, 6]; Fibroblasts (diploid, GM 1362, from patient with Lesch-Nyhan disease) [4]; Liver [3, 5]; Brain [5]; Heart [5]; Small intestine [5]; Kidney [5]; Lung [5]; Pancreas [5]; Stomach [5]; Testis [5]; Thigh muscle [5]; More (tissue distribution) [5]

Localization in source
Soluble [1, 2, 4, 6]

Purification
Human (partial) [1]; Rat (partial) [2]

Crystallization
–

Cloned
–

Renatured
–

---

5 STABILITY

pH
7–10 (stable) [2]; 8.8 (rapid inactivation below) [2]

Temperature (°C)
65 (10 min stable, heat treatment during purification) [1]; 70 (5 min stable, heat treatment during purification) [1]

Oxidation

Organic solvent

General stability information
Freeze-thawing inactivates [4]; Bovine serum albumin, glutathione, cysteine, DTT, EDTA, cystine or H_2O_2 does not stabilize during storage [2]

Storage
−80°C, crude, 4–6 months [4]; −70°C, in crude erythrocyte lysates, at least 30 days [6]; −20°C to +20°C, about 70% loss of activity overnight [2]; 4°C, crude, inactivation overnight [4]; Storage at room temperature, 4°C, or frozen, unstable upon [2]

---

6 CROSSREFERENCES TO STRUCTURE DATABANKS

PIR/MIPS code

Brookhaven code

---

7 LITERATURE REFERENCES

[1] Preiss, J., Handler, P.: J. Biol. Chem.,225,759–770 (1957)
[2] Lin, L.-F.H., Lan, S.J., Richardson, A.H., Henderson, L.V.M.: J. Biol. Chem.,247, 8016–8022 (1972)
[3] Dietrich, L.S., Muniz, O.: Biochemistry,11,1691–1695 (1972)
[4] Elliott, G.C., Rechsteiner, M.C.: Biochem. Biophys. Res. Commun.,104,996–1002 (1982)
[5] Shibata, K., Taguchi, H., Nishitani, H., Okumura, K., Shimabayashi, Y., Matsushita, N., Yamazaki, H.: Agric. Biol. Chem.,53,2283–2284 (1989)
[6] Rocchigiani, M., Micheli, V., Duley, J.A., Simmonds, H.A.: Anal. Biochem.,205, 334–336 (1992)

1 NOMENCLATURE

EC number
2.4.2.14

Systematic name
5-Phosphoribosylamine:pyrophosphate phospho-alpha-D-ribosyltransferase
(glutamate-amidating)

Recommended name
Amidophosphoribosyltransferase

Synonyms
Phosphoribosyldiphosphate 5-amidotransferase
Glutamine phosphoribosylpyrophosphate amidotransferase
Amidotransferase, phosphoribosyl pyrophosphate
alpha-5-Phosphoribosyl-1-pyrophosphate amidotransferase
5'-Phosphoribosylpyrophosphate amidotransferase
5-Phosphoribosyl-1-pyrophosphate amidotransferase
5-Phosphoribosylpyrophosphate amidotransferase
5-Phosphororibosyl-1-pyrophosphate amidotransferase
Glutamine 5-phosphoribosylpyrophosphate amidotransferase
Glutamine ribosylpyrophosphate 5-phosphate amidotransferase
Phosphoribose pyrophosphate amidotransferase
Phosphoribosyl pyrophosphate amidotransferase
Phosphoribosylpyrophosphate glutamyl amidotransferase

CAS Reg. No.
9031-82-7

---

2 REACTION AND SPECIFICITY

Catalysed reaction
5-Phospho-beta-D-ribosylamine + diphosphate + L-glutamate →
→ L-glutamine + 5-phospho-alpha-D-ribose 1-diphosphate + H_2O (mechanism [11])

Reaction type
Pentosyl group transfer

Natural substrates
L-Glutamine + 5-phospho-alpha-D-ribose 1-diphosphate + H_2O (first reaction
in de-novo pathway of purine biosynthesis [3, 10], rate-limiting [7], regulating [13] enzyme of purine biosynthetic pathway, regulatory enzyme in the
flow of recently fixed nitrogen from initial assimilation into amino acids via
purine biosynthesis [16]) [3, 7, 10, 13, 16]

Enzyme Handbook © Springer-Verlag Berlin Heidelberg 1996
Duplication, reproduction and storage in data banks are only
allowed with the prior permission of the publishers

Substrate spectrum

1 L-Glutamine + 5-phospho-alpha-D-ribose 1-diphosphate + H_2O (ir [6], reaction at 33% [14], 50% [16] or 70% (Bacillus subtilis) [11] the rate of NH_3-utilization, glutamine binding site distinct from NH_3-site [10], no substrate is carbamoyl phosphate [14]) [1–18]

2 NH_3 + 5-phospho-alpha-D-ribose 1-diphosphate + H_2O (NH_3-binding site distinct from glutamine-site [10], not NH_4^+ [11]) [3, 6, 10, 11, 14, 16]

Product spectrum

1 5-Phospho-beta-D-ribosylamine + L-glutamate + diphosphate [3, 6]

2 5-Phospho-beta-D-ribosylamine + ?

Inhibitor(s)

2-Mercaptoethanol (inactivation) [3]; Oxygen (inactivation, enzymes from Bacillus subtilis, Chinese hamster fibroblasts, murine erythroleukemia cells [10] or human placenta [10, 11], in vivo and in vitro, AMP protects [4], restorable by Fe^{2+}, S^{2-} and SH^- under anaerobic conditions (human placenta [10, 11], Bacillus subtilis [11]) [10, 11], phosphoribosyldiphosphate or purine nucleotides partially protect [10], effects of exposure to air is reversible by high concentrations of DTT [3]) [3, 4, 9–11]; Phosphate (at high concentrations [7], competitive to 5-phospho-alpha-D-ribose 1-diphosphate [10, 13, 18], reduces cooperativity in phosphoribosyl diphosphate-binding [10], potentiates [10] or slightly increases [7] nucleotide inhibition) [7, 10, 13, 18]; Sulfate [13]; Purine ribonucleotides (strong [1], allosteric inhibitors [11], all mammalian enzymes are subject to feed-back inhibition by purine nucleotides [10, 11], inhibition by nucleotides is a stable property of E. coli and Bacillus subtilis enzymes and a labile property of avian and yeast enzymes [11], nucleoside diphosphates or triphosphates inhibit to a lesser degree than monophosphates [8], diphosphates stronger than monophosphates [6], 5-phospho-alpha-D-ribose 1-diphosphate protects, not glutamine [10, 11], no inhibitors are pyrimidine nucleotides [3, 8, 13], purine [1, 13] or pyrimidine [13] nucleosides or bases [1, 13], 2'- or 3'-analogues, deoxyribosephosphate analogues [1], ribose 5-phosphate [1, 3] or 3',5'-cyclic derivatives of AMP or GMP [3]) [1, 3, 4, 6, 8, 10–14, 16–18]; AMP (most potent nucleotide inhibitor, kinetic study [13], competitive to 5-phospho-alpha-D-ribose 1-diphosphate (rat) [1], does not promote cooperativity in 5-phospho-alpha-D-ribose 1-diphosphate binding [16], synergism in combination with GMP [3, 8], stronger inhibition with NH_3 instead of glutamine [14], 5-phospho-alpha-D-ribose 1-diphosphate protects [7], not adenine or adenosine [13]) [1, 3, 6–8, 10–14, 16–18]; GMP (as strong as AMP [13], more effective than AMP [3], competitive to phosphoribosyldiphosphate (rat [1]) [1, 8], promotes cooperativity in phosphoribosyldiphosphate binding [16], synergism in combination with AMP [3, 8] or ADP [4], stronger inhibition with NH_3 instead of glutamine as substrate [14], 5-phospho-alpha-D-ribose 1-diphosphate protects [7], not guanine or guanosine [13]) [1, 3, 4, 6–8,

10–14, 16–18]; IMP (feed-back inhibition, more effective than AMP [3], pro-motes cooperativity in phosphoribosyldiphosphate binding [16], not inosine or ITP [13]) [1, 3, 13, 16]; XMP (does not promote cooperativity in 5-phos-pho-alpha-D-ribose 1-diphosphate-binding [16], not xanthine or xanthosine [13]) [3, 13, 16]; UMP (weak [13], not [8], not uracil [13]) [13]; CMP (not cytidine, CDP or CTP) [13]; OMP (not orotic acid or orotidine) [13]; TMP [13]; TTP (weak) [13]; XDP [13]; XTP [13]; Cyclic adenosine (weak [13], not [3]) [13]; Guanosine 3',5'-monophosphate (weak [13], not [3]) [13]; dCMP [13]; Methyl-dCMP and TMP (in combination, synergism) [13]; 6-Hydroxyp-urine nucleotides (in combination, synergism) [13]; 6-Aminopurine nu-cleotides (in combination, synergism) [13]; GDP (weak [1, 4, 6]) [1, 4, 6, 13]; ADP (weak [4], most potent [6]) [4, 6, 13]; 5',5'''-p^1,p^4-Diguanosine tetraphosphate [6]; ATP (rat, competitive to 5-phospho-alpha-D-ribose 1-diphosphate, loss of sensitivity towards ATP by heating at 60°C or gel fil-tration [1], not [3]) [1, 13]; GTP (weak) [13]; Adenylic acids (feed-back inhi-bition) [8]; Guanylic acids (feed-back inhibition) [8]; Azaserine (glutamine as substrate [10, 14], competitive to glutamine [14]) [10, 11, 14]; 1,10-Phe-nanthroline (human placenta [10], Bacillus subtilis [4]) [4, 10]; Mg^{2+} (above 10 mM [13], not [16]) [13]; SDS (t$_{1/2}$ at 4°C in phosphate buffer: more than 40 min, in Tris buffer about 2 min) [12]; Decrease of pH-value from 7.4 to 5 (at non-saturating glutamine concentrations) [14]; p-Substituted mercuri-benzoate (glutamine as substrate) [3]; p-Hydroxymercuribenzoate [1]; Urea [1]; N-Ethylmaleimide [1]; NH$_3$ (competitive to glutamine) [3, 14, 16]; 6-Diazo-5-oxo-L-norleucine (i.e. DON, glutamine analogue, chicken liver, E. coli [11], glutamine (not NH$_3$ [3]) as substrate [3, 10], irreversible inactiva-tion after two-stage incubation [14], complete inactivation by 1 equivalent per 57000 subunit, totally dependent on Mg-phosphoribosyldiphosphate [3]) [3, 10, 11, 14]; L-2-Amino-4-oxo-5-chloropentanoic acid (glutamine ana-logue, glutamine as substrate, not NH$_3$) [3]; Diphosphate (product inhibition, uncompetitive to phosphoribosyldiphosphate) [16]; Glutamate (product inhi-bition, competitive to glutamine) [16]; More (no inhibition by hypoxanthine, 6-mercaptopurine, uric acid [13], Mg-phosphoribosyldiphosphate-Mg-, gly-cine or HCO$_3^-$ [16]) [13, 16]

Cofactor(s)/prosthetic group(s)/activating agents

Mg-phosphoribosyldiphosphate (requirement, phosphoribosyldiphos-phate-Mg^3- is the reactive molecular species of phosphoribosyldiphosphate, K$_m$-value: 0.62 mM [16], no glutamine-binding site in the absence of Mg-phosphoribosyldiphosphate [3]) [3, 8, 16]; More (complex allosteric pro-tein whose activity is regulated by a series of conformational changes in-duced by a number of ligands [12]) [12, 18]

Metal compounds/salts

Mg^{2+} (requirement [2, 13, 16], activation [8], K$_m$-value: 0.65 mM, inhibitory above 10 mM [16]) [2, 3, 8, 13, 16]; Mn^{2+} (requirement, equally effective as

Mg^{2+}) [13]; Co^{2+} (activation, about 40% as effective as Mg^{2+} or Mn^{2+}) [13]; Ca^{2+} (activation, about 15% as effective as Mg^{2+} or Mn^{2+}) [13]; Fe (requirement, iron-sulfur protein [4], Bacillus subtilis and probably human placenta enzyme: diamagnetic [11] 4Fe-4S-cluster [4, 9, 10], iron is oxidized by O$_2$ to enzyme-bound Fe^{3+} [9], low temperature magnetic circular dichroism, electron paramagnetic resonance and resonance Raman spectroscopy [15], not E. coli enzyme [11]) [4, 9–11, 15]; S (requirement, iron-sulfur protein [4], Bacillus subtilis and probably human placenta enzyme: diamagnetic [11] 4Fe-4S-cluster [4, 9, 10], S^{2-} is oxidized by O$_2$ to a mixture of sulfur oxides bound as thiocystine and yet unidentified products [9], low temperature magnetic circular dichroism, electron paramagnetic resonance and resonance Raman spectroscopy [15], not avian liver [10] or E. coli enzyme [11]) [4, 9–11, 15]; More (no activation by Ba^{2+}, Cd^{2+}, Cu^{2+}, Fe^{2+}, Hg^{2+}, Ni^{2+}, Zn^{2+} [13], Fe-salts or sulfide [3]) [3, 13]

Turnover number (min^{-1})

Specific activity (U/mg)
More [6]; 0.00002 [17]; 0.001 (glutamine) [14]; 0.00125–0.0014 [8]; 0.00297 (NH$_3$) [14]; 0.0055 [13]; 0.043 (Bacillus subtilis) [11]; 1.82 [7]; 15.6 [16]; 17.2 (E. coli) [3, 11]

K$_m$-value (mM)
More (kinetic study [12, 16], effect of AMP on kinetic parameters [13], kinetic properties of glutamine and NH$_3$ utilization [14]) [12–14, 16]; 0.067–0.072 (5-phospho-alpha-D-ribose 1-diphosphate) [3, 11]; 0.14–0.48 (5-phospho-alpha-D-ribose 1-diphosphate, human, adenocarcinoma 755 [10]) [10, 13, 17]; 0.4 (5-phospho-alpha-D-ribose 1-diphosphate (+ NH$_3$ [6])) [6, 16]; 0.57–0.64 (5-phospho-alpha-D-ribose 1-diphosphate, rat [10]) [7, 10]; 0.7 (5-phospho-alpha-D-ribose 1-diphosphate (+ glutamine)) [6]; 1–4.5 (glutamine, human) [10, 13]; 1.24–1.5 (glutamine, rat) [7, 10]; 1.7 (glutamine) [3]; 1.8 (glutamine, adenocarcinoma 755) [10]; 2–3.8 (NH$_3$, human) [10]; 4.3 (glutamine) [11]; 8.8 (NH$_3$) [3]; 16 (NH$_3$) [16]; 18 (glutamine) [16]

pH-optimum
6–8 (broad, phosphate buffer) [13]; 6.5 (imidazole/HCl buffer) [13]; 6.5–8.5 (broad, Tris or phosphate buffer, constant activity) [7]; 6.6–7.7 (broad, glutamine-dependent activity, phosphate buffer) [3]; 6.8–7.4 (Tris/HCl buffer preferred) [8]; 7.5 (Tris/HCl buffer [13], unstable enzyme [17]) [13, 17]; 7.8–8.6 (broad, glutamine-dependent activity, Tris/HCl buffer) [3]; 8 (stable enzyme [17]) [16, 17]; 8.5 (NH$_3$-dependent activity, Tris buffer) [3]

pH-range
6–10 (about 70% of maximal activity at pH 6 and 10) [16]; 7–9 (about 65% (stable enzyme) or 70% (unstable enzyme) of maximal activity at pH 7 and

about 75% (stable enzyme) or 65% (unstable enzyme) of maximal activity at pH 9) [17]

Temperature optimum (°C)
25 (assay at) [16, 18]; 37 (assay at) [3, 6–8, 14]; 38 (assay at) [2]

Temperature range (°C)

---

3 ENZYME STRUCTURE

Molecular weight
More (enzymes from avian liver, human placenta, Chinese hamster fibroblasts and mouse liver exist in two molecular weight forms, the larger one is observed when incubated with purine nucleotides, the smaller one when incubated with phosphoribosyldiphosphate [10], in case of pigeon liver vice versa [10, 12], i.e. ligand-promoted interconversion [11], amino acid composition [3, 4]) [3, 4, 10–12]
93000 (Bacillus subtilis, sucrose density centrifugation, enzyme exists in equilibrium of tetrameric, dimeric and monomeric forms) [4]
102000 (pigeon, sedimentation equilibrium centrifugation, phosphate, phosphoribosyldiphosphate or purine mononucleotides influence the sedimentation profile) [12]
110000 (chicken, gel filtration or sedimentation equilibrium centrifugation, ligand-induced alteration of sedimentation coefficient and Stokes radius) [18]
127000 (mouse liver (small form)) [10]
133000 (human placenta (small form), large form is converted to small form by incubation with phosphoribosyldiphosphate) [10]
172000 (pigeon, sedimentation equilibrium centrifugation, in the presence of phosphoribosyldiphosphate, the sedimentation profile is influenced by phosphate, phosphoribosyldiphosphate or purine mononucleotides) [12]
180000 (Schizosaccharomyces pombe, mutant unstable enzyme, gel filtration) [17]
181000 (pigeon, sedimentation equilibrium centrifugation, in the presence of phosphate, the sedimentation profile is influenced by phosphate, phosphoribosyldiphosphate or purine mononucleotides) [12]
185000 (Bacillus subtilis, highly concentrated enzyme solution, sucrose density centrifugation, enzyme exists in equilibrium of tetramer, dimer and monomeric forms, conversion of dimer to tetramer within 10-fold increase in protein concentration) [4]
194000 (E. coli, sedimentation equilibrium centrifugation) [3, 11]
195000 (rat, sucrose density gradient centrifugation) [7]
200000 (Bacillus subtilis, highly concentrated enzyme solution, gel filtration, enzyme exists in equilibrium of tetrameric, dimeric and monomeric forms, conversion of dimer to tetramer within 10-fold increase in protein concentration [4], rat, PAGE [7], chicken in the presence of phosphoribosyldiphosphate, gel filtration and sedimentation equilibrium centrifugation, ligand-in-

duced alteration of sedimentation coefficient and Stokes radius [18]) [4, 7, 18]
215000 (rat, gel filtration [7, 10], only one MW enzyme form [10]) [7, 10]
224000 (E. coli, gel filtration) [3, 11]
270000 (human placenta (large form), small form is converted to large form by incubation with purine nucleotides, large form presumably catalytically inactive) [10]
292000 (mouse liver (large form)) [10]
360000 (Schizosaccharomyces pombe, wild-type stable enzyme, gel filtration) [17]

Subunits
? (2–4 × 50000, Bacillus subtilis, SDS-PAGE [4, 11], 3 or 4 × 56395, E. coli, calculated from deduced amino acid sequence [11], 3 or 4 × 57000, E. coli, SDS-PAGE [3, 11], x × 58000, chicken, SDS-PAGE [18]) [3, 4, 11, 18]

Glycoprotein/Lipoprotein
–

---

4 ISOLATION/PREPARATION

Source organism
E. coli (strains K-12 or B-96 (purine requiring strain) [3]) [3, 10, 11]; Bacillus subtilis [4, 5, 9–11, 15]; Artemia sp. (brine shrimp) [6]; Chicken [11, 18]; Chinese hamster [10]; Human [8, 10, 13, 14]; Mouse [10]; Pigeon [1, 2, 10, 12]; Rat [1, 7]; Schizosaccharomyces pombe (wild type strain 972 h- and mutant strain ade 2 h- (devoid of adenylosuccinate synthase)) [17]; Glycine max (soybean, Merr. cv. Williams) [16]

Source tissue
Cell [3–5, 9, 17]; Erythroleukemia cells (murine) [10]; Fibroblasts (hamster) [10]; Leukocytes (human) [10]; Liver (rat, mouse, bird) [1, 2, 7, 10–12, 18]; Lymphoblasts (diploid wil 2 line, tissue culture) [8]; Nauplii [6]; Placenta (human) [10, 13, 14]; Root nodules [16]

Localization in source
Soluble [1, 6, 7, 13]; Cytosol (little or no activity in mitochondria or microsomes) [13]

Purification
E. coli [3]; Bacillus subtilis (partial, ATP-agarose affinity chromatography [5]) [4, 5]; Artemia sp. (partial) [6]; Rat (partial) [1, 7]; Pigeon (partial) [1, 2, 12]; Human (partial) [13, 14]; Glycine max [16]; Schizosaccharomyces pombe (partial) [17]; Chicken [18]

Crystallization
–

Cloned

–

Renatured

–

---

5 STABILITY

pH

Temperature (°C)
More (AMP and GMP enhance thermal stability) [3]; 48 (wild-type stable enzyme from cells of mid-exponential growth phase: $t_{1/2}$: 34 min, from cells of stationary growth phase: $t_{1/2}$: 2–5 min, mutant enzyme: $t_{1/2}$: 2 min, in the presence of ADP or adenosine 3 min, and 4 min in the presence of IMP) [17]; 60 (6 min, about 25% loss of activity [1], at least 15 min stable [8], after 3–6 min loss of ATP-sensitivity [1]) [1, 8]

Oxidation
Oxygen-labile in vivo and in vitro, AMP protects [4]; Oxygen inactivates during purification [3]; Anaerobic conditions stabilize [3, 4]; DTT enhances aerobic inactivation [3]; O_2, rather than peroxide, superoxide, hydroxyl radical or singlet oxygen, inactivates, allosteric inhibitors, such as AMP, ADP, GMP or GDP modulate the rate of inactivation [9]

Organic solvent

General stability information
2-Mercaptoethanol stabilizes during purification (not [3]) [1]; AMP or GMP stabilizes dimeric enzyme form [4, 11]; GDP stabilizes tetrameric enzyme form [4, 11]; ADP or ADP and GMP, ratio 1:1, stabilize equilibrium between dimeric and tetrameric form [4]; AMP stabilizes against inactivation by O_2 [4, 9]; Phosphoribosyldiphosphate and other nucleotides antagonize stabilizing effect of AMP [9]; Stable during all stages of purification provided thiols and oxygen are avoided [3]; High concentrations of thiol reagents stabilize during purification [16]; Freezing inactivates, not stabilized by 30%, v/v, glycerol [16]; Prolonged dialysis against distilled water leads to precipitation [1]; Stability profile depends on growth rate and growth phase of cell culture [17]

Storage
2-Mercaptoethanol stabilizes during storage [1]; PMSF stabilizes during storage [17]; Phosphate stabilizes during storage [12, 13]; Mg^{2+} stabilizes during storage [13]; 2-Mercaptoethanol and phosphoribosyldiphosphate stabilize during storage at 4°C [12]; Storage in 25 mM Tris buffer inactivates with 26% or 4% residual activity after 6 or 10 days, respectively [12]; –20°C, crude, more than 50% loss of activity overnight [8]; 3°C, crude mutant en-

zyme extract, inactivation overnight, partially purified mutant enzyme: inactivation within 72 h, PMSF stabilizes [17]; 3°C, crude or partially purified wild-type enzyme, at least 72 h stable [17]; 4°C, in phosphate buffer, pH 7.4, 60 mM mercaptoethanol and phosphoribosyldiphosphate, at least 10 days [12]; 4°C, partially purified preparation in 50 mM phosphate buffer, pH 7.4, 5 mM Mg^{2+} and 60 mM mercaptoethanol, at least 4 weeks [13]; 4°C, 25% loss of activity per day [16]

6 CROSSREFERENCES TO STRUCTURE DATABANKS

PIR/MIPS code
PIR1:XQBS (Bacillus subtilis); PIR2:A38337 (chicken); PIR1:XQEC (Escherichia coli); PIR2:S01389 (Escherichia coli); PIR3:S38482 (fission yeast (Schizosaccharomyces pombe)); PIR2:A46088 (rat); PIR2:A22642 (yeast (Saccharomyces cerevisiae)); PIR2:A53342 (precursor human); PIR2:JC1414 (precursor human); PIR3:S47860 (fruit fly (Drosophila melanogaster)); PIR3:S43526 (yeast (Schizosaccharomyces pombe))

Brookhaven code
1GPH ((Bacillus Subtilis))

7 LITERATURE REFERENCES

[1] Caskey, C.T., Ashton, D.M., Wyngaarden, J.B.: J. Biol. Chem.,239,2570–2579 (1964)
[2] Hartman, S.C., Buchanan, J.M.: J. Biol. Chem.,233,451–455 (1958)
[3] Messenger, L.J., Zalkin, H.: J. Biol. Chem.,254,3382–3392 (1979)
[4] Wong, J.Y., Bernlohr, D.A., Turnbough, C.L., Switzer, R.L.: Biochemistry,20, 5669–5674 (1981)
[5] Wong, J.Y., Switzer, R.L.: Arch. Biochem. Biophys.,196,134–137 (1979)
[6] Liras, A., Argomaniz, L., Llorente, P.: Biochim. Biophys. Acta,1033,114–117 (1990)
[7] Tsuda, M., Katunuma, N., Weber, G.: J. Biochem.,85,1347–1354 (1979)
[8] Wood, A.W., Seegmiller, J.E.: J. Biol. Chem.,248,138–143 (1973)
[9] Bernlohr, D.A., Switzer, R.L.: Biochemistry,20,5675–5681 (1981)
[10] Holmes, E.W.: Adv. Enzyme Regul.,19,215–231 (1981) (Review)
[11] Zalkin, H.: Adv. Enzyme Regul.,21,225–237 (1983) (Review)
[12] Itoh, R., Holmes, E.W., Wyngaarden, J.B.: J. Biol. Chem.,251,2234–2240 (1976)
[13] Holmes, E.W., McDonald, J.A., McCord, J.M., Wyngaarden, J.B., Kelley, W.N.: J. Biol. Chem.,248,144–150 (1973)
[14] King, G.L., Boounous, C.G., Holmes, E.W.: J. Biol. Chem.,253,3933–3938 (1978)
[15] Oñate, Y.A., Vollmer, S.J., Switzer, R.L., Johnson, M.K.: J. Biol. Chem.,264, 18386–18391 (1989)
[16] Reynoldds, P.H.S., Blevins, D.G., Randall, D.D.: Arch. Biochem. Biophys.,229, 623–631 (1984)
[17] Nagy, M., Reichert, U., Ribet, A.-M.: Biochim. Biophys. Acta,370,85–95 (1974)
[18] Itoh, R., Gorai, I., Usami, C., Tsushima, K.: Biochim. Biophys. Acta,581,142–152 (1979)

1 NOMENCLATURE

EC number
2.4.2.15

Systematic name
Guanosine:orthophosphate D-ribosyltransferase

Recommended name
Guanosine phosphorylase

Synonyms
Phosphorylase, guanosine

CAS Reg. No.
9030-28-8

---

2 REACTION AND SPECIFICITY

Catalysed reaction
Guanosine + phosphate →
→ guanine + D-ribose 1-phosphate

Reaction type
Pentosyl group transfer

Natural substrates
Guanosine + phosphate (may play an important role in the utilization by bone marrow of the corresponding preformed base and nucleoside reaching this tissue via blood) [1]

Substrate spectrum
1 Guanosine + phosphate (r) [1]
2 Deoxyguanosine + phosphate [1]
3 More (activities at arsenate concentrations of 0.05 M and 0.1 M are 10% and 88% of the activities at comparable phosphate concentrations, respectively) [1]

Product spectrum
1 Guanine + D-ribose 1-phosphate [1]
2 ?
3 ?

Enzyme Handbook © Springer-Verlag Berlin Heidelberg 1996
Duplication, reproduction and storage in data banks are only
allowed with the prior permission of the publishers

Inhibitor(s)
  Tris buffer [1]; p-Chloromercuribenzoate [1]

Cofactor(s)/prosthetic group(s)/activating agents

Metal compounds/salts

Turnover number (min^{-1})

Specific activity (U/mg)
  4.27 [1]

K_m-value (mM)
  0.210 (deoxyguanosine) [1]; 0.216 (guanosine) [1]; 0.376 (phosphate
  (+ guanosine)) [1]

pH-optimum
  7.0 (deoxyguanosine) [1]; 7.0–7.4 (guanosine) [1]

pH-range
  5–8.5 (5.0: about 35% of activity maximum, 8.5: about 45% of activity maximum) [1]

Temperature optimum (°C)
  37 (assay at) [1]

Temperature range (°C)

3 ENZYME STRUCTURE

Molecular weight

Subunits

Glycoprotein/Lipoprotein
  –

4 ISOLATION/PREPARATION

Source organism
  Rabbit [1]; Trichomonas vaginalis [2]

Source tissue
  Bone marrow [1]

Localization in source

Purification
  Rabbit [1]

Crystallization
–

Cloned
–

Renatured
–

5 STABILITY

pH

Temperature (°C)

Oxidation

Organic solvent

General stability information

Storage
  –10°C, stable for 1 month [1]

6 CROSSREFERENCES TO STRUCTURE DATABANKS

PIR/MIPS code

Brookhaven code

7 LITERATURE REFERENCES

[1] Yamada, E.W.: J. Biol. Chem.,236,3043–3046 (1961)
[2] Heyworth, P.G., Gutteridge, W.E., Ginger, C.D.: FEBS Lett.,141,106–110 (1982)

1 NOMENCLATURE

EC number
2.4.2.16

Systematic name
Urate-ribonucleotide:orthophosphate D-ribosyltransferase

Recommended name
Urate-ribonucleotide phosphorylase

Synonyms
UAR phosphorylase [1]
Phosphorylase, urate ribonucleotide
Urate ribonucleotide phosphorylase

CAS Reg. No.
9030-29-9

2 REACTION AND SPECIFICITY

Catalysed reaction
Urate D-ribonucleoside + phosphate →
→ urate + D-ribose 1-phosphate

Reaction type
Pentosyl group transfer

Natural substrates
Uric acid ribonucleoside + phosphate [1]

Substrate spectrum
1 Uric acid ribonucleoside + phosphate (reverse reaction not demonstrated, arsenate can replace phosphate at a lower reaction rate) [1]

Product spectrum
1 Urate + D-ribose 1-phosphate [1]

Inhibitor(s)
p-Chloromercuribenzoate (reversed by cysteine) [1]; 1-Deoxyribose 5-methyluracil (i.e. thymidine) [1]; 2-Thiouracil [1]; Xanthine [1]; 4-Amino-5-imidazolecarboxamide [1]; Uracil [1]; 5-Methyluracil (i.e. thymine) [1]; 1-Ribosyluracil [1]; Colchicine [1]; Phenylbutazone [1]; 2-Thio-6-oxypurine [1]; Inosine (weak) [1]; Guanosine (weak) [1]

Cofactor(s)/prosthetic group(s)/activating agents

Metal compounds/salts

Turnover number (min^{-1})

Specific activity (U/mg)
 More [1]

K_m-value (mM)
 0.11 (uric acid ribonucleoside) [1]; 0.6 (inorganic phosphate) [1]; 11 (arsenate) [1]

pH-optimum
 7.8 (assay at) [1]

pH-range

Temperature optimum (°C)
 37 (assay at) [1]

Temperature range (°C)

---

3 ENZYME STRUCTURE

Molecular weight

Subunits

Glycoprotein/Lipoprotein
 –

---

4 ISOLATION/PREPARATION

Source organism
 Rat [1]; Pigeon [1]; Guinea pig [1]; Dog [1]

Source tissue
 Kidney [1]; Heart [1]; Small intestine mucosa [1]; Testis [1]; Spleen (low activity) [1]; Liver [1]; Skeletal muscle [1]

Localization in source

Purification
 Dog [1]

Crystallization
 –

Cloned
 –

Renatured

–

---

5 STABILITY

pH
  4.5 (0°C, 10 min, 30% loss of activity, partially purified enzyme) [1]

Temperature (°C)
  50 (pH 7.8, 5 min, 5% loss of activity, partially purified enzyme) [1];
  60 (pH 7.8, 5 min, 40% loss of activity, partially purified enzyme) [1];
  70 (pH 7.8, 5 min, 95% loss of activity, partially purified enzyme) [1];
  80 (pH 7.8, 5 min, 100% loss of activity, partially purified enzyme) [1]

Oxidation

Organic solvent

General stability information

Storage
  –10°C, partially purified enzyme is stable for at least 2 months [1]

---

6 CROSSREFERENCES TO STRUCTURE DATABANKS

PIR/MIPS code

Brookhaven code

---

7 LITERATURE REFERENCES

[1] Laster, L., Blair, A.: J. Biol. Chem.,238,3348–3357 (1963)

---

1 NOMENCLATURE

EC number
2.4.2.17

Systematic name
1-(5-Phospho-D-ribosyl)-ATP:pyrophosphate phospho-alpha-D-ribosyltransferase

Recommended name
ATP phosphoribosyltransferase

Synonyms
Phosphoribosyl-ATP pyrophosphorylase
Adenosine triphosphate phosphoribosyltransferase [10]
Phosphoribosyltransferase, adenosine triphosphate
Phosphoribosyl ATP synthetase
Phosphoribosyl ATP:pyrophosphate phosphoribosyltransferase
Phosphoribosyl-ATP:pyrophosphate-phosphoribosyl phosphotransferase
Phosphoribosyladenosine triphosphate pyrophosphorylase
Phosphoribosyladenosine triphosphate synthetase

CAS Reg. No.
9031-46-3

---

2 REACTION AND SPECIFICITY

Catalysed reaction
ATP + 5-phospho-alpha-D-ribose 1-diphosphate →
→ 1-(5-phospho-D-ribosyl)-ATP + diphosphate (sequential kinetic mechanism in biosynthetic direction, ordered bi-bi mechanism with ATP binding
first to free enzyme and phosphoribosyl-ATP dissociating last from en-
zyme-product complexes [10, 11], double displacement mechanism [2])

Reaction type
Pentosyl group transfer

Natural substrates
ATP + 5-phospho-alpha-D-ribose 1-diphosphate (first step in histidine
biosynthesis) [1, 3, 4, 10, 11]

Substrate spectrum
1 ATP + 5-phospho-alpha-D-ribose 1-diphosphate (r [3, 7, 10]) [1–11]
2 More (not: ribose 5-phosphate, AMP, UTP, CTP, GTP) [3]

Enzyme Handbook © Springer-Verlag Berlin Heidelberg 1996
Duplication, reproduction and storage in data banks are only
allowed with the prior permission of the publishers

Product spectrum
 1 1-(5-Phospho-D-ribosyl)-ATP + diphosphate [1, 3]
 2 ?

Inhibitor(s)
 beta,gamma-Methylene ATP (competitive to phosphoribosyl-ATP, noncompetitive to diphosphate) [11]; Histidine (feed-back inhibition [1, 4], reversed by: Hg^{2+} [1], p-hydroxymercuribenzoate [1], methylmercuric bromide [1], Ni^{2+} [1], inhibits reverse reaction cooperatively and completely [11]) [1–5, 11]; Zn^{2+} [1]; Ag^+ [1]; Cd^{2+} [1]; Ni^{2+} [1]; Co^{2+} [1]; Cu^{2+} [1]; 1-(5-Phospho-alpha-D-ribosyl)-ATP (product inhibition, competitive to both substrates) [10]; Guanosine 5'-diphosphate-3'-diphosphate (in presence of partially inhibiting concentrations of histidine guanosine 5'-diphosphate-3'-diphosphate becomes a potent inhibitor of the residual activity of ATP phosphoribosyltransferase, no inhibition in absence of histidine, inhibition is slowly reversible) [5]; Dicoumarol (competitive to ATP, inhibitor in both directions, diminishes yield of phosphoribosyladenosine triphosphate by acting as parasite substrate) [7]; Pentachlorophenol (competitive to ATP, inhibitor in both directions, diminishes yield of phosphoribosyladenosine triphosphate by acting as parasite substrate) [7]; Dinitrophenol (diminishes yield of phosphoribosyladenosine triphosphate by acting as parasite substrate) [7]; 5-Phospho-alpha-D-ribose 1-diphosphate (noncompetitive to both substrates in the reverse reaction) [11]; AMP (in the presence of histidine inhibition by AMP and ADP becomes positively cooperative and much more potent [6], competitive inhibitor of ATP [1, 2, 6], competitive inhibitor to 5-phospho-alpha-D-ribose 1-diphosphate [1, 6]) [1, 2, 6]; Diphosphate (competitive to both substrates) [10]; ADP (competitive to ATP [1, 6], in the presence of histidine inhibition by AMP and ADP becomes positively cooperative and much more potent [6]) [1, 6]; ATP (competitive to 5-phospho-alpha-D-ribose 1-diphosphate [1], competitive to ATP [10]) [1, 2, 10]; Adenine (competitive to ATP) [1]; Hg^{2+} [1]; p-Hydroxymercuribenzoate [1]; Methylmercuric bromide [1]; N-Ethylmaleimide [1]; More (not: carbonylcyanide m-chlorophenylhydrazone) [7]

Cofactor(s)/prosthetic group(s)/activating agents

Metal compounds/salts

Turnover number (min^{-1})

Specific activity (U/mg)
 More [1, 9]

K_m-value (mM)
 0.067 (5-phospho-alpha-D-ribose 1-diphosphate) [1]; 0.2 (ATP) [1]

pH-optimum
  8.5 (assay at) [3, 4]

pH-range

Temperature optimum (°C)
  25 (assay at) [3]

Temperature range (°C)

---

3 ENZYME STRUCTURE

Molecular weight
  210000–221000 (Salmonella typhimurium, equilibrium centrifugation with meniscus depletion method) [4]
  216000 (Salmonella typhimurium, ultracentrifugation analysis) [9]

Subunits
  ? (x × 36000, Salmonella typhimurium, viscometric methods, equilibrium centrifugation with meniscus depletion method after dialysis against 5.0 M guanidine-HCl and 0.143 M beta-mercaptoethanol) [4]
  Hexamer (6 × 33000, Salmonella typhimurium, SDS-PAGE) [9]

Glycoprotein/Lipoprotein
  –

---

4 ISOLATION/PREPARATION

Source organism
  Salmonella typhimurium (LT2 strain TA2165 [6, 10]) [1, 3–6, 9–11]; E. coli [2, 7, 8]

Source tissue
  Cell [1]

Localization in source

Purification
  Salmonella typhimurium [1, 3, 4, 9]; E. coli [2]

Crystallization
  –

Cloned
  –

Renatured
  –

---

5 STABILITY

pH
    More (overview: stability at various pH-values) [9]

Temperature (°C)
    47.5 (1.9 mg enzyme/ml, 80 min, 75% loss of activity without addition of
    AMP, with 0.05 mM AMP about 60% loss of activity, with 0.5 mM AMP
    about 40% loss of activity, with 5 mM AMP about 25% loss of activity) [8]; 4
    8 (3 mg enzyme/ml, 10 min, 15% remaining activity without addition of
    histidine, with 0.000086 mM histidine about 15% remaining activity after
    15 min, with 0.00069 mM histidine about 70% remaining activity after
    80 min, with 0.00345 mM histidine about 80% remaining activity after
    80 min, with 0.0069 mM histidine about 55% remaining activity after 50 min)
    [8]

Oxidation

Organic solvent

General stability information
    NaCl and 2-mercaptoethanol stabilize the very labile enzyme [4]; Histidine
    or AMP stabilizes the enzyme with respect to thermal inactivation [8];
    L-Histidine, 0.4 mM, stabilizes against heat inactivation, 0.04 mM does not
    stabilize against heat inactivation, at 1.33 mM and higher heat inactivation is
    greater than the control [1]; Slight stabilization by 10 mM $MgCl_2$ or $CaCl_2$ by
    1 mM $MnCl_2$ and by 1 mM histidine at pH-values above 7 [9]; Overview, sta-
    bility of the enzyme at various pH-values, salt concentrations and histidine
    concentrations [9]

Storage
    −15°C, several months [3]; 4°C, 0.01 M Tris, 0.10 M NaCl, 0.4 mM histidine,
    2.8 mM 2-mercaptoethanol, 0.5 mM EDTA, pH 7.5, 50% loss of activity after
    several days [4]; 4°C, HEPES buffer pH 7.5, final ammonium sulfate micro-
    crystalline enzyme in 60% saturated ammonium sulfate, 0.1 M NaCl, 0.01 M
    Tris, 0.5 mM EDTA, 10 mM DTT, stable for 1 month [9]; Storage in liquid ni-
    trogen of quick-frozen enzyme in HEPES buffer pH 7.5, 0.1 M NaCl, 0.01 M
    Tris, 0.5 mM EDTA, 1 mM DTT, 1 mM histidine, indefinitely stable [9]

6 CROSSREFERENCES TO STRUCTURE DATABANKS

PIR/MIPS code
    PIR1:XREC (Escherichia coli); PIR3:S19176 (Escherichia coli); PIR1:XREBT
    (Salmonella typhimurium); PIR2:S55497 (yeast (Candida albicans));
    PIR1:XRBY (yeast (Saccharomyces cerevisiae))

Brookhaven code

7 LITERATURE REFERENCES

[1] Martin, R.G.: J. Biol. Chem.,238,257–268 (1963)
[2] Tebar, A.R., Ballesteros, A.O.: Mol. Cell. Biochem.,11,131–136 (1976)
[3] Ames, B.N., Martin, R.G., Garry, B.J.: J. Biol. Chem.,236,2019–2026 (1961)
[4] Voll, M.J., Appella, E., Martin, R.G.: J. Biol. Chem.,242,1760–1767 (1967)
[5] Morton, D.P., Parsons, S.M.: Biochem. Biophys. Res. Commun.,74,172–177 (1977)
[6] Morton, D.P., Parsons, S.M.: Arch. Biochem. Biophys.,181,643–648 (1977)
[7] Dall-Larsen, T., Kryvi, H., Klungsoyr, L.: Eur. J. Biochem.,66,443–446 (1976)
[8] Kryvi, H.: Biochim. Biophys. Acta,317,123–130 (1973)
[9] Parsons, S.M., Koshland, D.E.: J. Biol. Chem.,249,4104–4109 (1974)
[10] Morton, D.P., Parsons, S.M.: Arch. Biochem. Biophys.,175,677–686 (1976)
[11] Kleeman, J.E., Parsons, S.M.: Arch. Biochem. Biophys.,175,687–693 (1976)

1 NOMENCLATURE

EC number
    2.4.2.18

Systematic name
    N-(5-Phospho-D-ribosyl)-anthranilate:pyrophosphate phospho-alpha-D-ribosyltransferase

Recommended name
    Anthranilate phosphoribosyltransferase

Synonyms
    Phosphoribosyl-anthranilate pyrophosphorylase
    Anthranilate-5-phosphoribosylphosphate phosphoribosyltransferase [5]
    PRT [11]
    Phosphoribosyltransferase, anthranilate
    Anthranilate 5-phosphoribosylpyrophosphate phosphoribosyltransferase
    Anthranilate phosphoribosylpyrophosphate phosphoribosyltransferase
    Phosphoribosylanthranilate pyrophosphorylase
    Phosphoribosylanthranilate transferase
    Anthranilate-PP-ribose-P phosphoribosyltransferase [2]
    More (in some organisms, this enzyme is part of a multifunctional protein together with one or more other components of the system for biosynthesis of tryptophan (EC 4.1.1.48, EC 4.1.3.27, EC 4.2.1.20, EC 5.3.1.24)) [11–13]

CAS Reg. No.
    9059-35-2

2 REACTION AND SPECIFICITY

Catalysed reaction
    Anthranilate + 5-phospho-alpha-D-ribose 1-diphosphate →
    → N-(5-phospho-D-ribosyl)-anthranilate + diphosphate (mechanism [7])

Reaction type
    Pentosyl group transfer

Natural substrates
    Anthranilate + 5-phospho-alpha-D-ribose 1-diphosphate (enzyme of tryptophan biosynthesis) [1–15]

Substrate spectrum
    1 Anthranilate + 5-phospho-alpha-D-ribose 1-diphosphate [1–15]

Product spectrum
1 N-(5-Phospho-D-ribosyl)-anthranilate + diphosphate [1–3]

Inhibitor(s)
Anthranilate (substrate inhibition above 0.008 mM) [7]; 3-Hydroxyanthranilate [7]; L-Tryptophan (transferase activity of component II is only inhibitable by L-tryptophan when the component is in the complex, this inhibition does not appear to depend upon the feedback-sensitive site of complex I [1], when phosphoribosyltransferase is not an aggregate with anthranilate synthase, it is not subject to tryptophan inhibition, inhibitor site is on the anthranilate synthase component [12]) [1, 3, 12]; N-(5-Phospho-D-ribosyl)-anthranilate [3]; Sodium diphosphate [3]

Cofactor(s)/prosthetic group(s)/activating agents

Metal compounds/salts
Mg^{2+} (required [3, 12], activates [7], K_m: 0.056 mM [7]) [3, 7, 12]

Turnover number (min^{-1})
20.4 (anthranilate + 5-phospho-alpha-D-ribose 1-diphosphate, Hansenula henricii) [6]; 174 (anthranilate + 5-phospho-alpha-D-ribose 1-diphosphate, Saccharomyces cerevisiae) [4]; 246 (anthranilate + 5-phospho-alpha-D-ribose 1-diphosphate, Salmonella typhimurium) [8, 9]; 264 (anthranilate + 5-phospho-alpha-D-ribose 1-diphosphate, E. coli) [5]

Specific activity (U/mg)
More [3, 12]; 1.54 [13]; 1.58 [4]; 0.4 [7]

K_m-value (mM)
0.0016 (anthranilate, Saccharomyces cerevisiae) [4]; 0.003 (anthranilate, Salmonella typhimurium TAX6trpR782) [3]; 0.004 (anthranilate, Salmonella typhimurium trpAB1653trpR782) [3]; 0.013 (5-phospho-alpha-D-ribose 1-diphosphate, Salmonella typhimurium TAX6trpR782) [3]; 0.0224 (5-phospho-alpha-D-ribose 1-diphosphate, Saccharomyces cerevisiae) [4]; 0.06 (5-phospho-alpha-D-ribose 1-diphosphate, Salmonella typhimurium trpAB1653trpR782) [3]; 0.1 (5-phospho-alpha-D-ribose 1-diphosphate, complex) [1]; 0.2 (5-phospho-alpha-D-ribose 1-diphosphate, component II) [1]

pH-optimum
6.9 [3]; 7.0 (assay at) [3]; 7.4–7.7 [7]; 7.5 (assay at) [4]

pH-range

Temperature optimum (°C)
25 (assay at) [4]; 37 (assay at) [7]

Temperature range (°C)

3 ENZYME STRUCTURE

Molecular weight

45000 (Serratia marcescens [10, 11], Enterobacter liquefaciens, Serratia marinorubra [11], gel filtration) [10, 11]

67000 (Erwinia carotovora, gel filtration [11, 15], Proteus vulgaris, Enterobacter hafniae, Proteus morganii, Aeromonas formicans, gel filtration [11]) [11, 15]

70000 (Hafnia alvei, gel filtration [14], Hansenula henricii, gel filtration [7], and 150000, Salmonella typhimurium, disc gel electrophoresis [13]) [7, 13, 14]

83000 (Saccharomyces cerevisiae, analytical ultracentrifugation) [4]

90000 (Aerobacter aerogenes, tryptophan auxotroph with no anthranilate synthase activity, sucrose gradient sedimentation) [12]

150000 (and 70000, Salmonella typhimurium, disc gel electrophoresis) [13]

170000 (Aerobacter aerogenes, sucrose gradient sedimentation, complex with anthranilate synthase activity) [12]

320000 (Salmonella typhimurium TAX6trpR782, gel filtration) [3]

220000 (and larger than 1000000, Salmonella typhimurium trpAB1653trpR782, gel filtration) [3]

Subunits

Monomer (1 × 43000, Serratia marcescens, SDS-PAGE) [10]

Dimer (2 × 40000, Erwinia carotovora, SDS-PAGE [15], 2 × 37000, Hafnia alvei, SDS-PAGE [14], 2 × 42000, Saccharomyces cerevisiae, SDS-PAGE [4]) [4, 14, 15]

? (x × 72000, Salmonella typhimurium trpAB1653trpR782, SDS-PAGE [3], Salmonella typhimurium enzyme exists in both monomeric and dimeric forms, SDS-PAGE [13]) [3, 13]

More (in some organisms, this enzyme is part of a multifunctional protein together with one or more other components of the system for biosynthesis of tryptophan (EC 4.1.1.48, EC 4.1.3.27, EC 4.2.1.20, EC 5.3.1.24) [11–13], Hansenula henricii: no aggregation with other enzymes [7], Enterobacteriaceae: anthranilate synthase-phosphoribosylanthranilate transferase complex exists in Citrobacter species [11], Enterobacter cloacae [11], Erwinia dissolvens [11], E. coli [11], Salmonella typhimurium [11, 13], Enterobacter aerogenes [11], Aerobacter aerogenes [12], in all other bacteria examined phosphoribosylanthranilate transferase and anthranilate synthase are separate enzyme molecules [11]) [7, 11–13]

Glycoprotein/Lipoprotein

4 ISOLATION/PREPARATION

Source organism
Enterobacter cloacae [11]; Enterobacter hafniae [11]; Enterobacter liquefaciens [11]; Citrobacter freundii [11]; Citrobacter ballerupensis [11]; Proteus vulgaris [11]; Proteus morganii [11]; Serratia marinorubra [11]; Aeromonas formicans [11]; Erwinia carotovora [11, 15]; Erwinia dissolvens [11]; Aerobacter aerogenes [12]; Hafnia alvei [14]; E. coli (anthranilate synthetase complex consisting of 2 separate subunits: component I and II [1]) [1, 5]; Neurospora crassa [2]; Salmonella typhimurium (mutant strains: TAX6trpR782 and trpAB1653trpR782 [3]) [3, 8, 9, 13]; Saccharomyces cerevisiae [4]; Hansenula henricii [6, 7]; Serratia marcescens [10, 11]

Source tissue
Cell [4, 14]

Localization in source

Purification
Erwinia carotovora [15]; Hafnia alvei [14]; Hansenula henricii [7]; Aerobacter aerogenes (partial, enzyme complex) [12]; Serratia marcescens [10]; Neurospora crassa [2]; Salmonella typhimurium (enzyme complex [3], unaggregated form [13]) [3, 13]; Saccharomyces cerevisiae [4]

Crystallization
[14]

Cloned
–

Renatured
–

---

5 STABILITY

pH

Temperature (°C)
46 (half-life: Salmonella typhimurium TAX6trpR782: 1.5 min, Salmonella typhimurium trpAB1653trpR782: 5 min) [3]

Oxidation

Organic solvent

General stability information
Freezing and thawing inactivates [12]; Dilution inactivates [12]; Glycerol, 10%, is essential for storage, but it must be replaced by polyethylene glycol to achieve crystals that are not severely temperature dependent and radiation sensitive [14]

Storage

4°C, 10% loss of activity after 2 months [12]; –25°C, 500 mM Tris-HCl, pH 7.5, 2 mM EDTA, 2 mM $MgCl_2$, 5% loss of activity after 2 weeks [7]

6 CROSSREFERENCES TO STRUCTURE DATABANKS

PIR/MIPS code

PIR1:NPKEDC (Acinetobacter calcoaceticus); PIR2:S17704 (Azospirillum brasilense); PIR2:JH0099 (Bacillus pumilus); PIR1:NPBS (Bacillus subtilis); PIR2:A49897 (Buchnera aphidicola); PIR2:A05003 (Erwinia carotovora (fragment)); PIR2:JS0340 (Lactobacillus casei); PIR2:S35126 (Lactococcus lactis subsp. lactis); PIR2:B35114 (Pseudomonas aeruginosa); PIR2:C35115 (Pseudomonas putida); PIR2:A05222 (Serratia marcescens (fragment)); PIR3:S34748 (Thermotoga maritima); PIR1:NPBY (yeast (Saccharomyces cerevisiae))

Brookhaven code

7 LITERATURE REFERENCES

[1] Ito, J., Yanofsky, C.: J. Bacteriol.,97,734–742 (1969)
[2] Wegman, J., DeMoss, J.A.: J. Biol. Chem.,240,3781–3788 (1965)
[3] Grieshaber, M.: Z. Naturforsch.,33c,235–244 (1978)
[4] Hommel, U., Lustig, A., Kirschner, K.: Eur. J. Biochem.,180,33–40 (1989)
[5] Gonzalez, J.E., Sommerville, R.L.: Biochem. Cell Biol.,64,681–691 (1986)
[6] Kane, J.F., Jensen, R.A.: J. Biol. Chem.,245,2384–2390 (1970)
[7] Bode, R., Birnbaum, D.: Z. Allg. Mikrobiol.,18,559–566 (1978)
[8] Henderson, E.J., Zalkin, H., Hwang, L.H.: J. Biol. Chem.,245,1424–1431 (1970)
[9] Grieshaber, M., Bauerle, R.: Biochemistry,13,373–383 (1974)
[10] Largen, M., Mills, S.E., Rowe, J., Yanofsky, C.: Eur. J. Biochem.,67,31–36 (1976)
[11] Largen, M., Belser, W.L.: J. Bacteriol.,121,239–249 (1975)
[12] Egan, A.F., Gibson, F.: Biochem. J.,130,847–859 (1972)
[13] Marcus, S.L., Balbinder, E.: Biochem. Biophys. Res. Commun.,47,438–444 (1972)
[14] Edwards, S.E., Kraut, J., Xuong, N., Ashford, V., Halloran, T.P., Mills, S.L.: J. Mol. Biol., 203,525–526 (1988)
[15] Largen, M., Mills, S.E., Rowe, J., Yanofsky, C.: J. Biol. Chem.,253,409–412 (1978)

1 NOMENCLATURE

EC number
2.4.2.19

Systematic name
Nicotinate-nucleotide:pyrophosphate phospho-alpha-D-ribosyltransferase (decarboxylating)

Recommended name
Nicotinate-nucleotide pyrophosphorylase (carboxylating)

Synonyms
Quinolinate phosphoribosyltransferase (decarboxylating)
Quinolinic acid phosphoribosyltransferase [5]
QAPRTase [13]
Pyrophosphorylase, nicotinate mononucleotide (carboxylating)
NAD pyrophosphorylase
Nicotinate mononucleotide pyrophosphorylase (carboxylating) (EC 2.4.2.19)
Quinolinic phosphoribosyltransferase

CAS Reg. No.
37277-74-0

---

2 REACTION AND SPECIFICITY

Catalysed reaction
Nicotinate D-ribonucleotide + diphosphate + CO_2 →
→ pyridine-2,3-dicarboxylate + 5-phospho-alpha-D-ribose 1-diphosphate
(ternary complex of the enzyme, quinolinate and 5-phosphoribosyl 1-diphosphate [7, 20], forms a binary complex (enzyme-quinolinic acid or enzyme-5-phosphoribose 1-diphosphate) [12])

Reaction type
Pentosyl group transfer

Natural substrates
Quinolinate + 5-phospho-alpha-D-ribose 1-diphosphate (step in biosynthesis of the pyridine nucleotides from tryptophan [2], enzyme is probably involved in regulation of nicotine biosynthesis in tobacco [5], enzyme is involved in the de novo synthesis of pyridine nucleotides [7], intermediate enzyme in the de novo NAD biosynthesis [9, 14, 16]) [2, 5, 7, 9, 14, 16]

Substrate spectrum

1 Quinolinate + 5-phospho-alpha-D-ribose 1-diphosphate (i.e pyridine-2,3-dicarboxylate, strictly specific for quinolinate and 5-phospho-alpha-D-ribose 1-diphosphate [6, 16]) [1–21]

Product spectrum

1 Nicotinic acid mononucleotide + diphosphate + CO_2 (i.e. beta-niacin mononucleotide [6, 16]) [1–3, 6, 7, 11, 16, 20]

Inhibitor(s)

2,4-Pyridine dicarboxylic acid (weak) [7, 9]; Nicotinic acid mononucleotide [21]; ADP (weak [9, 18]) [6, 9, 18]; Picolinic acid [7]; 2-Hydroxynicotinate [7]; NAD^+ (slight) [7]; 2,6-Pyridine dicarboxylic acid [7, 9]; 3,4-Pyridine dicarboxylic acid (weak [9]) [7, 9]; 4-Hydroxyquinolinic acid [7]; Methyl-3-carboxypyridine-2-carboxylate [7]; 3-Cyanopyridine-2-carboxylate [7]; 3-Amidopyridine-2-carboxylate [7]; Hg^{2+} [16, 21]; Co^{2+} (can partially replace Mg^{2+} in activation, inhibitory at 1 mM [16]) [16, 21]; Sr^{2+} [16]; Ba^{2+} [16]; Ca^{2+} [16]; Cr^{3+} [16]; Ag^+ [16]; Mn^{2+} (can partially replace Mg^{2+}, inhibitory at 1 mM [16], activation [2, 12, 20]) [16]; Cd^{2+} (can partially replace Mg^{2+} in activation, inhibitory at 1 mM [16]) [16, 21]; 5-Phospho-alpha-D-ribose 1-diphosphate (inhibition of crystalline enzyme at alkaline pH and physiological pH (pH 7.4), but not at acidic pH, competitive to quinolinate, in presence of 30% glycerol inhibition even at acidic pH) [8]; $H_2PO_4^-$ [20]; Maleic acid [12, 19]; Fumaric acid [12, 19]; L-Malic acid [12, 19]; Citric acid [12, 19]; ADP (weak) [9, 18]; GDP [9]; GTP [9, 21]; CDP (weak) [9]; CTP [9, 21]; UDP (weak) [9]; UTP (weak [21]) [9, 21]; IDP [9, 21]; ITP [9]; dTTP [21]; Inorganic diphosphate [18, 21]; Deamino-NAD (weak) [9]; Pyridine 2,5-dicarboxylic acid (weak) [9]; N-Ethylmaleimide [9]; Dithiobis(2-nitrobenzoic acid) [9]; Monoiodoacetic acid [6, 9]; Glutamic acid [12, 19]; 2-Oxoglutaric acid [13, 19]; Aspartic acid [12, 19]; Succinic acid [12, 19]; Oxaloacetic acid [19]; Lactic acid [19]; Acetic acid [19]; Formic acid [19]; Fe^{2+} [4, 16]; Ni^{2+} [4, 16, 21]; Zn^{2+} (can partially replace Mg^{2+} in activation, inhibitory at 1 mM [16]) [4, 16, 21]; Tris (not [7]) [2]; Al^{3+} [4, 16]; Fe^{2+} [4, 6, 16]; ATP (and other nucleotides [4], inhibition removed by rising Mg^{2+} concentration [9, 21]) [4, 6, 9, 21]; Cu^{2+} [6, 16, 21]; Phthalic acid [6, 9, 17]; PCMB [6, 9]; Glycerol (30%, 87% inhibition at pH 8.0, 85% inhibition at pH 7.0, 73% inhibition at pH 6.1, 53% inhibition at pH 5.0 [12], activation at pH 6.1 and 6.5, inhibition at pH above 7.0, inhibition is strongest at pH 9.0 and decreases above [21]) [12, 21]; More (activity is inhibited by modifying reagents for residues of lysine, histidine and arginine, these results and the effect of preincubation with the substrates on chemical modifications suggest that residues of lysine, histidine and cysteine may be important to the binding site of quinolinate [6], inhibition by carboxylic acids is reversed by increasing Mg^{2+} ion concentration [12, 19], activity inhibited by increase in ionic strength [20], not: nicotinic acid [9], nicotinamide [9], amino acids [9, 21], carboxylic acids [21]) [6, 9, 12, 19–21]

Cofactor(s)/prosthetic group(s)/activating agents
Glycerol (activation at pH 6.1 and 6.5, inhibition at pH above 7.0, inhibition is strongest at pH 9.0 and decreases above) [21]

Metal compounds/salts
Mg^{2+} (absolute requirement for divalent metal ion [2, 3, 6, 7, 12, 16, 19, 20], maximal activity with Mg^{2+} [6, 16, 19, 20], K_m: 0.23 mM [3], optimal concentration: 0.1–0.2 mM [2], 1 mM [6, 16, 18], 1.5 mM [20], optimal concentrations are 4, 6 and 10 mM for 20, 40 and 100 mM sodium acetate/acetic acid buffer, pH 5.5, respectively [19]) [1–3, 6, 7, 12, 16, 18–20]; Mn^{2+} (one-third as effective as Mg^{2+} at equimolar concentration [7], four-fifth as effective as Mg^{2+} at 1 mM, Mn^{2+} is most effective at 2 mM and the activity is 80% of that with 1.5 mM Mg^{2+} [20], Mn^{2+} is one-half as effective as Mg^{2+} at concentrations below 0.2 mM [2], Mn^{2+} can fully replace Mg^{2+} [12], can partially replace Mg^{2+}, inhibitory at 1 mM [16]) [2, 7, 12, 16, 20]; Co^{2+} (can partially replace Mg^{2+}, inhibitory at 1 mM [16], inhibits [21]) [16]; Cd^{2+} (can partially replace Mg^{2+}, inhibitory at 1 mM [16], inhibits [21]) [16]; Zn^{2+} (can partially replace Mg^{2+}, inhibitory at 1 mM [16], inhibits [4, 21]) [16]

Turnover number (min^{-1})

Specific activity (U/mg)
0.013 (human liver) [18]; 0.018 (human brain) [18]; 0.9 [13]; 0.052 [6]; 0.89 [1]; 0.75 [7]; 0.05 [14, 15]; 0.0123 [17]; More (assay [5]) [5, 8]

K_m-value (mM)
0.001 (quinolinate) [2]; 0.0051 (quinolinate) [5]; 0.011 (quinolinate) [3]; 0.021 (5-phospho-alpha-D-ribose 1-diphosphate, root) [5]; 0.023 (5-phospho-alpha-D-ribose 1-diphosphate) [3]; 0.05 (5-phospho-alpha-D-ribose 1-diphosphate) [2]; 0.12 (quinolinate) [6, 16]; 0.18 (5-phospho-alpha-D-ribose 1-diphosphate) [6, 16]; More [7, 11, 13, 17–21]

pH-optimum
4.4 [19]; 6.1 [6, 16]; 6.2 [2]; 6.5 [3]; 6.5–7.0 [17]; 6.5–7.7 [7]; 9.5 [20]

pH-range
5.2–7.8 (50% of activity maximum at pH 5.2 and 7.8) [16]; 5.3–7.3 (5.3: about 45% of activity maximum, 7.3: about 55% of activity maximum) [2]

Temperature optimum (°C)
30 (assay at) [5, 13]; 37 (assay at) [6, 14, 15, 18–20]; 50 [4]; 60 [20]

Temperature range (°C)

3 ENZYME STRUCTURE

Molecular weight

68000–72000 (Ricinus communis, gel filtration, sucrose density gradient centrifugation) [7]

72000 (Salmonella typhimurium, gel filtration) [13]

160000 (Lentinus edodes, gel filtration [3], rat, gel filtration, sucrose density gradient centrifugation [17]) [3, 17]

167000–170000 (human, gel filtration, sucrose density gradient centrifugation) [18]

172000–173000 (pig liver, gel filtration, sedimentation velocity method) [14, 15]

178000 (Pseudomonas sp., sedimentation and diffusion data, amino acid analysis) [1]

202000–210000 (pig, gel filtration, sedimentation equilibrium) [6, 12]

210000–220000 (Alcaligenes eutrophus, gel filtration, ultracentrifugation) [20]

220000 (pig kidney, gel filtration) [19]

More (amino acid sequence of Salmonella typhimurium enzyme) [13]

Subunits

Dimer (2×35000, Ricinus communis, SDS-PAGE [7], 2×35000, Salmonella typhimurium, SDS-PAGE [13]) [7, 13]

Pentamer (5×34000–35000, pig liver, SDS-PAGE, sedimentation equilibrium analysis in guanidine-HCl [14, 15], 5×32000, rat, SDS-PAGE [17], 5×34000, human, SDS-PAGE [18]) [14, 15, 17, 18]

Hexamer (6×33500–34200, pig, SDS disc gel electrophoresis, sedimentation equilibrium in guanidine-HCl [6, 12], 6×35000, pig kidney, SDS-PAGE [19]) [6, 12, 19]

Octamer (8×27500, Alcaligenes eutrophus, SDS-PAGE) [20]

? ($x \times 53000$, unidentified bacterium ATCC 23269, SDS-PAGE) [11]

Glycoprotein/Lipoprotein

Glycoprotein (sugar content 1% [12]) [12, 19]

---

4 ISOLATION/PREPARATION

Source organism

Human [18]; Lentinus edodes (Shiitake mushroom) [3, 4]; Pseudomonas sp. [1]; Bovine [2]; Nicotiana tabacum (var. Samsun) [5]; Unidentified bacterium ATCC 23269 [11]; Pig [6, 8–10, 12, 14–16, 19]; Ricinus communis [7]; Salmonella typhimurium [13]; Rat [17]; Alcaligenes eutrophus (nov. subsp. quinolinicus) [20, 21]

Source tissue
   Cells [1]; Kidney [8, 19]; Brain [17, 18]; Liver [2, 6, 8–10, 12, 14–18]; Root
   [5]; Stem [5]; Leaf lamina (weak activity) [5]; Suspension cultured cells
   (weak) [5]; Seedlings (etiolated, endosperm) [7]

Localisation in source

Purification
   Pseudomonas sp. [1]; Pig [6, 8, 14–16, 19]; Ricinus communis [7]; Rat [17];
   Bovine [2]; Lentinus edodes [3, 4]; Unidentified bacterium ATCC 23269
   [11]; Salmonella typhimurium [13]; Human [18]; Alcaligenes eutrophus [20]

Crystallization
   [1, 6, 10, 14, 16, 19]

Cloned
   –

Renatured
   –

---

5 STABILITY

pH
   3.0–3.5 (abrupt denaturation) [19]; 4.5–9.5 (37°C, 30 min, stable) [19];
   5.5–10.0 (stable) [6]

Temperature (°C)
   80 (3 min, 50% loss of activity) [17]; More (quinolinic acid, 0.8 mM, 50%
   protection against heat inactivation) [7]

Oxidation

Organic solvent

General stability information
   Stabilizing agent not required during purification [6]; Quinolinic acid,
   0.8 mM, 50% protection against heat inactivation [7]; Complete inactivation
   when frozen at –20°C in 0.05 M potassium phosphate buffer, pH 7.0, 0.01 M
   2-mercaptoethanol [7]; Not stable to freezing [6]

Storage
   0°C, 0.05 M potassium phosphate buffer, pH 7.0, containing ammonium sulfate at 30% saturation, stable for 2 years [6]; –90°C, 0.05 M potassium
   phosphate buffer, pH 7.0, 50% w/v sucrose, 0.01 M dithiothreitol, stable for
   several months [7]; 0–4°C, crystalline conditions, stable for at least 2 years
   [19]

6 CROSSREFERENCES TO STRUCTURE DATABANKS

PIR/MIPS code

Brookhaven code

---

7 LITERATURE REFERENCES

[1]   Packman, P.M., Jacoby, W.B.: J. Biol. Chem.,240, PC4107-PC4108 (1965)
[2]   Gholson, R.K., Ueda, I., Ogasawara, N., Henderson, L.M.: J. Biol. Chem.,239, 1208–1214 (1964)
[3]   Taguchi, H., Iwai, K.: J. Nutr. Sci. Vitaminol.,20,269–281 (1974)
[4]   Taguchi, H., Iwai, K.: J. Nutr. Sci. Vitaminol.,20,283–291 (1974)
[5]   Wagner, R., Wagner, K.G.: Phytochemistry,23,1881–1883 (1984)
[6]   Iwai, K., Taguchi, H.: Methods Enzymol.,66,96–101 (1980) (Review)
[7]   Mann, D.F., Byerrum, R.U.: J. Biol. Chem.,249,6817–6823 (1974)
[8]   Shibata, K., Iwai, K.: Agric. Biol. Chem.,44,2785–2791 (1980)
[9]   Taguchi, H., Iwai, K.: Agric. Biol. Chem.,40,385–389 (1976)
[10] Musick, W.D.L.: J. Mol. Biol.,117,1101–1107 (1977)
[11] Kalikin, L., Calvo, K. C.: Biochem. Biophys. Res. Commun.,152,559–564 (1988)
[12] Iwai, K., Shibata, K., Taguchi, H.: Agric. Biol. Chem.,43,351–355 (1979)
[13] Hughes, K.T., Dessen, A., Gray, J.P., Grubmeyer, C.: J. Bacteriol.,175,479–486 (1993)
[14] Iwai, K., Taguchi, H.: Biochem. Biophys. Res. Commun.,56,884–891 (1974)
[15] Taguchi, H., Iwai, K.: Agric. Biol. Chem.,39,1493–1500 (1975)
[16] Taguchi, H., Iwai, K.: Agric. Biol. Chem.,39,1599–1604 (1975)
[17] Okuno, E., Schwarcz, R.: Biochim. Biophys. Acta,841,112–119 (1985)
[18] Okuno, E., White, R.J., Schwarcz, R.: J. Biochem.,103,1054–1059 (1988)
[19] Shibata, K., Iwai, K.: Biochim. Biophys. Acta,611,280–288 (1980)
[20] Iwai, K., Shibata, K., Taguchi, H., Itakura, T.: Agric. Biol. Chem.,43,345–350 (1979)
[21] Shibata, K., Iwai, K.: Agric. Biol. Chem.,44,119–123 (1980)

1 NOMENCLATURE

EC number
2.4.2.20

Systematic name
2,4-Dioxotetrahydropyrimidine-nucleotide:pyrophosphate phospho-alpha-D-ribosyltransferase

Recommended name
Dioxotetrahydropyrimidine phosphoribosyltransferase

Synonyms
Dioxotetrahydropyrimidine-ribonucleotide pyrophosphorylase
Phosphoribosyltransferase, dioxotetrahydropyrimidine
Dioxotetrahydropyrimidine phosphoribosyl transferase
Dioxotetrahydropyrimidine ribonucleotide pyrophosphorylase

CAS Reg. No.
37277-75-1

---

2 REACTION AND SPECIFICITY

Catalysed reaction
A 2,4-dioxotetrahydropyrimidine D-ribonucleotide + diphosphate →
→ a 2,4-dioxotetrahydropyrimidine + 5-phospho-alpha-D-ribose 1-diphosphate

Reaction type
Pentosyl group transfer

Natural substrates

Substrate spectrum
1 Xanthine + diphosphate [1]
2 Uric acid + diphosphate [1]
3 More (enzyme also synthesizes a number of pyrimidine ribonucleotides from the corresponding base (e.g. uracil, orotic acid, thymine, 6-azathymine, 6-azauracil, 5-fluorouracil or 5-iodouracil) and pyrophosphorylribose phosphate) [1]

Product spectrum
1 3-Ribosylxanthine 5'-phosphate + ? [1]
2 3-Ribosyluric acid 5-phosphate + ? [1]
3 ?

Enzyme Handbook © Springer-Verlag Berlin Heidelberg 1996
Duplication, reproduction and storage in data banks are only
allowed with the prior permission of the publishers

Inhibitor(s)
UMP (xanthine and uracil pyrophosphorylase) [1]

Cofactor(s)/prosthetic group(s)/activating agents

Metal compounds/salts
Mg^{2+} (required at 2 mM or higher for maximal activity) [1]

Turnover number (min^{-1})

Specific activity (U/mg)
0.497 [1]

K_m-value (mM)

pH-optimum
More [1]

pH-range

Temperature optimum (°C)
37 (assay at) [1]

Temperature range (°C)

3 ENZYME STRUCTURE

Molecular weight

Subunits

Glycoprotein/Lipoprotein
–

4 ISOLATION/PREPARATION

Source organism
Bovine [1]

Source tissue
Erythrocytes [1]

Localization in source

Purification
Bovine (partial) [1]

Crystallization
–

Cloned

–

Renatured

–

---

5 STABILITY

pH

Temperature (°C)
50 (pH 6.0, 40 min, 12–22% loss of activity) [1]

Oxidation

Organic solvent

General stability information

Storage

---

6 CROSSREFERENCES TO STRUCTURE DATABANKS

PIR/MIPS code

Brookhaven code

---

7 LITERATURE REFERENCES

[1] Hatfield, D., Wyngaarden, J.B.: J. Biol. Chem.,239,2580–2586 (1964)

1 NOMENCLATURE

EC number
2.4.2.21

Systematic name
Nicotinate-nucleotide:dimethylbenzimidazole phospho-D-ribosyltransferase

Recommended name
Nicotinate-nucleotide-dimethylbenzimidazole phosphoribosyltransferase

Synonyms
Phosphoribosyltransferase, nicotinate mononucleotide-dimethylbenzimidazole
Nicotinate mononucleotide-dimethylbenzimidazole phosphoribosyltransferase
Nicotinate ribonucleotide:benzimidazole (adenine) phosphoribosyltransferase [2]

CAS Reg. No.
37277-76-2

---

2 REACTION AND SPECIFICITY

Catalysed reaction
beta-Nicotinate D-ribonucleotide + dimethylbenzimidazole →
→ nicotinate + N^1-(5-phospho-alpha-D-ribosyl)-5,6-dimethylbenzimidazole
(single displacement mechanism [2])

Reaction type
Pentosyl group transfer

Natural substrates

Substrate spectrum
1 beta-Nicotinate D-ribonucleotide + 5,6-dimethylbenzimidazole (ir [1]) [1–4]
2 beta-Nicotinate D-ribonucleotide + 5,6-dichlorobenzimidazole [2]
3 beta-Nicotinate D-ribonucleotide + 5(6)-nitrobenzimidazole [2]
4 Nicotinic acid mononucleotide + benzimidazole (ir [1], highly specific for nicotinic acid mononucleotide [2]) [1, 2, 4]
5 NMN + benzimidazole (ir) [1]
6 Nicotinamide nucleoside + benzimidazole (ir) [1]
7 Nicotinic acid nucleoside + benzimidazole (ir) [1]
8 Nicotinic acid mononucleotide + adenine [2, 3]
9 More (not: NAD, ribose 1-phosphate, ribose 5-phosphate, 2-methyl-benzimidazole) [1]

Enzyme Handbook © Springer-Verlag Berlin Heidelberg 1996
Duplication, reproduction and storage in data banks are only
allowed with the prior permission of the publishers

Product spectrum

1 Nicotinate + N^1-(5-phospho-alpha-D-ribosyl)-5,6-dimethylbenzimidazole [1]
2 ?
3 ?
4 1-alpha-D-Ribofuranosylbenzimidazole 5'-phosphate + benzimidazole (alpha-ribazole 5'-phosphate [4]) [1, 2]
5 ?
6 ?
7 ?
8 7-alpha-D-Ribofuranosyladenine 5'-phosphate + ? [2, 3]
9 ?

Inhibitor(s)

Cofactor(s)/prosthetic group(s)/activating agents

Metal compounds/salts

Turnover number (min^{-1})

Specific activity (U/mg)

11.6 [1]; 2.3 [2]; More [4]

K_m-value (mM)

0.000016 (dimethylbenzimidazole) [4]; 0.01 (adenine) [2]; 0.083 (nicotinic acid mononucleotide) [4]; 0.3 (nicotinic acid mononucleotide (+ adenine)) [2]; 0.5 (benzimidazole) [2]; 0.7 (nicotinic acid mononucleotide (+ benzimidazole)) [2]; 2 (benzimidazole) [1]

pH-optimum

8.5–9.4 [1]; 8.6 (assay at) [2]; 8.6–9.2 (nicotinic acid mononucleotide + adenine or benzimidazole) [2]

pH-range

7.3–9.7 (7.3: about 60% of activity maximum, 9.7: about 90% of activity maximum) [1]

Temperature optimum (°C)

30 (assay at) [1, 4]; 37 (assay at) [2]

Temperature range (°C)

3 ENZYME STRUCTURE

Molecular weight

71000 (Pseudomonas denitrificans, gel filtration) [4]

Subunits
    Dimer (2 × 35000, Pseudomonas denitrificans, SDS-PAGE) [4]

Glycoprotein/Lipoprotein
    –

---

4 ISOLATION/PREPARATION

Source organism
    Propionibacterium shermanii [1]; Clostridium sticklandii [2, 3]; Pseudomonas denitrificans [4]

Source tissue
    Cell [1, 2, 4]

Localization in source

Purification
    Propionibacterium shermanii (partial) [1]; Clostridium sticklandii [2]; Pseudomonas denitrificans [4]

Crystallization
    –

Cloned
    –

Renatured
    –

---

5 STABILITY

pH

Temperature (°C)

Oxidation

Organic solvent

General stability information

Storage
    –15°C, 0.01 M potassium phosphate buffer, pH 6.8, 0.005 M DTT, 10% ethylene glycol, several months [2]

---

6 CROSSREFERENCES TO STRUCTURE DATABANKS

PIR/MIPS code

Brookhaven code

7 LITERATURE REFERENCES

[1] Friedmann, H.C.: J. Biol. Chem.,240,413–418 (1965)
[2] Fyfe, J.A., Friedmann, H.C.: J. Biol. Chem.,244,1659–1666 (1969)
[3] Friedmann, H.C., Fyfe, J.A.: J. Biol. Chem.,244,1667–1671 (1969)
[4] Cameron, B., Blanche, F., Rouyez, M.-C., Bisch, D., Famechon, A., Couder, M., Cauchois, L., Thibaut, D., Debussche, L., Crouzet, J.: J. Bacteriol.,173,6066–6073 (1991)

1 NOMENCLATURE

EC number
2.4.2.22

Systematic name
5-Phospho-alpha-D-ribose-1-diphosphate:xanthine phospho-D-ribosyltrans-ferase

Recommended name
Xanthine phosphoribosyltransferase

Synonyms
Xan phosphoribosyltransferase [3]
Phosphoribosyltransferase, xanthine
Xanthosine 5'-phosphate pyrophosphorylase
Xanthylate pyrophosphorylase
Xanthylic pyrophosphorylase
XMP pyrophosphorylase

CAS Reg. No.
9023-10-3

2 REACTION AND SPECIFICITY

Catalysed reaction
5-Phospho-alpha-D-ribose 1-diphosphate + xanthine →
→ (9-D-ribosylxanthine)-5'-phosphate + diphosphate

Reaction type
Peritosyl group transfer

Natural substrates

Substrate spectrum
1   5-Phospho-alpha-D-ribose 1-diphosphate + xanthine (specific for xanthine [1–3], 2 different enzymes or enzyme forms catalyze phosphoribosyl transfer to hypoxanthine, guanine and xanthine, with one of these hypo-xanthine is the most efficient phosphoribosyl acceptor whereas with the other guanine is the most efficient acceptor, no distinct xanthine phos-phoribosyltransferase detected [1]) [1–4]
2   5-Phospho-alpha-D-ribose 1-diphosphate + hypoxanthine (weak activity) [4]
3   5-Phospho-alpha-D-ribose 1-diphosphate + adenine (weak activity) [4]
4   4,6-Dihydroxypyrazolo[3,4-d]pyrimidine + 5-phospho-alpha-D-ribose 1-diphosphate (weak activity) [4]

Product spectrum
1 (9-D-Ribosylxanthine)-5'-phosphate + diphosphate
2 ?
3 ?
4 ?

Inhibitor(s)
ATP (weak) [4]; CDP (weak) [4]; UDP (weak) [4]; TDP (weak) [4]; CTP
(weak) [4]; UTP (weak) [4]; TTP (weak) [4]; GMP [4]; GDP [4]; GTP [4]; XMP
[4]; XDP [4]; XTP [4]; Xanthine [4]; 8-Mercaptoxanthine [4]; 2-Thioxanthine
[4]; 6-Thioxanthine [4]; Guanine [1]; More (not: dinitrofluorobenzene, p-chloromercuribenzoate, N-ethylmaleimide, diisopropylfluorophosphate, 2-mer-
captoethanol) [4]

Cofactor(s)/prosthetic group(s)/activating agents

Metal compounds/salts
Mn^{2+} (no activity in dialyzed preparations unless activated by divalent cation, Mn^{2+} optimally efficient) [2]; Mg^{2+} (no activity in dialyzed preparations
unless activated by divalent cation, low activation by Mg^{2+}) [2]; Zn^{2+} (no activity in dialyzed preparations unless activated by divalent cation, low activation by Zn^{2+}) [2]; Co^{2+} (no activity in dialyzed preparations unless activated by divalent cation, Co^{2+} is equal to Mn^{2+} for activation of enzyme from
Leishmania mexicana and L. braziliensis, much less efficient for enzyme
from L. donovani and L. tarentolae) [2]

Turnover number (min^{-1})

Specific activity (U/mg)
3.65 [4]

K_m-value (mM)
0.001 (xanthine, Lactobacillus casei) [1]; 0.020 (xanthine) [4]; 0.053
(5-phospho-alpha-D-ribose 1-diphosphate) [4]

pH-optimum
7.4–8.8 [4]

pH-range
6.9–9.1 (about 80% of activity maximum at pH 6.9 and 9.1) [4]

Temperature optimum (°C)
37 (assay at) [4]; 38 (assay at) [1]

Temperature range (°C)

3 ENZYME STRUCTURE

Molecular weight
42000 (Streptococcus faecalis, gel filtration) [4]
54000–62000 (Leishmania donovani, gel filtration) [3]

Subunits

Glycoprotein/Lipoprotein
–

---

4 ISOLATION/PREPARATION

Source organism
Streptococcus faecalis (ATCC 8043) [4]; Lactobacillus casei [1]; E. coli (2 different enzymes or enzyme forms catalyze phosphoribosyl transfer to hypoxanthine, guanine and xanthine, with one of these hypoxanthine is the most efficient phosphoribosyl acceptor whereas with the other guanine is the most efficient acceptor, no distinct xanthine phosphoribosyltransferase detected) [1]; Leishmania mexicana (enzymes catalyzing the transribosylation from 5-phospho-alpha-D-ribose 1-diphosphate to guanine, hypoxanthine and xanthine are inseparable) [2]; Leishmania donovani [2, 3]; Leishmania braziliensis [2]; Leishmania tarentolae [2]; Streptococcus faecalis [4]

Source tissue
Promastigotes [2]

Localization in source

Purification
Streptococcus faecalis (ATCC 8043) [4]

Crystallization
–

Cloned
–

Renatured
–

---

5 STABILITY

pH
5.6–10 (25°C, 30 min) [4]

Temperature (°C)
60 (5 min, complete inactivation, crude enzyme extract, Lactobacillus casei) [1]

Oxidation

Organic solvent

General stability information
Increasing the potassium phosphate buffer concentration from 1 to 50 mM increases stability [4]; XMP stabilizes [4]; 5-Phospho-alpha-D-ribose 1-diphosphate stabilizes [4]; 6-Oxo-substituted purine nucleotides are good stabilizers [4]; Stable to freezing, when protected by nonspecific cell protein (40000 g supernatant as opposed to 100000 g supernatant) [2]

Storage
−70°C, 6 months stable in intact cells [4]; −70°C, ammonium sulfate fraction, stored in 10 mM potassium phosphate, pH 7.4 [4]; −70°C, Sepharose-phenylguanine column eluate in either 10 mM potassium phosphate, pH 7.4, 1 mM XMP + 0.1 mM Na_4−5-phospho-alpha-D-ribose 1-diphosphate or 9 mM Tris-HCl, pH 7.4, 0.2 mM $MgSO_4$ and 0.1 mM Na_4−5-phospho-alpha-D-ribose 1-diphosphate [4]

---

6 CROSSREFERENCES TO STRUCTURE DATABANKS

PIR/MIPS code
PIR1:RTECGX (Escherichia coli)

Brookhaven code

---

7 LITERATURE REFERENCES

[1] Krenitsky, T.A., Neil, S.M., Miller, R.L.: J. Biol. Chem.,245,2605–2611 (1970)
[2] Kidder, G.W., Nolan, L.L.: J. Protozool.,29,405–409 (1982)
[3] Tuttle, J.V., Krenitsky, T.A.: J. Biol. Chem.,255,909–916 (1980)
[4] Miller, R.L., Adamczyk, D.L., Fyfe, J.A., Elion, G.B.: Arch. Biochem. Biophys., 165,349–358 (1974)

1 NOMENCLATURE

EC number
2.4.2.23

Systematic name
Deoxyuridine:orthophosphate deoxy-D-ribosyltransferase

Recommended name
Deoxyuridine phosphorylase

Synonyms
Phosphorylase, deoxyuridine

CAS Reg. No.
37277-77-3

---

2 REACTION AND SPECIFICITY

Catalysed reaction
Deoxyuridine + phosphate →
→ uracil + deoxy-D-ribose 1-phosphate

Reaction type
Pentosyl group transfer

Natural substrates

Substrate spectrum
1 Deoxyuridine + phosphate (r) [1]

Product spectrum
1 Uracil + deoxy-D-ribose 1-phosphate [1]

Inhibitor(s)
5-Azauracil (reaction is inhibited in the direction of deoxyuridine synthesis more than in the direction of their cleavage at the same concentration of 5-azauracil) [1]

Cofactor(s)/prosthetic group(s)/activating agents

Metal compounds/salts

Turnover number (min^{-1})

Specific activity (U/mg)

K$_m$-value (mM)

pH-optimum

pH-range

Temperature optimum (°C)

Temperature range (°C)

3 ENZYME STRUCTURE

Molecular weight

Subunits

Glycoprotein/Lipoprotein

–

4 ISOLATION/PREPARATION

Source organism
  Mouse [1]

Source tissue
  Liver [1]

Localization in source

Purification

Crystallization

–

Cloned

–

Renatured

–

5 STABILITY

pH

Temperature (°C)

Oxidation

Organic solvent

General stability information

Storage

---

6 CROSSREFERENCES TO STRUCTURE DATABANKS

PIR/MIPS code

Brookhaven code

---

7 LITERATURE REFERENCES

[1] Cihak, A., Sorm, F.: Biochim. Biophys. Acta,80,672–674 (1964)

1 NOMENCLATURE

EC number
2.4.2.24

Systematic name
UDP-D-xylose:1,4-beta-D-xylan 4-beta-D-xylosyltransferase

Recommended name
1,4-beta-D-Xylan synthase

Synonyms
EC 2.4.1.72 (formerly)
Xylosyltransferase, uridine diphosphoxylose-1,4-beta-xylan
1,4-beta-Xylan synthase
Synthase, 1,4-beta-xylan
Xylan synthase
Xylan synthetase

CAS Reg. No.
37277-73-9

---

2 REACTION AND SPECIFICITY

Catalysed reaction
UDP-D-xylose + (1,4-beta-D-xylan)$_n$ →
→ UDP + (1,4-beta-D-xylan)$_{n+1}$

Reaction type
Pentosyl group transfer

Natural substrates
UDP-D-xylose + (1,4-beta-D-xylan)$_n$ (formation of hemicellulose of cell wall [6], catalyzes synthesis of the xylan main chain during the biogenesis of the plant cell wall [7]) [6, 7]

Substrate spectrum
1 UDP-D-xylose + (1,4-beta-D-xylan)$_n$ [1–3, 5–9]

Product spectrum
1 UDP + (1,4-beta-D-xylan)$_{n+1}$ (with high substrate concentration or prolonged incubation, three polysaccharides other than neutral xylan are synthesized, one of these is apparently a glycolipid, and the other two are apparently glucuronoxylans [3]) [1, 3, 7]

Inhibitor(s)
   UDP-D-glucuronic acid [9]; UMP [1, 8]; UDP [1, 8]; UTP [1, 8]; AMP [1];
   GMP [1, 8]; CMP [8]; CDPcholine [8]; ADP [8]

Cofactor(s)/prosthetic group(s)/activating agents
   More (no requirement for detergent) [7]

Metal compounds/salts
   Mg^{2+} (stimulates, Mg^{2+} more effective than Mn^{2+} [7], no requirement for divalent metal ion [5]) [7]; Mn^{2+} (stimulates, Mg^{2+} more effective than Mn^{2+}
   [7], no requirement for divalent metal ion [5]) [7]

Turnover number (min^{-1})

Specific activity (U/mg)

K_m-value (mM)
   0.4 (UDP-D-xylose) [7]

pH-optimum
   6.5–7.2 (crude extract) [1]

pH-range

Temperature optimum (°C)
   25 (assay at) [1]

Temperature range (°C)

---

3 ENZYME STRUCTURE

Molecular weight

Subunits

Glycoprotein/Lipoprotein
   –

---

4 ISOLATION/PREPARATION

Source organism
   Pisum sativum [9]; Acer pseudoplatanus (sycamore) [7, 8]; Zea mays (corn)
   [1]; Zinnia elegans [2]; Avena sativa [3]; Phaseolus vulgaris (L., cv. Canadian wonder [6]) [4, 6]; Mung bean [5]; Populus robusta (poplar) [8]

Source tissue
   Seedlings [3]; Xylem cells (differentiated [7, 8], differentiating [8]) [7, 8];
   Cobs (immature) [1]; Mesophyll cells (differentiating into tracheary ele-
   ments) [2]; Suspension cells [6]; Shoots [5]; Cambial cells [8]; Epicotyl [9]

Localization in source
  Membrane (bound [6, 8], associated [7]) [6–8]

Purification

Crystallization
  –

Cloned
  –

Renatured
  –

---

5 STABILITY

pH

Temperature (°C)

Oxidation

Organic solvent

General stability information
  Sucrose, 0.4 M, stabilizes enzyme in crude extracts [1]

Storage
  0°C, inactivation of particulate system after 2 days [7]; –15°C, 50% inactivation of particulate system after 4–5 days [7]; Storage in liquid N_2 keeps activity of particulate system constant over a period of 30 days [7]

---

6 CROSSREFERENCES TO STRUCTURE DATABANKS

PIR/MIPS code

Brookhaven code

---

7 LITERATURE REFERENCES

[1] Bailey, R.W., Hassid, W.Z.: Proc. Natl. Acad. Sci. USA,56,1586–1593 (1966)
[2] Suzuki, K., Ingold, E., Suguyama, M., Komamine, A.: Plant Cell Physiol.,32,303–306
    (1991)
[3] Ben-Arie, R., Ordin, L., Kindinger, J.I.: Plant Cell Physiol.,14,427–434 (1973)
[4] Bolwell, G.P., Northcote, D.H.: Biochem. J.,210,497–507 (1983)
[5] Odzuck, W., Kauss, H.: Phytochemistry,11,2489–2494 (1972)
[6] Bolwell, G.P., Northcote, D.H.: Biochem. J.,210,509–515 (1983)
[7] Dalessandro, G., Northcote, D.H.: Planta,151,53–60 (1981)
[8] Dalessandro, G., Northcote, D.H.: Planta,151,61–67 (1981)
[9] Baydoun, E. A.-H., Waldron, K.W., Brett, C.T.: Biochem. J.,257,853–858 (1989)

---

1 NOMENCLATURE

EC number
2.4.2.25

Systematic name
UDPapiose:7-O-beta-D-glucosyl-5,7,4'-trihydroxyflavone
beta-D-apiofuranosyltransferase

Recommended name
Flavone apiosyltransferase

Synonyms
Apiosyltransferase, uridine diphosphoapiose-flavone
UDPapiose:7-O-(beta-D-glucosyl)-flavone apiosyltransferase [1]

CAS Reg. No.
37332-49-3

2 REACTION AND SPECIFICITY

Catalysed reaction
UDPapiose + 7-O-beta-D-glucosyl-5,7,4'-trihydroxyflavone →
→ UDP + 7-O-(beta-D-apiofuranosyl-1,2-beta-D-gluco-
syl)-5,7,4'-trihydroxyflavone

Reaction type
Pentosyl group transfer

Natural substrates
UDPapiose + 7-O-(beta-D-glucosyl)-apigenin (biosynthesis of apiin) [1]

Substrate spectrum
1 UDPapiose + 7-O-(beta-D-glucosyl)-apigenin (specific for UDPapiose as
  glycosyl donor) [1]
2 UDPapiose + biochanin A (i.e. 5,7-dihydroxy-4'-methoxyisoflavone-7-
  glucoside) [1]
3 More (overview: 7-O-beta-glucosides of a number of flavonoids and of
  4-substituted phenols can act as acceptors, not: flavonol-3-glucoside,
  flavonol-7-glucoside, apigenin-8-C-glucoside, aglycones of flavonoids,
  glucose) [1]

Product spectrum
1 UDP + 7-O-beta-D-apiofuranosyl-1,2-beta-D-glucosyl-apigenin (i.e. apiin) [1]
2 ?
3 ?

Inhibitor(s)
Apiin [1]; Apigenin-7-glucoside [1]; Chrysoeriol-7-glucoside [1]; Mn^{2+} [1]; Co^{2+} [1]; Ca^{2+} [1]; Mg^{2+} [1]; p-Chloromercuribenzoate [1]; Iodoacetamide (reversed by cysteine) [1]; UDP [1]; More (no effect: NH_4^+, K^+) [1]

Cofactor(s)/prosthetic group(s)/activating agents
Dithioerythritol (stimulates, optimum concentration: 6 mM) [1]

Metal compounds/salts
More (higher apiin yield in Tris-HCl than in phosphate buffer) [1]

Turnover number (min^{-1})

Specific activity (U/mg)
More [1]

K_m-value (mM)
0.066 (apigenin-7-glucoside) [1]

pH-optimum
7.0 [1]

pH-range

Temperature optimum (°C)
30 (assay at) [1]

Temperature range (°C)

3 ENZYME STRUCTURE

Molecular weight
50000 (Petroselinum hortense, gel filtration) [1]

Subunits

Glycoprotein/Lipoprotein
–

4 ISOLATION/PREPARATION

Source organism
Petroselinum hortense [1]

Source tissue
Cell suspension culture [1]

Localization in source
Soluble [1]

Purification
  Petroselinum hortense [1]

Crystallization
  –

Cloned
  –

Renatured
  –

---

5 STABILITY

pH
  6.0–8.0 (highest stability) [1]

Temperature (°C)

Oxidation

Organic solvent

General stability information
  Solution of enzyme after hydroxylapatite chromatography in sodium
  phosphate buffer, pH 7.0, rapidly loses activity upon freezing [1]; Albumin,
  0.1 mg, stabilizes [1]

Storage
  –20°C, Tris-HCl, dithiothreitol, protein concentration 0.5–1 mg/ml [1]

---

6 CROSSREFERENCES TO STRUCTURE DATABANKS

PIR/MIPS code

Brookhaven code

---

7 LITERATURE REFERENCES

[1] Ortmann, R., Sutter, A., Grisebach, H.: Biochim. Biophys. Acta,289,293–302 (1972)

---

1 NOMENCLATURE

EC number
2.4.2.26

Systematic name
UDP-D-xylose:protein beta-D-xylosyltransferase

Recommended name
Protein xylosyltransferase

Synonyms
UDP-D-xylose:core protein xylosyltransferase [2]
UDP-D-xylose:core protein beta-D-xylosyltransferase [2, 5]
UDP-D-xylose:proteoglycan core protein beta-D-xylosyltransferase [3]
Xylosyltransferase, uridine diphosphoxylose-core protein beta-
UDP-xylose-core protein beta-D-xylosyltransferase
Uridine diphosphoxylose-protein xylosyltransferase

CAS Reg. No.
55576-38-0

2 REACTION AND SPECIFICITY

Catalysed reaction
Transfers a beta-D-xylosyl residue from UDP-D-xylose to the serine hydroxyl group of an acceptor protein substrate (most probable mechanism: ordered single displacement with UDPxylose as the leading substrate and the xylosylated peptide as the first product released [5])

Reaction type
Pentosyl group transfer

Natural substrates
More (involved in the biosynthesis of the linkage region of proteochondroitin sulfate) [1]

Substrate spectrum

1 UDP-D-xylose + acceptor protein substrate (transfers a beta-D-xylosyl residue from UDP-D-xylose to the serine hydroxyl group of an acceptor protein substrate: exogenous protein acceptor obtained by Smith degradation of bovine chondroitin sulfate-protein complex [1], Smith-degraded or HF-treated cartilage proteoglycan [3, 4], silk fibroin from Bombyx mori (consists in large part of the repeating hexapeptide: Ser-Gly-Ala-Gly-Ala-Gly) [3], silk sequence hexapeptide Ser-Gly-Ala-Gly-Ala-Gly [3], peptide fragments derived by digestion of deglycosylated core protein with Staphylococcus aureus V8 protease [5], tryptic and chymotryptic fragments from fibroin [3]) [1, 3–5]

Product spectrum

1 UDP + acceptor protein substrate with xyloserine

Inhibitor(s)

Cofactor(s)/prosthetic group(s)/activating agents

Metal compounds/salts

Turnover number (min^{-1})

Specific activity (U/mg)

More (enzyme assay [4]) [1, 4]

K$_m$-value (mM)

0.019 (UDPxylose (+ Smith-degraded proteoglycan), value on serine basis) [4]; 0.18 (UDPxylose) [5]; More [5]

pH-optimum

6.5 (assay at) [4]

pH-range

Temperature optimum (°C)

37 (assay at) [1, 4]

Temperature range (°C)

3 ENZYME STRUCTURE

Molecular weight

95000–100000 (chicken) [2]
110000–120000 (chicken, gel filtration) [1]

Subunits
Tetramer (2 × 23000 + 2 × 27000, chicken) [2]

Glycoprotein/Lipoprotein
–

---

4 ISOLATION/PREPARATION

Source organism
Chicken [1–4]; Rat [5]

Source tissue
Brain [1]; Epiphyseal cartilage (highest activity [1], embryonic [2–4]) [1–4];
Chondrosarcoma cells [5]

Localization in source
Soluble [1]

Purification
Chicken (partial [1]) [1, 2]

Crystallization
–

Cloned
–

Renatured
–

---

5 STABILITY

pH

Temperature (°C)

Oxidation

Organic solvent

General stability information

Storage

---

6 CROSSREFERENCES TO STRUCTURE DATABANKS

PIR/MIPS code

Brookhaven code

7 LITERATURE REFERENCES

[1] Stoolmiller, A.C., Horwitz, A.L., Dorfman, A.: J. Biol. Chem.,247,3525–3532 (1972)
[2] Schwartz, N.B., Roden, L.: Carbohydr. Res.,37,167–180 (1974)
[3] Campbell, P., Jacobsson, I., Benzing-Purdie, L., Roden, L., Fessler, J.H.: Anal. Biochem.,137,505–516 (1984)
[4] Sandy, J.D.: Biochem. J.,177,569–574 (1979)
[5] Kearns, A.E., Campbell, S.C., Westley, J., Schwartz, N.B.: Biochemistry,30, 7477–7483 (1991)

1 NOMENCLATURE

EC number
   2.4.2.27

Systematic name
   dTDP-L-dihydrostreptose:streptidine-6-phosphate dihydrostreptosyltransferase

Recommended name
   dTDPdihydrostreptose-streptidine-6-phosphate dihydrostreptosyltransferase

Synonyms
   Dihydrostreptosyltransferase, thymidine diphosphodihydrostreptose-strepti-
   dine 6-phosphate

CAS Reg. No.
   73699-20-4

2 REACTION AND SPECIFICITY

Catalysed reaction
   dTDP-L-dihydrostreptose + streptidine 6-phosphate →
   → dTDP + O-1,4-alpha-L-dihydrostreptosyl-streptidine 6-phosphate

Reaction type
   Pentosyl group transfer

Natural substrates
   dTDP-L-dihydrostreptose + streptidine 6-phosphate (biosynthesis of strepto-
   mycin) [1]

Substrate spectrum
   1 dTDP-L-dihydrostreptose + streptidine 6-phosphate [1]
   2 More (not: streptidine, 2-deoxystreptamine, 4-deoxystreptamine) [1]

Product spectrum
   1 dTDP + O-alpha-L-dihydrostreptosyl-1,4-streptidine 6-phosphate [1]
   2 ?

Inhibitor(s)
   dTDP [1]; dTTP [1]

Cofactor(s)/prosthetic group(s)/activating agents

Metal compounds/salts
Mn^{2+} (Mn^{2+} or Mg^{2+} required, maximal activity with Mn^{2+} at 3 mM) [1]; Mg^{2+} (Mn^{2+} or Mg^{2+} required, maximal activity with Mg^{2+} at 10 mM) [1]; Co^{2+} (as effective as Mg^{2+}) [1]

Turnover number (min^{-1})

Specific activity (U/mg)
More [1]

K_m-value (mM)

pH-optimum
8.2 (glycine/NaOH buffer) [1]; More (no distinct pH-optimum observed in Tris-HCl buffer between pH 7.3 and 9) [1]

pH-range

Temperature optimum (°C)
0 (assay at) [1]

Temperature range (°C)

3 ENZYME STRUCTURE

Molecular weight
63000 (Streptomyces griseus, gel filtration) [1]

Subunits
Dimer (2×35000, Steptomyces griseus, SDS-PAGE in presence of 2-mercaptoethanol) [1]

Glycoprotein/Lipoprotein
−

4 ISOLATION/PREPARATION

Source organism
Streptomyces griseus [1]

Source tissue

Localization in source

Purification
Streptomyces griseus [1]

Crystallization

–

Cloned

–

Renatured

–

---

5 STABILITY

pH

Temperature (°C)

Oxidation

Organic solvent

General stability information
Addition of streptidine is important for stabilization of enzyme activity during purification, purified transferase is also stable in absence of streptidine [1]

Storage
4°C, 50 mM Tris-HCl, pH 7.8 at 0°C, containing 5 mM mercaptoethanol, 1 mM EDTA, 2.5 mM streptidine, 10% v/v glycerol, stable for at least 4 months [1]

---

6 CROSSREFERENCES TO STRUCTURE DATABANKS

PIR/MIPS code

Brookhaven code

---

7 LITERATURE REFERENCES

[1] Kniep, B., Grisebach, H.: Eur. J. Biochem.,105,139–144 (1980)

1 NOMENCLATURE

EC number
2.4.2.28

Systematic name
5'-Methylthioadenosine:orthophosphate methylthio-D-ribosyltransferase

Recommended name
5'-Methylthioadenosine phosphorylase

Synonyms
5'-Methylthioadenosine nucleosidase [2]
5'-Deoxy-5'-methylthioadenosine phosphorylase [4, 10, 18]
MTA phosphorylase [4]
MeSAdo phosphorylase [7]
MeSAdo/Ado phosphorylase [8]
Phosphorylase, methylthioadenosine
Methylthioadenosine nucleoside phosphorylase

CAS Reg. No.
61970-06-7

---

2 REACTION AND SPECIFICITY

Catalysed reaction
5'-Methylthioadenosine + phosphate →
→ adenine + 5-methylthio-D-ribose 1-phosphate (sequential mechanism [5], phosphorolytic mechanism [1, 2], equilibrium-ordered reaction, 5'-methylthioadenosine is the first substrate to bind and 5-methylthioribose 1-phosphate is the first product to be released [3], ordered bisubstrate biproduct reaction with methylthioadenosine the first substrate to add and adenine the last product to leave the enzyme [11])

Reaction type
Pentosyl group transfer

Natural substrates
5'-Methylthioadenosine + phosphate (physiological significance of 5'-methylthioadenosine cleavage is probably related to removal of the thioether which in turn exerts a significant inhibition on methyl transfer reactions [2, 14] and an antiproliferative effect on stimulated human lymphocytes and virally transformed mouse fibroblasts [14], involved in salvage of adenine and methionine from 5'-methylthioadenosine [2, 7, 14]) [2, 7, 14]

Substrate spectrum

1 5'-Methylthioadenosine + phosphate (r [13, 17]) [1–3, 8, 9, 11–15, 17]
2 5'-Deoxyadenosine + phosphate [9, 12]
3 5'-n-Butylthioadenosine + phosphate (93.3% of the reaction with 5'-methylthioadenosine) [2, 14]
4 5'-Isobutylthioadenosine + phosphate (97% of the reaction with 5'-methylthioadenosine) [2, 14]
5 5'-Isobutylthioinosine + phosphate (8.1% of the reaction with 5'-methylthioadenosine) [2, 14]
6 5'-Methylthioinosine + phosphate (8.8% [2], 8.9% [15] of the reaction with 5'-methylthioadenosine) [2, 14, 15]
7 5'-Deoxy-5'-methylthioadenosine + phosphate [4, 10, 12, 18]
8 2'-Deoxyadenosine + phosphate (substrate for trypanosomal but not for mammalian enzyme [8]) [8, 9]
9 3'-Deoxyadenosine + phosphate (substrate for trypanosomal but not for mammalian enzyme) [8]
10 2',3'-Dideoxyadenosine + phosphate (substrate for trypanosomal but not for mammalian enzyme) [8]
11 6-Methylpurine 2'-deoxyribonucleoside + phosphate (substrate for trypanosomal but not for mammalian enzyme) [8]
12 5'-Methylselenoadenosine + phosphate (95% of the activity with 5'-methylthioadenosine) [15]
13 5'-Ethylthioadenosine + phosphate (60% of the activity with 5'-methylthioadenosine [15]) [15, 16]
14 5'-n-Propylthioadenosine + phosphate [16]
15 More (specificity in both directions of nucleoside cleavage and nucleoside synthesis [6], substrate specificity [9], not: S-adenosylhomocysteine [4]) [4, 6, 9]

Product spectrum

1 Adenine + 5-methylthio-D-ribose 1-phosphate [2, 3, 12, 13, 15, 16]
2 Adenine + 5'-deoxyribose 1-phosphate [12]
3 ?
4 ?
5 ?
6 ?
7 ?
8 ?
9 ?
10 ?
11 ?
12 ?
13 ?
14 ?
15 ?

Inhibitor(s)

SH-group blocking compounds [4, 17]; 2-Halogenated derivatives of
9-(2-hydroxyethoxymethyl)adenine (chloro-, bromo- and iodo-congeners) [7];
Halogenated derivatives of 9-(1,3-dihydroxy-2-propoxymethyl)adenine
(strong) [7]; 9-[(1-Hydroxy-3-iodo-2-propoxy)methyl]adenine [7]; 9-(Phospho-
noalkyl)adenines [7]; Adenine arabinoside [9]; Inosine [9]; 2'-Deoxyinosine
[9]; Guanosine [9]; 2'-Deoxyguanosine [9]; 5'-Deoxy-5'-methylthiotubercidin
[10]; Adenine [3, 10]; 5'-Deoxy-5'-chloroformycin [12]; 5'-Methylthiotuberci-
din [15, 16]; 5'-Dimethylthioadenosine [15]; O_2 (partial inactivation) [15];
Chloromercuribenzoic acid (partially reversed by dithiothreitol) [15]; Iodoacetamide (reversal by dithiothreitol) [15]; 5'-Ethylthioadenosine [16];
5'-n-Propylthioadenosine [16]; L-Methionine (weak) [3]; Ribose 1-phosphate
[3]; 2-Deoxyribose 1-phosphate [3]; Fructose 1-phosphate [3]; Guanine [3];
More (not: alkylating, mercaptide-forming or oxidizing thiol reagents [2],
EDTA [3], putrescine [3], cadaverine [3]) [2, 3]

Cofactor(s)/prosthetic group(s)/activating agents

Putrescine (increases activity [4], no effect [3]) [4]; Spermidine (increases
activity) [4]; Spermine (increases activity) [4]; Thiols (activate, maximal activation: 1 mM dithiothreitol) [15]; SH-group reducing agents (requirement)
[4, 17]

Metal compounds/salts

More (no requirement for metal ion (Mg^{2+}, Mn^{2+} or Ca^{2+} [1])) [1, 3]

Turnover number (min^{-1})

Specific activity (U/mg)

10.2 [17]; 10.3 [18]; 0.0283 [1]; 0.023 [16]; 2.4 [2]; 2.46 [14]; 0.131 [15]

K_m-value (mM)

0.00001 (5'-deoxy-5'-methylthioadenosine) [4]; 0.005 (5'-methylthioadeno-
sine) [11, 17]; 0.008 (5-methylthioribose 1-phosphate) [17]; 0.023 (adenine)
[17]; 0.026 (5'-deoxy-5'-methylthioadenosine) [10]; 0.095 (5'-methyl-
thioadenosine) [2, 14]; 0.3 (5'-methylthioadenosine) [1]; 0.320 (phosphate)
[17]; 3.5 (phosphate) [12]; 6.1 (phosphate) [2, 14]; 7.5 (PO_4^{3-}) [10]; 13.5
(phosphate) [11]; More [13, 15, 16]

pH-optimum

7.2 [2, 4]; 7.2–7.6 [16]; 7.4–8.0 [3]; 7.5 [1]

pH-range

6.0–9.0 (50% of activity maximum at pH 6.0 and 9.0) [2]; 6.2–8.6 (50% of
activity maximum at pH 6.2 and 8.6) [17]; 6.5–8.5 (considerable decrease of
activity below pH 6.5 and above pH 8.5) [1]

Temperature optimum (°C)
  37 (assay at) [1, 3, 16–18]; 47 [10]; 70 (assay at [2, 14]) [2, 4, 14]; 95 [2, 14]; 120 [5]

Temperature range (°C)
  86–95 (86°C: 50% of activity maximum, 95°C: activity maximum, no activity at 40°C) [2, 14]

---

3 ENZYME STRUCTURE

Molecular weight
  55000 (human, gel filtration) [6]
  64000 (rat, gel filtration) [4]
  89000 (Leishmania donovani, gel filtration) [9]
  90000 (rat, gel filtration) [16]
  98000 (bovine, sedimentation equilibrium [18], human, gel filtration [17]) [17, 18]

Subunits
  Dimer (2 × ?, rat, SDS-PAGE [4], 2 × 30300, human, SDS-PAGE [6]) [4, 6]
  Trimer (3 × 32500, human, SDS-PAGE with addition of 8 M urea and 1% 2-mercaptoethanol [17], 3 × 32000, bovine, SDS-PAGE [18]) [17, 18]

Glycoprotein/Lipoprotein
  –

---

4 ISOLATION/PREPARATION

Source organism
  Trypanosoma brucei brucei [8]; Leishmania donovani [9]; Rat [1, 3, 4, 16]; Caldariella acidophila [2, 14]; Sulfolobus solfataricus [5]; Human [6, 7, 10, 15, 17]; Mouse [7, 12]; Drosophila melanogaster [11, 13]; Bovine [18]

Source tissue
  Liver [3, 6, 16, 18]; Lung [3]; Prostate [1, 15]; Spleen [3, 4]; Testis [3]; Promastigotes [9]; Peripheral lymphocytes [10]; Heart (low activity) [3]; Kidney [3]; Sarcoma 180 cells (murine [7]) [7, 12]; HL-60 human promyelocytic leukemia cells [7]; Full-term placenta [17]

Localization in source
  Soluble [1, 3]

Purification
  Sulfolobus solfataricus [5]; Rat (partial [1, 3], lung [3]) [1, 3, 4, 16]; Caldariella acidophila [2, 14]; Human [6, 10, 15]; Leishmania donovani [9]; Drosophila melanogaster [13]; Bovine [18]

Crystallization
–

Cloned
–

Renatured
–

---

5 STABILITY

pH

Temperature (°C)
40–55 (15 min, stable) [17]; 65 (rapid inactivation) [15]; 70 (15 min, complete loss of activity) [17]; 100 (1 h, stable [2], stable for at least 2 h [5])
[2, 5]; 130 (half-life: 15 min) [5]; 132 (10 min, melting temperature) [5]; 140
(half-life: 5 min) [5]; More (high degree of thermal stability [4], resistance to
thermal inactivation is increased remarkably by addition of 5'-methyl-
thioadenosine or phosphate [17]) [4, 17]

Oxidation

Organic solvent

General stability information
Stable to freeze-thawing [15]; Rapid inactivation in absence of reducing
agents: 50% inactivation within 24 h at both 4°C and –20°C [17]; Resistance
to thermal inactivation is increased remarkably by addition of 5'-methyl-
thioadenosine or phosphate [17]

Storage
–20°C, stable for at least 3 weeks [1]; –70°C, stable for at least 4 months
[3]; –20°C, stable for several months [14]; –20°C, 5 mM DTT, 50 mM potassium phosphate, pH 7.4, less than 10% loss of activity after 1 month [17]

---

6 CROSSREFERENCES TO STRUCTURE DATABANKS

PIR/MIPS code

Brookhaven code

---

7 LITERATURE REFERENCES

[1] Pegg, A.E., Williams-Ashman, H.G.: Biochem. J.,115,241–247 (1969)
[2] Carteni'-Farina, M., Oliva, A., Romeo, G., Napolitano, G., De Rosa, M.,
    Gambacorta, A., Zappia, V.: Eur. J. Biochem.,101,317–324 (1979)

---

[3] Garbers, D.L.: Biochim. Biophys. Acta,523,82–93 (1978)
[4] Lee, S.H., Cho, Y.D.: Korean Biochem. J.,26,433–439 (1993)
[5] Cacciapuoti, G., Porcelli, M., Bertoldo, C., Zappia, V.: Life Chem. Rep.,10,75–81 (1992)
[6] Toorchen, D., Miller, R.L.: Biochem. Pharmacol.,41,2023–2030 (1991)
[7] Savarese, T.M., Harrington, S., Nakamura, C., Chen, Z.H., Kumar, P., Mikkilineni, A., Abushanab, E., Chu, S.H., Parks, R.E.: Biochem. Pharmacol.,40,2465–2471 (1990)
[8] Ghoda, L.Y., Savarese, T.M., Northup, C.H., Parks, R.E., Garofalo, J., Katz, L., Ellenbogen, B.B., Bacchi, C.J.: Mol. Biochem. Parasitol.,27,109–118 (1988)
[9] Koszalka, G.W., Krenitsky, T.A.: Adv. Exp. Med. Biol.,195B, Purine Pyrimidine Metab. Man5, Pt. B,559–563 (1986)
[10] White, M.W., Vandenbark, A.A., Barney, C.L., Ferro, A.J.: Biochem. Pharmacol.,31, 503–507 (1982)
[11] Shugart, L., Mahoney, L., Chastain, B.: Int. J. Biochem.,13,559–564 (1981)
[12] Savarese, T.M., Crabtree, G.W., Parks, R.E.: Biochem. Pharmacol.,30,189–199 (1981)
[13] Shugart, L., Tancer, M., Moore, J.: Int. J. Biochem.,10,901–904 (1979)
[14] Zappia, V., Carteni-Farina, M., Romeo, G., De Rosa, M., Gambacorta, A.: Methods Enzymol.,94,355–361 (1983) (Review)
[15] Zappia, V., Oliva, A., Cacciapuoti, G., Galletti, P., Mignucci, G., Carteni-Farina, M.: Biochem. J.,175,1043–1050 (1978)
[16] Ferro, A.J., Wrobel, N.C., Nicolette, J.A.: Biochim. Biophys. Acta,570,65–73 (1979)
[17] Della Ragione, F., Carteni-Farina, M., Gragnaniello, V., Schettino, M.I., Zappia, V.: J. Biol. Chem.,261,12324–12329 (1986)
[18] Della Ragione, F., Oliva, A., Gragnaniello, V., Russo, G.L., Palumbo, R., Zappia, V.: J. Biol. Chem.,265,6241–6246 (1990)

1 NOMENCLATURE

EC number
2.4.2.29

Systematic name
tRNA-guanine:queuine tRNA-D-ribosyltransferase

Recommended name
Queuine tRNA-ribosyltransferase

Synonyms
tRNA-guanine transglycosylase
TGT [3]
Guanine insertion enzyme
tRNA transglycosylase [1]
tRNA guanine transglycosylase [3]
Queuine tRNA ribosyltransferase [6]
Q-insertase [6, 8]
Ribosyltransferase, queuine transfer ribonucleate
Transfer ribonucleate glycosyltransferase
tRNA guanine transglycosidase
Guanine, queuine-tRNA transglycosylase [7]

CAS Reg. No.
72162-89-1

---

2 REACTION AND SPECIFICITY

Catalysed reaction
tRNAguanine + queuine →
→ tRNAqueuine + guanine

Reaction type
Pentosyl group transfer

Natural substrates
tRNAguanine + queuine (post-transcriptional modification of tRNA [1, 2],
queuine or some queuine derivatives may be the actual substrate in the
biosynthesis of queuosine in tRNA of mammalian cells [9]) [1, 2, 9]

Enzyme Handbook © Springer-Verlag Berlin Heidelberg 1996
Duplication, reproduction and storage in data banks are only
allowed with the prior permission of the publishers

Substrate spectrum

1 tRNAguanine + queuine (catalyzes the incorporation of bases of queuosine precursors into E. coli undermodified tRNA^{Asn} and tRNA^{Tyr}, these bases are inserted in the first position of the anticodon of tRNA^{Asn} and tRNA^{Tyr}, replacing guanine originally located at that position [1], exchange of queuine as well as its precursors and guanine for guanine originally located in the first position of the anticodon of undermodified tRNA^{Tyr}, tRNA^{His}, tRNA^{Asn} and tRNA^{Asp} from an E. coli mutant or rat ascites hepatoma cells [9], catalyzes the exchange of queuine precursors such as 7-(aminomethyl)-7-deazaguanine and 7-cyano-7-deazaguanine, but not queuine itself [1], tRNA(Asn, Asp, His or Tyr) [3], the enzyme also catalyzes the exchange of precursors of queuine and of guanine itself for guanine located in the first position of certain tRNA anticodons, specificity [5, 8]: virtually all purines and pteridines that are inhibitors or substrates have an amino nitrogen at the 2 position, in addition the 9 position and the oxygen at the 6 position may be important for recognition by the enzyme [5], saturation of the double bond in the cyclopentenediol moiety of queuine reduces substrate activity and queuine analogs lacking the cyclopentenediol moiety are relatively poor substrates (e.g. 7-deazaguanine or 7-aminomethyl-7-deazaguanine) [5], exchange of guanine for queuine in position 34 of tRNAs having a GUN anticodon, where N (position 36) stands for A, U, C or G [6], influence of nucleosides in position 36, 37 and 38 of an anticodon loop on conversion of G-34 to Q-34 [6]) [1–9]

2 Guanosine-tRNA^{Tyr} + 7-(aminomethyl)-7-deazaguanine [1]

Product spectrum

1 tRNAqueuine + guanine [1]

2 ?

Inhibitor(s)

8-Bromoguanine [5]; 7-Methylguanine [5]; 6-Methylmercaptoguanine [5]; 6-Thioguanine [5]; 8-Azaguanine [5]; 3-Deazaguanine [5]; Isocytosine [5]; 2,4-Diamino-6-hydroxypyrimidine [5]; Pterin [5]; Folic acid [5]; 7-Deazaguanine [5]; Pterin 6-carboxylic acid [5]; Tetrahydrobiopterin [5]; Zn^{2+} [7]; Mn^{2+} [7]; Cd^{2+} [7]; Co^{2+} [7]; Ni^{2+} [7]; Pb^{2+} [7]; Cu^{2+} [7]; Neoplanocin A (poor) [5]; Bisimidoester [4]; Dimethylsuberimidate (tRNA protects) [4]; More (virtually all purines and pteridines that are inhibitors or substrates have an amino nitrogen at the 2 position, in addition the 9 position and the oxygen at the 6 position may be important for recognition by the enzyme) [5]

Cofactor(s)/prosthetic group(s)/activating agents

Metal compounds/salts
  Na^+ (monovalent or divalent cation required, Na^+ and Mg^{2+} are more effective activators than other cations, optimal concentration of NaCl: 114 mM,
  order of effectiveness in stimulation: $Na^+ > Li^+ > K^+ > Cs^+ > Rb^+$) [7]; Mg^{2+}
  (monovalent or divalent cation required, Na^+ and Mg^{2+} are more effective
  activators than other cations [7], no stimulation [9], loss of activity following
  dialysis can be restored in part by the addition of Mg^{2+} and to a lesser degree with K^+, activity is highest when K^+ and Mg^{2+} are both present, optimal
  Mg^{2+} concentration: 5 mM [7]) [7]; Sr^{2+} (stimulation) [7]; Ba^{2+} (stimulation)
  [7]; Li^+ (order of effectiveness in stimulation: $Na^+ > Li^+ > K^+ > Cs^+ > Rb^+$)
  [7]; K^+ (order of effectiveness in stimulation: $Na^+ > Li^+ > K^+ > Cs^+ > Rb^+$,
  loss of activity following dialysis can be restored in part by the addition
  of Mg^{2+} and to a lesser degree with K^+, activity is highest when K^+ and
  Mg^{2+} are both present) [7]; Cs^+ (order of effectiveness in stimulation:
  $Na^+ > Li^+ > K^+ > Cs^+ > Rb^+$) [7]; Rb^+ (order of effectiveness in stimulation:
  $Na^+ > Li^+ > K^+ > Cs^+ > Rb^+$) [7]

Turnover number (min^{-1})

Specific activity (U/mg)
  More [2, 4, 7, 9]

K_m-value (mM)
  0.0000033 (guanine-accepting tRNA) [2]; 0.000014 (7-(aminomethyl)-7-
  deazaguanine) [1]; 0.000053 (guanine) [1]; 0.00006 (guanine) [1]; 0.00015
  (guanine) [2]; 0.00029 (queuine) [9]; 0.00083 (guanine) [9]; 0.00165 (guanine) [3]; 0.0021 (7-(aminomethyl)-7-deazaguanine) [9]; More [4]

pH-optimum
  7.3 [9]; 7.6 [7]; 8.0 [3]

pH-range
  7.0–7.5 (more than 80% of activity maximum at pH 7.0 and 7.5) [9]

Temperature optimum (°C)
  37 [3]

Temperature range (°C)

3 ENZYME STRUCTURE

Molecular weight
  80000 (rat, gel filtration) [9]
  104000 (rabbit, gel filtration) [2]
  140000 (wheat, gel filtration) [7]
  255000 (E. coli, native PAGE) [4]

Subunits
Dimer (2 × 68000, wheat, SDS-PAGE [7], 1 × 60000 + 1 × 43000, rabbit, denaturing electrophoresis of enzyme denatured at 100°C in 5 M urea and 0.01 M mercaptoethanol [2]) [2, 7]
Hexamer (6 × 42500, E. coli, SDS-PAGE, cross-linking data) [4]

Glycoprotein/Lipoprotein
–

---

4 ISOLATION/PREPARATION

Source organism
E. coli [1, 4]; Wheat [7]; Rat [9]; Rabbit [2, 5]; Salmonella typhimurium [3]; Xenopus leavis [6, 8]

Source tissue
Erythrocytes [2]; Reticulocytes [5]; Germ [7]; Oocytes [6, 8]; Liver [9]

Localization in source
Cytosol [2]

Purification
Rabbit [2]; E. coli [4]; Wheat [7]; Rat [9]

Crystallization
–

Cloned
[4]

Renatured
–

---

5 STABILITY

pH

Temperature (°C)

Oxidation

Organic solvent

General stability information
Degassed HEPES/DTT buffer system is optimal for stability [4]

Storage
–80°C, 10% glycerol, 0.5 M KCl, 50 mM Tris, pH 7.6, 10 mM beta-mercapto-ethanol, 25% loss of activity per month [2]; –20°C, 50% glycerol, stable for 10 months [9]

6 CROSSREFERENCES TO STRUCTURE DATABANKS

PIR/MIPS code
PIR2:C38530 (Escherichia coli plasmid PR20)

Brookhaven code

7 LITERATURE REFERENCES

[1] Okada, N., Noguchi, S., Kasai, H., Shindo-Okada, N., Ohgi, T., Goto, T., Nishimura, S.: J. Biol. Chem.,254,3067–3073 (1979)
[2] Howes, N.K., Farkas, W.R.: J. Biol. Chem.,253,9082–9087 (1978)
[3] Gunduz, U., Kacar, Y.: Period. Biol.,90,321–326 (1988)
[4] Garcia, G.A., Koch, K.A., Chong, S.: J. Mol. Biol.,231,489–497 (1993)
[5] Farkas, W.R., Jacobson, K.B., Katze, J.R.: Biochim. Biophys. Acta,781,64–75 (1984)
[6] Haumont, E., Droogmans, L., Grosjean, H.: Eur. J. Biochem.,168,219–225 (1987)
[7] Walden, T.L., Howes, N., Farkas, W.R.: J. Biol. Chem.,257,13218–13222 (1982)
[8] Carbon, P., Haumont, E., Fournier, M., de Henau, S., Grosjean, H.: EMBO J.,2, 1093–1097 (1983)
[9] Shindo-Okada, N., Okada, N., Ohgi, T., Goto, T., Nishimura, S.: Biochemistry,19, 395–400 (1980)

1 NOMENCLATURE

EC number
2.4.2.30

Systematic name
NAD+:poly(adenine-diphosphate-D-ribosyl)-acceptor ADP-D-ribosyltransferase

Recommended name
NAD+ ADP-ribosyltransferase

Synonyms
Poly(ADP-ribose) synthase
ADP-ribosyltransferase (polymerizing)
(Adenosine diphosphoribose)transferase, nicotinamide adenine dinucleotide-protein
Adenosine diphosphate ribosyltransferase
ADP-ribosyltransferase
C_3 exoenzyme
Exoenzyme C_3
Exoenzyme S
NAD-protein ADP-ribosyltransferase
Poly(ADP-ribose) transferase
Poly(ADP-ribosyl)transferase
Poly(ADP-ribose) polymerase [12]

CAS Reg. No.
58319-92-9

2 REACTION AND SPECIFICITY

Catalysed reaction
NAD+ + (ADP-D-ribosyl)$_n$-acceptor →
→ nicotinamide + (ADP-D-ribosyl)$_{n+1}$-acceptor (mechanism [8])

Reaction type
Pentosyl group transfer

Natural substrates
NAD+ + (ADP-D-ribosyl)$_n$-acceptor (involved in posttranslational covalent modification of cellular proteins by attaching one or more ADP-ribosyl residues, involved in DNA-synthesis and cellular proliferation, differentiation and development [3], may be involved in DNA-repair [8], enzyme induced by single-strand breaks of DNA [9]) [3, 8, 9]

Substrate spectrum

1 NAD+ + (ADP-D-ribosyl)$_n$-acceptor (r [13], the ADP-D-ribosyl-group is transferred to an acceptor carboxyl group on a histone [1, 4, 9] or the enzyme itself (automodification, leads to inactivation [9], presumably mono-ADP-ribosylation [13]) [1, 4, 9, 13], and further ADP-groups are transferred to the 2'-position of the terminal adenosine moiety, building up a polymer with an average chain length of 20–30 units [1], amino acids at position 2, 4 and 116 of C- and N-terminal positions are poly(ADP-ribosyl)ated [4], catalyzes initiation, elongation and branching of (ADP-ribose)$_n$-chain [11], no ribosylation of free arginine [10], acceptors are DNA-binding proteins, chromatin, Ca^{2+}-Mg^{2+}-endonuclease [4], core histones H2A and H2B [4], histone H1 [1, 4], DNA-ligase (ADP-ribosylation leads to inactivation), high-mobility-group (i.e. HMG) proteins [9], terminal deoxynucleotidyltransferase [9], DNA-polymerase (alpha and beta [4]), topoisomerase I and II (ADPribosylation leads to inactivation [9]) [4, 9], human neutrophil GTP-binding protein G22K (related to ras GTP-binding proteins, not G24K) [15], human platelet membrane proteins, low molecular weight eukaryotic GTP-binding proteins [2, 13, 14], e.g. from pig brain [14] or recombinant proteins rhoA or rhoB [13]) [1–16]

Product spectrum

1 Nicotinamide + (ADP-D-ribosyl)$_{n+1}$-acceptor [1, 3–14]

Inhibitor(s)

Benzamide (strong [12]) [11, 12, 16]; 3-Monosubstituted benzamides (strong, kinetics) [11]; 3-Hydroxybenzamide (most potent of benzamides [11], 0.001–0.1 mM [5]) [5, 11]; 3-Aminobenzamide (less effective than benzamide [11]) [1, 3, 11, 12]; 3-Guanidinobenzamide (0.01–0.1 mM) [5]; 3-Methoxybenzamide (0.001–1 mM, stimulating at 50 nM [5]) [3, 5, 11, 12]; 3-Acetamidobenzamide [11]; 3-Methylbenzamide [11]; 3-Fluorobenzamide [11]; 3-Chlorobenzamide [11]; 2-Methoxybenzamide [11]; 2-Aminobenzamide [12]; 4-Aminobenzamide [12]; 1,8-Naphthalimide derivatives (strong, kinetics) [11]; 6-Phenanthridinone derivatives (strong, kinetics) [11]; Isoquinoline derivatives (strong, kinetics) [11]; Carsalam derivatives (stronger inhibition in the presence of Mg^{2+}, kinetics) [11]; Phthalazine derivatives (kinetics) [11]; Chromone derivatives (kinetics) [11]; Metabolites of tryptophan (moderate inhibitors, e.g. xanthurenic acid (kinetics, stronger inhibition in the presence of Mg^{2+}), nicotinamide, kynurenic acid, 1-methylnicotinamide (weak)) [11]; 1,4-Naphthoquinone derivatives (e.g. vitamine K1) [11]; Thymidine (1 mM [5]) [3, 5, 7, 12, 16]; Theophylline (1 mM [5]) [3, 5, 12, 16]; Theobromine [3, 12]; Caffeine (1 mM [5], weak [3]) [3, 5, 12]; Juglone [11]; Lawsone [11]; Plumbagin [11]; 1,5-Dihydroxy-4-phthalazinone (0.0001–0.01 mM) [5]; 3-Aminophthalhydrazide (0.001–0.1 mM) [5]; Nicotinamide (human enzyme [1], 0.0001–1 mM [5]) [1–3, 5–7, 11, 12, 16]; 1-Methylnicotinamide [12]; 8-Methylnicotinamide [12]; 5-Methylnicotinamide [12, 16]; Pyrazinamide [12]; alpha-NAD$^+$ (0.5 mM) [5]; NAD$^+$ [2]; NADP$^+$ [10]; Hypoxanthine [12]; 3-Isobutyl-1-methylxanthine [12]; Isonicotinate hydrazide [12]; Unsaturated fatty acids [11]; Novobiocin (stronger inhibition in the presence of Mg^{2+}) [11]; Chlorthenoxazin (kinetics) [11]; 5-Nitrouracil [11]; 5-Bromodeoxyuridine [12]; GTP-gamma-S (in the presence of Mg^{2+}, stimulates in its absence [6], with protein G22K as substrate [15]) [6, 15]; 3-Aminobenzoic acid (weak, human enzyme) [1]; Nicotinic acid (weak, human enzyme [1], not [12]) [1]; Thiocyanide (0.3 M) [10]; Thiocyanate (at higher concentrations) [10]; 1,10-Phenanthroline (in the absence of Mg^{2+}, stimulation in its presence) [11]; EDTA (stronger inhibition in the presence of Mg^{2+}) [11]; DTT (up to 2.5 mM, reversible with 5–10 mM DTT [10], not [1]) [10]; GSH [10]; Cysteine (in the absence of sodium nitroprusside) [10]; PCMB (human enzyme) [1]; N-Ethylmaleimide [7, 10]; NaCl (0.3 M) [10]; KCl (MW 90000 enzyme protein) [5]; Cu^{2+} (human enzyme) [1]; Hg^{2+} (human enzyme) [1]; Zn^{2+} (human enzyme [1]) [1, 12]; Cd^{2+} (human enzyme) [1]; Co^{2+} (human enzyme) [1]; Mn^{2+} (human enzyme) [1]; SDS (with protein rhoB as substrate) [13]; Dimethylsulfoxide (weak) [11]; Anti-exoenzyme C$_3$ antibodies [13]; More (in vivo and in vitro inhibitor studies [12], no inhibitors: methanol [3], ATP [7], benzoate, 3- or 4-aminobenzoate [12]) [3, 7, 12]

Cofactor(s)/prosthetic group(s)/activating agents
DNA (requirement, induced by single-strand breaks [4, 9], only together with histone [3], DNAse I-treated [5]) [1, 3–5, 9, 11, 12, 16]; Histone (activation [1, 3, 5], histone H1 [5], substrate and activator for automodification [1], in the presence of DNA [1, 3]) [1, 3, 5]; 3-Methoxybenzamide (stimulation at 50 nM, MW 116000, not MW 90000 enzyme protein, inhibits at 0.001–1 mM) [5]; ATP (requirement [2], stimulation at 5 mM and above [7], not [6, 10, 14]) [2, 7]; GTP (activation, in the absence of divalent cations) [6, 14]; GTP-gamma-S (activation, in the absence of divalent cations) [6, 14]; GDP (activation, in the absence of divalent cations) [6, 14]; GDP-beta-S (activation, in the absence of divalent cations [14], not [6]) [14]; 1,10-Phenanthroline (activation, in the presence of Mg^{2+}, inhibitory in its absence) [11]; p-Phthalic acid (activation, in the presence of Mg^{2+}) [11]; Harmaline (activation, in the presence of Mg^{2+}) [11]; SDS (activation, 0.01%, with rhoA protein as substrate) [13]; Cytosolic factor (activation, heat-labile and trypsin-sensitive) [15]; More (no activation by DTT [1], bovine serum albumin [3], GMP, ATP-beta-S [6, 14]) [1, 3, 6, 14]

Metal compounds/salts
Mg^{2+} (requirement, 1 mM [6], 5 mM [7], activation [1], in the presence of DNA, not [10]) [1, 6, 7, 16]; Ba^{2+} (activation, can replace Mg^{2+}) [1]; Sr^{2+} (activation, can replace Mg^{2+} with 90% efficiency) [1]; Ca^{2+} (activation [1, 6, 16], can replace Mg^{2+}, with about 80% efficiency [1]) [1, 6, 16]; Mn^{2+} (activation, 1 mM [6], 5 mM [7], can replace Mg^{2+} to some extent [7]) [6, 7]; Divalent cations (requirement) [2]; Nitric oxide (and agents releasing nitric

oxide, e.g. 3-morpholinosydnonimine (i.e. SIN-1) or sodium nitroprusside, activation, in the presence of DTT, GSH or cysteine) [10]; Phosphate (slight stimulation at high concentrations) [10]; More (no activation by Cd^{2+}, Cu^{2+}, Hg^{2+}, Zn^{2+} [1], NaCl or KCl [7]) [1, 7]

Turnover number (min^{-1})

Specific activity (U/mg)
  0.0007 [5]; 0.0031 (Clostridium limosum) [13]; 0.0064 (Clostridium botulinum) [13]; 0.007 [6]; 0.0545 [3]; 0.2–0.5 [12]; 0.325 [9]; 1.02 (human) [1]

K_m-value (mM)
  More (kinetic study [7]) [7, 9]; 0.0003 (NAD+, Clostridium botulinum) [13]; 0.0004 (NAD+, Clostridium limosum) [13]; 0.002 (NAD+) [6]; 0.02 (NAD+, MW 90000 enzyme protein) [5]; 0.0267 (NAD+) [3]; 0.0617 (NAD+, human) [1]; 0.077 (NAD+, MW 116000 enzyme protein) [5]

pH-optimum
  More (pI: 10.0, human [1], pI: 10.3 (Clostridium limosum), pI: 10.6 (Clostridium botulinum) [13]) [1, 13]; 8 [3, 5, 7]; 8.5 (0.1 M glycine-NaOH buffer, human) [1]; 8.7 (0.1 M Tris-HCl buffer, human) [1]

pH-range
  6.8–8.6 (about half-maximal activity at pH 6.8 and 8.6, MW 116000 enzyme protein) [5]; 7.5–9.6 (about 80% of maximal activity at pH 7.5 and about half-maximal activity at pH 9.6, MW 90000 enzyme protein) [5]

Temperature optimum (°C)
  6–10 [5]; 17.5 [3]; 25 [7]; 30 (assay at) [15]; 37 (assay at) [6, 10–13]

Temperature range (°C)
  0–22 (about 75% of maximal activity at 0°C and about half-maximal activity at 22°C, MW 90000 enzyme protein, about 88% of maximal activity at 0°C and about half-maximal activity at 22°C, MW 116000 enzyme protein) [5]

3 ENZYME STRUCTURE

Molecular weight
  More (two catalytically active enzyme proteins: MW 90000 and MW 116000, Dictyostelium discoideum) [5]

Subunits
  ? (x × 25000, Clostridium botulinum [6, 13], Clostridium limosum [13], SDS-PAGE [6, 13], x × 75000, Helix pomatia, SDS-PAGE [3], x × 116000, human, SDS-PAGE [1, 9]) [1, 3, 6, 9, 13]

Glycoprotein/Lipoprotein

4 ISOLATION/PREPARATION

Source organism
Bovine (adult [11], calf [1, 11, 12]) [1, 11, 12]; Hamster (baby) [7]; Human [1, 4, 8–10]; Mouse [1]; Helix pomatia [3]; Dictyostelium discoideum (Ax-2 wild-type) [5]; Crypthecodinium cohnii (dinoflagellate, strain GC) [16]; Clostridium botulinum (type C strain 4/12 [6], type D [15]) [6, 13–15]; Clostridium limosum [13]; Legionella pneumophila [2]

Source tissue
Fibroblasts (from healthy and patients with Fanconi's anemia or Cockaye syndrome [9]) [4, 9]; Foot [3]; HeLa-cells [1]; Kidney (BHK-21/C13 cells) [7]; Liver [3]; Lymphoblastoids (cell line GM06315A) [8]; Placenta (human) [1]; Platelets [10]; Testis (mouse) [1]; Thymus (calf [1], C3H10T1/2 cells [12]) [1, 11, 12]; Amoebae [5]; Cell [16]; Cell suspension culture [7, 9, 12]; Cell culture filtrate [6, 13–15]

Localization in source
Nucleus (distribution [4]) [4, 7]; Nucleolus (HEP-2 cells [4]) [4, 7]; Cytosol (associated with polysomes [4, 9]) [4, 9, 10]; Extracellular [6, 13–15]

Purification
Human (affinity chromatography on 3-aminobenzamide-agarose, very close to homogeneity, amino acid analysis [1]) [1, 9]; Bovine (affinity chromatography on 3-aminobenzamide-agarose) [1]; Helix pomatia (partial) [3]; Dictyostelium discoideum (affinity chromatography on poly(A)-oligo(dT)-cellulose) [5]; Crypthecodinium cohnii (partial) [16]; Clostridium botulinum [6]; Clostridium limosum [13]; Legionella pneumophila (partial) [2]

Crystallization
–

Cloned
(human fibroblast enzyme) [4, 9]

Renatured
–

---

5 STABILITY

pH

Temperature (°C)
More (in the presence of GTP-gamma-S, GTP, GDP, GDP-gamma-S or GMP the enzyme from Clostridium botulinum is extremely stable against heat inactivation) [6]; 25 (in the absence of Mg^{2+}, $t_{1/2}$: 95 min, Mg^{2+} enhances thermal stability) [7]; 37 (in the absence of Mg^{2+}, $t_{1/2}$: 20 min, Mg^{2+} enhances thermal stability) [7]

Oxidation

Organic solvent

General stability information
Freeze-thawing, unstable to [5]; Mg^{2+} enhances thermal stability [7]

Storage

6 CROSSREFERENCES TO STRUCTURE DATABANKS

PIR/MIPS code
PIR3:S31735 (African clawed frog); PIR2:PN0495 (African clawed frog (fragment)); PIR2:JS0428 (bovine); PIR2:PN0494 (cherry salmon (fragment)); PIR2:JH0581 (chicken); PIR2:A47474 (fruit fly (Drosophila melanogaster)); PIR2:S04200 (mouse); PIR2:S21163 (rat (fragment)); PIR2:S05343 (Rhodospirillum rubrum); PIR2:A41021 (C3 precursor Clostridium botulinum); PIR2:S08407 (C3 precursor Clostridium botulinum phage (type D) (fragment)); PIR2:A29725 (nuclear human)

Brookhaven code

7 LITERATURE REFERENCES

[1] Ushiro, H., Yokoyama, Y., Shizuta, Y.: J. Biol. Chem.,262,2352–2357 (1987)
[2] Belyi, Yu.F., Tartakovskii, I.S., Vertiev, Yu.V., Prozorovskii, S.V.: Biomed. Sci. (London) ,2,169–174 (1991)
[3] Burtscher, H.J., Klocker, H., Schneider, R., Auer, B., Hirsch-Kauffmann, M., Schweiger, M.: Biochem. J.,248,859–864 (1987)
[4] De Murcia, G., Huletsky, A., Poirier, G.G.: Biochem. Cell Biol.,66,626–635 (1987) (Review)
[5] Kofler, B., Wallraff, E., Herzog, H., Schneider, R., Auer, B., Schweiger, M.: Biochem. J.,293,275–281 (1993)
[6] Aktories, K., Rösener, S., Blaschke, U., Chhatwal, G.S.: Eur. J. Biochem.,172, 445–450 (1988)
[7] Furneaux, H.M., Pearson, C.K.: Biochem. J.,187,91–103 (1980)
[8] Satoh, M.S., Lindahl, T.: Nature,356,356–358 (1992)
[9] Schweiger, M., Auer, B., Burtscher, H.J., Hirsch-Kauffmann, M., Klocker, H., Schneider, R.: Eur. J. Biochem.,165,235–242 (1987) (Review)
[10] Brüne, B., Lapetina, E.G.: Arch. Biochem. Biophys.,279,286–290 (1990)
[11] Banasik, M., Komura, H., Shimoyama, M., Ueda, K.: J. Biol. Chem.,267,1569–1575 (1992)
[12] Rankin, P.W., Jacobson, E.L., Benjamin, R.C., Moss, J., Jacobson, M.K.: J. Biol. Chem.,264,4312–4317 (1989)
[13] Just, I., Mohr, C., Schallehn, G., Menard, L., Didsbury, J.R., Vandekerckhove, J., van Damme, J., Aktories, K.: J. Biol. Chem.,267,10274–10280 (1992)
[14] Braun, U., Habermann, B., Just, I., Aktories, K., Vandekerckhove, J.: FEBS Lett., 243,70–76 (1989)
[15] Bokoch, G.M., Parkos, C.A., Mumby, S.M.: J. Biol. Chem.,163,16744–16749 (1988)
[16] Werner, E., Sohst, S., Gropp, F., Simon, D., Wagner, H., Kröger, H.: Eur. J. Biochem., 139,81–86 (1984)

1 NOMENCLATURE

EC number
2.4.2.31

Systematic name
NAD⁺:L-arginine ADP-D-ribosyltransferase

Recommended name
NAD(P)⁺-arginine ADP-ribosyltransferase

Synonyms
ADP-ribosyltransferase
Mono(ADPribosyl)transferase
(Adenosine diphosphoribose)transferase, nicotinamide adenine dinucleotide-arginine
NAD-arginine ADP-ribosyltransferase
Arginine specific ADP-ribosyltransferase
Arginine specific mono-ADP-ribosyltransferase
NAD-arginine mono-ADP-ribosyltransferase B [3]
More (cf. EC 2.4.2.36)

CAS Reg. No.
81457-93-4

---

2 REACTION AND SPECIFICITY

Catalysed reaction
NAD⁺ + L-arginine →
→ nicotinamide + N^2-(ADP-D-ribosyl)-L-arginine

Reaction type
Pentosyl group transfer

Natural substrates
Nicotinamide + arginine containing protein (involved in posttranslational, covalent modification of proteins [7], may be involved in phagocytosis, secretion and migration [5]) [5, 7]
Nicotinamide + E. coli RNA-polymerase (involved in coliphage T4-induced alteration and modification) [14]

Substrate spectrum

1 NAD⁺ + L-arginine (beta-NAD⁺ [1], NADP⁺ can act as donor [1, 4], NAD⁺ in preference to NADP⁺ [4], arginine residues in proteins can act as acceptors) [1, 4]

2 NAD⁺ + L-arginine methyl ester (better substrate than arginine [4]) [1, 3, 4, 12]

3 NAD⁺ + agmatine (better substrate than arginine [4]) [1, 4, 7–9, 11, 13]

4 NAD⁺ + poly-(L-arginine) (poor substrate of modifying ADPribosyltransferase [14]) [3, 4, 6, 11, 14, 15]

5 NAD⁺ + poly-(L-lysine) (poor substrate [3], not [14]) [3]

6 NAD⁺ + guanidine (and derivatives: guanidinobutyrate or guanidinopropionate) [1]

7 NAD⁺ + guanyltyramine [4]

8 NAD⁺ + guanylhydrazone-diethylamino-(benzylidineamino)guanidine [12]

9 NAD⁺ + guanylhydrazones (e.g. p-nitrobenzylidine aminoguanidine or methylglyoxal bis-(guanylhydrazone) dihydrochloride monohydrate) [16]

10 NAD⁺ + histones (altering ADPribosyltransferase, histone F1 [15]) [2, 3, 10, 15]

11 NAD⁺ + arginine containing protein (e.g. lysozyme (from egg-white [15], less effective than poly-(L-arginine) [6], not modifying ADPribosyltransferase [14, 15]) [3, 6, 15], (altering [15]) ADPribosyltransferase itself (i.e. auto-ADPribosylation) [12, 15], non-muscle beta/gamma-actin, skeletal muscle alpha-actin, smooth muscle gamma-actin, casein [5], soluble and particulate proteins from bovine thymus [6], less effective substrates than poly-(L-arginine): bovine plasma albumin, beta-lactoglobulin, ovalbumin, human alpha-, beta- or gamma-globulin, DNase I, trypsin inhibitor [6], poor substrates are ovalbumin or bovine serum albumin [3]) [3, 5, 6, 12, 15]

12 NAD⁺ + E. coli RNA-polymerase (i.e. EC 2.7.7.6, ir [14], holo enzyme, core enzyme or free alpha-subunit [14] or the other subunits (altering ADPribosyltransferase, at a low level) [15]. Modifying ADPribosyltransferase has high target specificity in vivo and in vitro: complete and exclusive mono-ADPribosylation of arginine residue 265 of E. coli RNA-polymerase alpha-subunit [14]. Altering transferase ADPribosylates arginine residues 191 or 195 and 265 of one alpha-subunit, possibly due to steric hindrance [15]. No substrate is T4-modified RNA-polymerase core enzyme [14]) [14, 15]

13 More (catalyzes auto-ADPribosylation [12] or hydrolysis of NAD⁺ to nicotinamide and ADPribose either in the absence of acceptor [4] or at saturating agmatine concentrations [8], catalyzes NAD⁺-dependent activation of EC 4.6.1.1, some bacterial endotoxins possess similar activities) [4, 8, 12]

Product spectrum

1 Nicotinamide + alpha-ADP-D-ribose-L-arginine [1, 4]
2 Nicotinamide + alpha-ADP-D-ribose-L-arginine methyl ester [1, 4]
3 Nicotinamide + ADPribose-agmatine [4, 7]
4 ?
5 ?
6 ?
7 Nicotinamide + ADPribose-guanyltyramine [4]
8 ?
9 Nicotinamide + ADPribose-guanylhydrazones [16]
10 ?
11 Nicotinamide + ADPribose-protein [4]
12 Nicotinamide + ADPribose-E. coli-RNA-polymerase (i.e T4-modified RNA-polymerase) [14, 15]
13 ?

Inhibitor(s)

Thymidine (strong [10]) [2, 10]; 5-Bromodeoxyuridine (strong) [10]; 3-Methoxybenzamide [10]; 3-Aminobenzamide [10]; Benzamide (ADPribosyltransferase A, A', B, C or C') [10]; $ZnCl_2$ [10]; Nicotinamide (ir [14], product inhibition [8], not "altering"-ADPribosyltransferase [15]) [2, 8, 10, 14, 15]; 8-Methylnicotinamide (weak) [10]; Theophylline [2, 10]; Chaotropic salts [3]; DTT (poly(L-Arg) as substrate, NAD+ protects to some extent) [12]; T4-modified RNA-polymerase (product inhibition) [14]; Increasing ionic strength (strong decrease from 7 to 30 mM NH_4Cl, above 30 mM moderate decrease (up to 120 mM)) [14]; ATP (inhibition or activation depends on protein substrate) [6]; AMP [14]; More (no inhibition by benzoate, 3-aminobenzoate, 2-aminobenzamide, 4-aminobenzamide, nicotinate, 1-methylnicotinamide, 5-methylnicotinamide, pyrazinamide, isonicotinate hydrazide, caffeine, 3-isobutyl-1-methylxanthine) [10]

Cofactor(s)/prosthetic group(s)/activating agents

Histone (activation, of auto-ADPribosylation [12], 0.01 mg/ml, in the absence of salts [4], converts inactive transferase to active protomeric species [4, 7], transferase A' or C: no activation [9], not [3]) [4, 7, 12]; Nucleoside triphosphates (activation, only at low NAD+-concentrations with lysozyme as substrate, with descending efficiency: ATP, ITP (GTP as good as ITP), CTP (UTP as good as CTP)) [6]; App(NH)p (activation, less efficient than ATP) [6]; Tripolyphosphates (activation, less efficient than ATP) [6]; Tetrapolyphosphates (activation, less efficient than ATP) [6]; Lysolecithin (rapid activation, selective: low molecular weight guanidino proteins preferred substrates, reversible, not as effective as chaotropic salts or histone, molecules containing long-chain fatty acids more effective than short-chain fatty acids lysolecithins) [7]; Lysophosphatidylcholine (activation, ADPribosyltransferase A, no activation by phosphatidylcholine or lysophospholipids containing choline)

[4]; Triton X-100 (activation, ADPribosyltransferase A [4]) [4, 7]; Triton X-114 or X-305 (activation) [7]; Tween 20 (activation) [7]; CHAPS (activation, ADP-ribosyltransferase A [4]) [4, 7]; More (no activation by DNA [2], ADP, AMP, cAMP or diphosphate [6], alpha-glycerophosphocholine, choline, lysophosphatidic acid, lysophosphatidylserine, phosphatidylserine, lysophosphatidylglycerol, lysophosphatidylethanolamine, phosphatidylethanolamine, lecithin phosphatidic acid [7]) [2, 6, 7]

Metal compounds/salts

NaCl (activation of arginine methylester ADPribosylation, variable effects on protein ADPribosylation depending on protein substrate) [4]; Chaotropic salts (activation, e.g. SCN$^-$, Br$^-$, Cl$^-$, F$^-$, PO$_4^{3-}$ (with descending efficiency), ADPribosyltransferase A [4]) [4]; More (no Mg^{2+}-requirement) [14]

Turnover number (min^{-1})

Specific activity (U/mg)

0.00011 (modifying ADPribosyltransferase) [15]; 0.014 (altering ADPribosyl transferase) [15]; 0.156 (transferase A) [10]; 0.215 [11]; 1.9 (alpha-type enzyme) [12]; 6 (beta-type enzyme) [12]; 7.6 (transferase C) [9]; 14 [13]

K$_m$-value (mM)

More (kinetic study) [8]; 0.0025 (alpha-actin) [5]; 0.007 (NAD$^+$, hydrolysis, turkey enzyme) [8]; 0.01 (gamma-actin) [5]; 0.014–0.0143 (NAD$^+$, lysolecithin assay [7], (+ E. coli RNA-polymerase), 20°C [14, 15]) [7, 14, 15]; 0.015 (beta/gamma-actin) [5]; 0.025 (NAD$^+$, NaCl assay) [7]; 0.05 (NAD$^+$, 15°C) [15]; 0.1 (NAD$^+$ (+ poly(L-Arg))) [11]; 0.56 (NAD$^+$) [12]; 1.0 (agmatine) [7]; 1.1 (NAD$^+$, hydrolysis, cholera toxin) [8]; 1.2 (L-arginine methyl ester) [12]; 1.3 (arginyl methyl ester, at optimal NaCl concentration) [4]; 2 (agmatine, ADPribosyl transferase C) [9]; 35 (agmatine, cholera toxin) [7]

pH-optimum

7.5 [14, 15]; 8.5 (higher activity in Tris-HCl buffer than in phosphate buffer [11]) [5, 11]

pH-range

6–9 (about half-maximal activity at pH 6 and about 80% of maximal activity at pH 9) [14]; 7.3–10 (about half-maximal activity at pH 7.3 and about 60% of maximal activity at pH 10) [5]; 8–10 (about 80% of maximal activity at pH 8 and about half-maximal activity at pH 10) [11]

Temperature optimum (°C)

15–20 [15]; 20 (modifying-ADPribosyltransferase) [14]; 30 (assay at, ADP-ribosyltransferase A [4]) [1, 2, 4, 6, 8–10, 12]; 37 (assay at, ADPribosyltransferase B [4]) [3, 4, 7, 11]

Temperature range (°C)
20–37 (maximal activity at 20°C and about 30% of maximal activity at 37°C)
[14]

---

3 ENZYME STRUCTURE

Molecular weight
24300 (turkey transferase A, gel filtration) [9]
25300 (turkey transferase A, gel filtration in the presence of NaCl, in the absence of salt: inactive, aggregated form of high molecular weight of above MW 200000, activation upon dissociation) [4]
25500 (turkey transferase A', gel filtration) [9]
26000 (turkey transferase C [9], modifying enzyme of coliphage-T4-infected E. coli [14, 15], gel filtration [9, 14, 15]) [9, 14, 15]
32000 (turkey transferase B, gel filtration) [4]
32700 (turkey transferase B, gel filtration) [9]
39000 (rabbit, SDS-PAGE/zymographic in situ assay of reactivated enzyme) [12]
61000 (rabbit, HPLC-gel filtration) [13]
61000–70000 (coliphage T4) [15]

Subunits
? (x × 38000, rabbit, SDS-PAGE [13], x × 38500–39000, rabbit alpha- and beta-form, SDS-PAGE [12]) [12, 13]
Monomer (1 × 28000, turkey transferase A, SDS-PAGE [3, 4], 1 × 28300, turkey, SDS-PAGE [2], 1 × 32000, turkey transferase B, SDS-PAGE [3, 4]) [2–4]

Glycoprotein/Lipoprotein
Glycoprotein [12]

---

4 ISOLATION/PREPARATION

Source organism
Turkey [1–4, 6–10, 16]; Rabbit [11–13, 16]; Chicken [5]; E. coli (strains B [14, 15] or K12 [15], infected with coliphage T4 [15] or coliphage T4D$^+$ mutants [14]) [14, 15]; Vibrio cholerae [8, 16]; T-even coliphages (T2, T4 or T6) [15]

Source tissue
Virions [15]; Cells (Vibrio cholerae [8, 16]) [8, 14–16]; Peripheral polymorphonuclear leukocytes (heterophils) [5]; Erythrocytes (turkey) [1–4, 6–10, 16]; Skeletal muscle (rabbit) [11–13, 16]

Localization in source
    Soluble (transferase A and B [9], choleragen (i.e. toxin from Vibrio cholerae)
    [8, 16]) [1, 3–5, 7–9, 16]; Granules [5]; Membrane-bound (transferase C [9],
    integral membrane protein [12]) [7, 9, 11–13]; Nucleus (transferase A') [9];
    Sarcoplasmic reticulum [11]; Microsomes [12]; More (turkey ADPribosyl-
    transferases: multiple forms of different intracellular localization and regula-
    tory properties) [9]

Purification
    Turkey (partial [1], transferase A and B [4]) [1–4, 9]; Rabbit (partial, solubili-
    zed with trypsin [11] or 0.3% w/v deoxycholate (two forms, alpha and beta,
    separable by concavalin A agarose chromatography, possibly modifications
    of the same enzyme form [12])) [11, 12]; Coliphage T4 (altering transferase)
    [15]; Coliphage-infected E. coli (modifying transferase) [14, 15]

Crystallization
    –

Cloned
    (rabbit, cloned to and expressed in E. coli) [13]

Renatured
    –

---

5 STABILITY

pH

Temperature (°C)
    More (histone, lysolecithin or non- or zwitter-ionic detergents enhance stabil-
    ity of ADPribosyl transferase A at 30°C [4], detergents or lysolecithin stabi-
    lize against thermal inactivation [7]) [4, 7]; 0–30 (at least 10 min stable, 30%
    glycerol) [1]; 60 ($t_{1/2}$: 10 min) [1]; 75 ($t_{1/2}$: 10 min, 30% glycerol) [1]

Oxidation

Organic solvent

General stability information
    Glycerol stabilizes during storage [1]; Ovalbumin stabilizes dilute enzyme
    solutions [1]; Propylene glycol stabilizes during purification [2]; NaCl stabi-
    lizes during purification [2]; Detergents or lysolecithin stabilize against ther-
    mal inactivation [7]; NAD+ stabilizes [12]; Rapid inactivation in 0.3% w/v so-
    dium deoxycholate, $t_{1/2}$: 1 day at 4°C, dialysis against detergent-free buffer
    prevents [12]; Sulfhydryl reagents, such as 2-mercaptoethanol stabilize [14];
    Modifying ADPribosyltransferase is unstable at low ionic strength or upon
    velocity gradient centrifugation [14, 15]; Histone, lysolecithin or non- or zwit-
    ter-ionic detergents enhances stability of ADPribosyl transferase A at 30°C
    [4]

Storage
-20°C, up to 4 weeks [12]; 0°C, 0.05 M phosphate buffer, 1 M NaCl, 50%
glycerol, 10% loss of activity within 16 days [1]; 0–4°C, ADPribosyltransferase B, 0.05 M sodium phosphate, pH 7.1, 0.1 M NaCl, several months [4];
0–4°C, ADPribosyltransferase A, 50% propylene glycol, several months [4]

6 CROSSREFERENCES TO STRUCTURE DATABANKS

PIR/MIPS code
PIR2:S52910 (chicken (fragment)); PIR2:A47239 (rabbit); PIR2:A55461 (AT1
chicken); PIR2:B55461 (AT2 chicken)

Brookhaven code

7 LITERATURE REFERENCES

[1] Moss, J., Stanley, S.J., Oppenheimer, N.J.: J. Biol. Chem.,254,8891–8894 (1979)
[2] Moss, J., Stanley, S.J., Watkins, P.A.: J. Biol. Chem.,255,5838–5840 (1980)
[3] Yost, D.A., Moss, J.: J. Biol. Chem.,258,4926–4929 (1983)
[4] Moss, J., Vaughan, M.: Methods Enzymol.,106,430–437 (1984) (Review)
[5] Terashima, M., Mishima, K., Yamada, K., Wakutani, T., Shimoyama, M.: Eur. J. Biochem.,204,305–311 (1992)
[6] Watkins, P.A., Moss, J.: Arch. Biochem. Biophys.,216,74–80 (1982)
[7] Moss, J., Osborne, J.C., Stanley, S.J.: Biochemistry,23,1353–1357 (1984)
[8] Osborne, J.C., Stanley, S.J., Moss, J.: Biochemistry,24,5235–5240 (1985)
[9] West, R.E., Moss, J.: Biochemistry,25,8057–8062 (1986)
[10] Rankin, P.W., Jacobson, E.L., Benjamin, R.C., Moss, J., Jacobson, M.K.: J. Biol. Chem.,264,4312–4317 (1989)
[11] Taniguchi, M., Tanigawa, Y., Tsuchiya, M., Mishima, K., Obara, S., Yamada, K., Shimoyama, M.: Biochem. Biophys. Res. Commun.,164,128–133 (1989)
[12] Peterson, J.E., Larew, J.S.-A., Graves, D.J.: J. Biol. Chem.,265,17062–17069 (1990)
[13] Zolkiewska, A., Nightingale M.S., Moss, J.: Proc. Natl. Acad. Sci. USA,89, 11352–11356 (1992)
[14] Skórko, R., Zillig, W., Rohrer, H., Mailhammer, R.: Eur. J. Biochem.,79,55–66 (1977)
[15] Goff, C.G.: Methods Enzymol.,106,418–429 (1984) (Review)
[16] Soman, G., Miller, J.F., Graves, D.: Methods Enzymol.,106,403–410 (1984) (Review)

1 NOMENCLATURE

EC number
2.4.2.32

Systematic name
UDP-D-xylose:dolichyl-phosphate D-xylosyltransferase

Recommended name
Dolichyl-phosphate D-xylosyltransferase

Synonyms

CAS Reg. No.

---

2 REACTION AND SPECIFICITY

Catalysed reaction
UDP-D-xylose + dolichyl phosphate →
→ GDP + dolichyl D-xylosyl phosphate

Reaction type
Pentosyl group transfer

Natural substrates
UDP-D-xylose + dolichyl phosphate (involved in glycoprotein biosynthesis)
[1]

Substrate spectrum
1 UDP-D-xylose + dolichyl phosphate [1]

Product spectrum
1 GDP + dolichyl D-xylosyl phosphate [1]

Inhibitor(s)
EDTA (strong) [1]; UDP [1]

Cofactor(s)/prosthetic group(s)/activating agents

Metal compounds/salts
Mn^{2+} (requirement) [1]

Turnover number (min^{-1})

Specific activity (U/mg)

K_m-value (mM)
0.00055 (UDP-D-xylose) [1]

Enzyme Handbook © Springer-Verlag Berlin Heidelberg 1996
Duplication, reproduction and storage in data banks are only
allowed with the prior permission of the publishers

pH-optimum
  6–7 [1]

pH-range
  5.2–7.8 (about half-maximal activity at pH 5.2 and 7.8) [1]

Temperature optimum (°C)
  37 (assay at) [1]

Temperature range (°C)

---

3 ENZYME STRUCTURE

Molecular weight

Subunits

Glycoprotein/Lipoprotein
  –

---

4 ISOLATION/PREPARATION

Source organism
  Chicken (hen) [1]

Source tissue
  Oviduct [1]

Localization in source
  Membrane-bound [1]

Purification

Crystallization
  –

Cloned
  –

Renatured
  –

---

5 STABILITY

pH

Temperature (°C)

Oxidation

2

Organic solvent

General stability information

Storage

---

6 CROSSREFERENCES TO STRUCTURE DATABANKS

PIR/MIPS code

Brookhaven code

---

7 LITERATURE REFERENCES

[1] Waechter, C.J., Lucas, J.J., Lennarz, W.J.: Biochem. Biophys. Res. Commun.,56, 343–350 (1974)

---

1 NOMENCLATURE

EC number
2.4.2.33

Systematic name
Dolichyl-D-xylosyl-phosphate:protein D-xylosyltransferase

Recommended name
Dolichyl-xylosyl-phosphate-protein xylosyltransferase

Synonyms

CAS Reg. No.

---

2 REACTION AND SPECIFICITY

Catalysed reaction
Dolichyl D-xylosyl phosphate + protein →
→ dolichyl phosphate + D-xylosylprotein

Reaction type
Pentosyl group transfer

Natural substrates
Dolichyl D-xylosyl phosphate + protein (involved in glycoprotein bio-synthesis) [1]

Substrate spectrum
1 Dolichyl D-xylosyl phosphate + protein [1]

Product spectrum
1 Dolichyl phosphate + D-xylosylprotein [1]

Inhibitor(s)
More (no inhibition by EDTA) [1]

Cofactor(s)/prosthetic group(s)/activating agents

Metal compounds/salts
Mn^{2+} (requirement) [1]

Turnover number (min^{-1})

Specific activity (U/mg)

K_m-value (mM)

pH-optimum

---

pH-range

Temperature optimum (°C)

Temperature range (°C)
    37 (assay at) [1]

---

3 ENZYME STRUCTURE

Molecular weight

Subunits

Glycoprotein/Lipoprotein
    –

---

4 ISOLATION/PREPARATION

Source organism
    Chicken (hen) [1]

Source tissue
    Oviduct [1]

Localization in source
    Membrane-bound [1]

Purification

Crystallization
    –

Cloned
    –

Renatured
    –

---

5 STABILITY

pH

Temperature (°C)

Oxidation

Organic solvent

General stability information

Storage

6 CROSSREFERENCES TO STRUCTURE DATABANKS

PIR/MIPS code

Brookhaven code

7 LITERATURE REFERENCES

[1] Waechter, C.J., Lucas, J.J., Lennarz, W.J.: Biochem. Biophys. Res. Commun.,56, 343–350 (1974)

1 NOMENCLATURE

EC number
2.4.2.34

Systematic name
UDP-L-arabinose:indol-3-ylacetyl-myo-inositol L-arabinosyltransferase

Recommended name
Indolylacetylinositol arabinosyltransferase

Synonyms
Arabinosylindolylacetylinositol synthase
Arabinosyltransferase, uridine diphosphoarabinose-indolylacetylinositol
UDP-arabinose:indol-3-ylacetyl-myo-inositol arabinosyl transferase [1]

CAS Reg. No.
84720-96-7

2 REACTION AND SPECIFICITY

Catalysed reaction
UDP-L-arabinose + indol-3-ylacetyl-myo-inositol →
→ UDP + indol-3-ylacetyl-myo-inositol L-arabinoside

Reaction type
Pentosyl group transfer

Natural substrates
UDP-L-arabinose + indol-3-ylacetyl-myo-inositol (involved in biosynthesis of
low molecular weight esters of indol-3-ylacetic acid in maize kernels) [1]

Substrate spectrum
1 UDP-L-arabinose + indol-3-ylacetyl-myo-inositol [1]

Product spectrum
1 UDP + indol-3-ylacetyl-myo-inositol L-arabinoside [1]

Inhibitor(s)

Cofactor(s)/prosthetic group(s)/activating agents

Metal compounds/salts

Turnover number (min⁻¹)

Specific activity (U/mg)

K_m-value (mM)

pH-optimum

pH-range

Temperature optimum (°C)
    37 (assay at) [1]

Temperature range (°C)

---

3 ENZYME STRUCTURE

Molecular weight

Subunits

Glycoprotein/Lipoprotein
    –

---

4 ISOLATION/PREPARATION

Source organism
    Zea mays (sweet corn) [1]

Source tissue
    Kernels (immature) [1]

Localization in source

Purification

Crystallization
    –

Cloned
    –

Renatured
    –

5 STABILITY

pH

Temperature (°C)

Oxidation

Organic solvent

General stability information

Storage

6 CROSSREFERENCES TO STRUCTURE DATABANKS

PIR/MIPS code

Brookhaven code

7 LITERATURE REFERENCES

[1] Curcuera, L.J., Bandurski, R.S.: Plant Physiol.,70,1664–1666 (1982)

---

1 NOMENCLATURE

EC number

2.4.2.35

Systematic name

UDP-D-xylose:flavonol-3-O-glycoside D-xylosyltransferase

Recommended name

Flavonol-3-O-glycoside xylosyltransferase

Synonyms

Xylosyltransferase, uridine diphosphoxylose-flavonol 3-glycoside

UDP-xylose:flavonol 3-glycoside xylosyltransferase

CAS Reg. No.

83380-90-9

---

2 REACTION AND SPECIFICITY

Catalysed reaction

UDP-D-xylose + flavonol 3-O-glycoside →

→ UDP + flavonol 3-O-D-xylosylglycoside

Reaction type

Pentosyl group transfer

Natural substrates

UDP-D-xylose + flavonol 3-O-glycoside (involved in flavonoid metabolism, pathway of flavonol 3-O-triglycoside biosynthesis) [1]

Substrate spectrum

1 UDP-D-xylose + flavonol 3-O-glycoside [1]

2 UDP-D-xylose + quercetin 3-O-galactoside (best substrate) [1]

3 UDP-D-xylose + kaempferol 3-O-glucoside [1]

4 UDP-D-xylose + quercetin 3-O-glucoside (i.e. isoquercitrin) [1]

5 UDP-D-xylose + isorhamnetin 3-O-glucoside [1]

6 UDP-D-xylose + isorhamnetin 3,7-O-glucoside (poor substrate) [1]

7 UDP-D-xylose + quercetin 3-O-arabinoglucoside (poor substrate) [1]

8 UDP-D-xylose + myricetin 3-O-rhamnoside (poor substrate) [1]

9 UDP-D-xylose + kaempferol 3-O-rhamnoside (poor substrate) [1]

10 UDP-D-xylose + quercetin 3-O-rhamnoside (poor substrate) [1]

11 UDP-D-xylose + quercetin 3-O-glucoside (poor substrate) [1]

12 UDP-D-xylose + quercetin 3-O-rhamnosylglucoside (i.e. rutin, poor sub-strate) [1]

13 UDP-D-xylose + flavonol 3-O-diglycoside [1]

14 More (aglycones are no substrates) [1]

Product spectrum
1 UDP + flavonol 3-O-D-xylosylglycoside [1]
2 UDP + quercetin 3-O-xylosylgalactoside
3 UDP + kaempferol 3-O-xylosylglucoside
4 UDP + quercetin 3-O-xylosylglucoside
5 UDP + isorhamnetin 3-O-xylosylglucoside
6 UDP + ?
7 UDP + quercetin 3-O-xylosylarabinoglucoside
8 UDP + myricetin 3-O-xylosylrhamnoside
9 UDP + kaempferol 3-O-xylosylrhamnoside
10 UDP + quercetin 3-O-xylosylrhamnoside
11 UDP + quercetin 3-O-D-xylosylglucoside [1]
12 UDP + quercetin 3-O-D-xylosylrhamnosylglucoside [1]
13 UDP + flavonol 3-O-triglycoside [1]
14 ?

Inhibitor(s)
Mn^{2+} [1]; PCMB [1]

Cofactor(s)/prosthetic group(s)/activating agents
2-Mercaptoethanol (slight stimulation) [1]; Dithioerythritol (slight stimulation) [1]; Glutathione (slight stimulation) [1]; Sucrose (slight stimulation) [1]; More (no stimulation by bovine serum albumin) [1]

Metal compounds/salts
Ca^{2+} (activation) [1]; NH_4^+ (activation) [1]; More (no activation by Mg^{2+}) [1]

Turnover number (min^{-1})

Specific activity (U/mg)

K_m-value (mM)

pH-optimum
8.5–9 [1]

pH-range

Temperature optimum (°C)

Temperature range (°C)

3 ENZYME STRUCTURE

Molecular weight
30000 (Tulipa, gel filtration) [1]

Subunits

Glycoprotein/Lipoprotein
–

2

4 ISOLATION/PREPARATION

Source organism
Tulipa (tulip, cv. Apeldoorn) [1]

Source tissue
Anthers (tapetum and pollen, at the stage of middle postmeiotic pollen ripening) [1]

Localization in source

Purification
Tulipa (partial) [1]

Crystallization
–

Cloned
–

Renatured
–

5 STABILITY

pH

Temperature (°C)

Oxidation

Organic solvent

General stability information

Storage

6 CROSSREFERENCES TO STRUCTURE DATABANKS

PIR/MIPS code

Brookhaven code

7 LITERATURE REFERENCES

[1] Kleinehollenhorst, G., Behrens, H., Pegels, G., Srunk, N., Wiermann, R.: Z. Natur-forsch.,37c,587–599 (1982)

1 NOMENCLATURE

EC number
2.4.2.36

Systematic name
NAD+:peptide-diphthamide N-(ADP-D-ribosyl)transferase

Recommended name
NAD+-diphthamide ADP-ribosyltransferase

Synonyms
ADP-ribosyltransferase
Mono(ADPribosyl)transferase
(Adenosine diphosphoribose)transferase, nicotinamide adenine dinucleo-
tide-elongation factor 2
NAD-diphthamide ADP-ribosyltransferase NAD-elongation factor 2 ADP-
ribosyltransferase
NAD:elongation factor 2-adenosine diphosphate ribose-transferase
More (cf. EC 2.4.2.31)

CAS Reg. No.
52933-21-8

---

2 REACTION AND SPECIFICITY

Catalysed reaction
NAD+ + peptide diphthamide →
→ nicotinamide + peptide N-(ADP-D-ribosyl)diphthamide

Reaction type
Pentosyl group transfer

Natural substrates

Substrate spectrum
1 NAD+ + elongation factor 2 (i.e. EF2, of pyBHK-cells [1], pig [2] or rat [3]
   liver, or wheat germ [3], r (in the presence of excess nicotinamide and
   bacterial exotoxin) [1], without acceptor substrate: NAD+-glycohydrolase
   activity [3]) [1–3]

Product spectrum
1 Nicotinamide + ADPribose-elongation factor 2 [1]

Enzyme Handbook © Springer-Verlag Berlin Heidelberg 1996
Duplication, reproduction and storage in data banks are only
allowed with the prior permission of the publishers

Inhibitor(s)
Histamine (not bacterial toxins) [1]; Cytoplasmic extract of pyBHK-cells (not fragment A) [1]; Elastase (from Pseudomonas) [2]; More (cellular ADPribosyltransferase: no inhibition by anti-fragment A-antiserum) [1]

Cofactor(s)/prosthetic group(s)/activating agents
DTT (enhanced activation together with urea) [3]; Urea (enhanced activation together with DTT) [3]; Guanidine hydrochloride (enhanced activation together with DTT, cysteine, 2-mercaptoethanol or sulfite) [3]; SDS (enhanced activation together with DTT, cysteine, 2-mercaptoethanol or sulfite) [3]; Cysteine (activation, together with SDS or guanidine hydrochloride) [3]; 2-Mercaptoethanol (activation, together with SDS or guanidine hydrochloride) [3]; More (the exotoxin is synthesized in a catalytically inactive, proenzyme form, activation correlates with unfolding of the toxin molecule) [3]

Metal compounds/salts
Sulfite (activation, together with SDS or guanidine hydrochloride) [3]

Turnover number (min^{-1})

Specific activity (U/mg)

K$_m$-value (mM)

pH-optimum
6.6 (assay at, nicotinamide + ADPribose-elongation factor 2) [1]; 8 (assay at, NAD+ + elongation factor 2) [1, 2]

pH-range

Temperature optimum (°C)
22 (assay at) [1]; 25 (assay at) [2]

Temperature range (°C)

---

3 ENZYME STRUCTURE

Molecular weight

Subunits
Monomer (1 × 72000, Pseudomonas aeruginosa exotoxin A, SDS-PAGE) [2]

Glycoprotein/Lipoprotein
–

---

4 ISOLATION/PREPARATION

Source organism
Pseudomonas aeruginosa [1–3]; Hamster (baby) [1]; Bovine [1]

Source tissue
Kidney (polyoma virus transformed cells, i.e. pyBHK-cells) [1]; Culture supernatant [2, 3]; Liver [1]

Localization in source
Extracellular (exotoxin A [2, 3], fragment A of diphtheria toxin [1]) [1–3]

Purification

Crystallization
–

Cloned
–

Renatured
–

---

5 STABILITY

pH

Temperature (°C)

Oxidation

Organic solvent

General stability information

Storage

---

6 CROSSREFERENCES TO STRUCTURE DATABANKS

PIR/MIPS code
PIR1:DOCGA (precursor corynephage beta); PIR1:DOCGPO (precursor corynephage omega)

Brookhaven code

---

7 LITERATURE REFERENCES

[1] Lee, H., Iglewski, W.J.: Proc. Natl. Acad. Sci. USA,81,2703–2707 (1984)
[2] Sanai, Y., Morihara, K., Tsuzuki, H., Homma, J.Y., Kato, I.: FEBS Lett.,120,131–134 (1980)
[3] Leppla, S.H., Martin, O.C., Muehl, L.A.: Biochem. Biophys. Res. Commun.,81, 532–538 (1978)

---

1 NOMENCLATURE

EC number
2.4.2.37

Systematic name
NAD+:[dinitrogen reductase] (ADP-D-ribosyl)transferase

Recommended name
NAD+-dinitrogen-reductase ADP-D-ribosyltransferase

Synonyms
NAD-azoferredoxin (ADPribose)transferase
(Adenosine diphosphoribose)transferase, nicotinamide adenine dinucleo-
tide-azoferredoxin
Azoferredoxin ADP-ribosyltransferase
Dinitrogenase reductase ADP-ribosyltransferase

CAS Reg. No.
117590-45-1

---

2 REACTION AND SPECIFICITY

Catalysed reaction
NAD+ + [dinitrogen reductase] →
→ nicotinamide + ADP-D-ribosyl-[dinitrogen reductase]

Reaction type
Pentosyl group transfer

Natural substrates
NAD+ + [dinitrogen reductase] (involved in regulation of nitrogenase (EC
1.18.6.1) activity [1, 2] through reversible ADP-ribosylation of one of the two
identical subunits of dinitrogenase reductase (i.e. component II or iron protein) [2]. Controls level of activity of nitrogenase together with EC 3.2.2.24)
[1, 2]

Substrate spectrum
1 NAD+ + [dinitrogen reductase] (high specificity for acceptor substrate,
the only acceptor is dinitrogen reductase (from Rhodcspirillum rubrum
[1–3], Azotobacter vinelandii [1, 3], Klebsiella pneumoniae [1, 3, 5], and
Clostridium pasteurianum [1]), donor substrates besides NAD+ are ethe-
no-NAD+, nicotinamide hypoxanthine dinucleotide and nicotinamide guanine dinucleotide [1], ADP-ribose attaching site: Arg-101 residue of dinitrogen reductase [4], no acceptor substrates are lysozyme, bovine serum

albumin, ovalbumin, histone, polyarginine, H_2O, arginine, agmatine, arginine methyl ester, arginine dansyl chloride, benzylarginine ethyl ester, tosylarginine methyl ester, Gly-Arg-Gly-Val-Ile-Thr [1], no donor substrates are 3-aminopyridine adenine nucleotide, acetopyridine adenine dinucleotide, nicotinic acid adenine dinucleotide, NADP+ or NADH [1]) [1–7]

Product spectrum
1 Nicotinamide + ADP-D-ribosyl-[dinitrogen reductase] [1]

Inhibitor(s)
NADH [1]; NADP+ [1]; Nicotinamide [1]; 3-Aminobenzamide [1]; NaCl (reversible) [1]; KCl (reversible) [1]; NaBr (strong) [1]; ATP (in the presence of ADP) [3]; Mg-ATP (in the presence of Mg-ADP) [3]

Cofactor(s)/prosthetic group(s)/activating agents
Mg-ADP (stimulation by binding to dinitrogen reductase (not from Azotobacter vinelandii), no activation by ATP, GDP, IDP, etheno-ADP, AMP-CH_2-P, 8-bromo-ADP) [3]; ADP (activation) [6]; ADP-beta-S (i.e. adenosine 5'-O-(2-thiodiphosphate), activation) [3]; 2'-Deoxy-ADP (activation) [3]

Metal compounds/salts
Mg-ADP (stimulation by binding to dinitrogen reductase (not from Azotobacter vinelandii)) [3]; Mg^{2+} (activation) [6]

Turnover number (min^{-1})

Specific activity (U/mg)
0.004–0.015 [3]; 0.077 [1]

K_m-value (mM)
0.7 (NAD+ (+ ADP-D-ribosyl-[dinitrogen reductase] from Azotobacter vinelandii), in the absence of Mg-ADP) [3]; 2 (NAD+ (+ native ADP-D-ribosyl-[dinitrogen reductase]), in the presence of Mg-ADP [3]) [1, 3]; 2.5 (NAD+ (+ ADP-D-ribosyl-[dinitrogen reductase] from Klebsiella pneumoniae), in the absence of Mg-ADP) [3]

pH-optimum
7 [1]

pH-range

Temperature optimum (°C)
30 (assay at) [1, 3, 6]

Temperature range (°C)

3 ENZYME STRUCTURE

Molecular weight
30900 (Rhodospirillum rubrum, gel filtration) [1]

Subunits
Monomer (1 × 29000, Rhodospirillum rubrum, SDS-PAGE) [1]

Glycoprotein/Lipoprotein
–

4 ISOLATION/PREPARATION

Source organism
Azospirillum brasilense (strain Sp7 [6]) [6, 7]; Azospirillum lipoferum [5];
Rhodospirillum rubrum (strain UR2 [2]) [1–4, 6, 7]

Source tissue
Cell [1–7]

Localization in source
Soluble [1]

Purification
Rhodospirillum rubrum (to near homogeneity) [1]

Crystallization
–

Cloned
(Azospirillum brasilense (expressed in Rhodospirillum rubrum mutants
UR212-UR215 via plasmid pYPZ106) [7], Azospirillum lipoferum (cloned to
plasmid pUC19 yielding pHAF102, transformed into E. coli CAG2041 via expression vector pHAF210, yielding E. coli strain UQ790) [5], Rhodospirillum
rubrum (expressed in E. coli and Klebsiella pneumoniae [4]) [2, 4]) [2, 4, 5, 7]

Renatured
–

5 STABILITY

pH

Temperature (°C)

Oxidation

Organic solvent

General stability information
ADP and NaCl stabilize during purification and storage [1]; DTT stabilizes during purification [1]; Extremely unstable in crude extract or partially purified preparation [1]; One freeze-thawing cycle leads to 50% loss of activity, ADP stabilizes [1]; Desalting procedures during purification inactivate [1]; Bovine serum albumin increases stability towards freeze thawing in the absence of ADP [3]

Storage
$-80°C$, in liquid N_2 [1]; $0°C$, in 0.2 M NaCl and 1 mM ADP, 18 h [1]

---

6 CROSSREFERENCES TO STRUCTURE DATABANKS

PIR/MIPS code

Brookhaven code

---

7 LITERATURE REFERENCES

[1] Lowery, R.G., Ludden, P.W.: J. Biol. Chem.,263,16714–16719 (1988)
[2] Fitzmaurice, W.P., Saari, L.L., Lowery, R.G., Ludden, P.W., Roberts, G.P.: Mol. Gen. Genet.,218,340–347 (1989)
[3] Lowery, R.G., Ludden, P.W.: Biochemistry,28,4956–4961 (1989)
[4] Fu, H.A., Wirt, H.J., Burris, R.H., Roberts, G.P.: Gene,85,153–160 (1989)
[5] Fu, H.A., Fitzmaurice, W.P., Roberts, G.P., Burris, R.H.: Gene,86,95–98 (1990)
[6] Fu, H.A., Hartman, A., Lowery, R.G., Fitzmaurice, W.P., Roberts, G.P., Burris, R.H.: J. Bacteriol.,171,4679–4685 (1989)
[7] Zhang, Y., Burris, R.H., Roberts, G.P.: J. Bacteriol.,174,3364–3369 (1992)

1 NOMENCLATURE

EC number
2.4.99.1

Systematic name
CMP-N-acetylneuraminate:beta-D-galactosyl-1,4-N-acetyl-beta-D-glucos-amine alpha-2,6-N-acetylneuraminyltransferase

Recommended name
beta-Galactoside alpha-2,6-sialyltransferase

Synonyms
Sialyltransferase, cytidine monophosphoacetylneuraminate-galactosylglyco-protein
alpha2–6 Sialyltransferase
beta-Galactoside alpha-2,6-sialyltransferase
Antigens, CD75
CMP-acetylneuraminate-galactosylglycoprotein sialyltransferase
CMP-acetylneuraminate-glycoprotein sialyltransferase
CMP-N-acetylneuraminic acid-glycoprotein sialyltransferase
Cytidine monophosphoacetylneuraminate-galactosylglycoprotein sialyltrans-ferase
Sialotransferase
Sialyltransferase

CAS Reg. No.
9075-81-4

---

2 REACTION AND SPECIFICITY

Catalysed reaction
CMP-N-acetylneuraminate + beta-D-galactosyl-1,4-N-acetyl-beta-D-glucos-amine →
→ CMP + alpha-N-acetylneuraminyl-2,6-beta-D-galactosyl-1,4-N-acetyl-beta-D-glucosamine (mechanism [9])

Reaction type
Glycosyl group transfer

Natural substrates
CMP-N-acetylneuraminate + beta-D-galactosyl-1,4-N-acetyl-beta-D-glucos-aminyl-R (R: glycoprotein or glycopeptide, one of a group of glycoslytrans-ferases which act to assemble the carbohydrate units of thyroglobulin [2], final step in synthesis of serum-type glycoproteins [7]) [2, 7]

Substrate spectrum

1 CMP-N-acetylneuraminate + beta-D-galactosyl-1,4-N-acetyl-beta-D-glucosaminyl-R (R: glycoprotein or glycopeptide [1, 2, 9], the non-reducing terminal galactosyl residue is essential for activity [1], transfers sialic acid to terminal positions on N-linked glycans [3, 4, 11, 13, 14]. High molecular weight substrates are more efficient acceptors than low molecular weight substrates [7, 10], specificity [9, 15], acceptors are lacto-N-neotetraose [3], asialo-ceruloplasmin [1, 11], asialo-alpha$_1$-acid glycoprotein [3, 4], asialo-prothrombin [3, 6], asialo-Tamm-Horsfall glycoprotein [6], asialo-orosomucoid (best substrate [7]) [6, 7], asialo-apoceruloplasmin (best substrate) [1], asialo-human chorionic gonadotropin, asialo-fibrinogen, asialo-thyroglobulin [1], asialo-fetuin (best substrate [10]) [1, 2, 5, 6, 9, 10], asialo-immunoglobulin G or M [9], asialo-transferrin (poor substrate [1]) [1, 15], fetuin (about 30% as effective as asialofetuin [10]) [9, 10] or p-nitrophenyl-D-galactoside [12]. No substrates are the corresponding sialylated native glycoproteins [1], antifreeze glycoprotein [3, 4], prothrombin, native alpha$_1$-acid glycoprotein or transferrin, asialo/agalacto-alpha$_1$-acid glycoprotein, asialo/agalacto-prothrombin [3], asialo/agalacto-fetuin [10], bovine [1, 6], porcine [6] or ovine [2–4] submaxillary asialo-mucin (no transfer to terminal N-acetyl-D-galactosamine [2]), galactose [2, 10], beta-methyl-L-arabinopyranoside [9], N-acetyl-D-galactosamine or N-acetyl-D-glucosamine [10], galactosylhydroxylysine and its derivatives, or galactose-containing oligosaccharides linked to Ser/Thr-glycopeptide of earthworm cuticle collagen [2]) [1–15]
2 CMP-N-acetylneuraminate + beta-D-galactosyl-1,4-N-acetyl-beta-D-glucosamine (i.e. N-acetyllactosamine, most active disaccharide substrate [7], best substrate [9], bovine or goat enzyme [7], poor substrate [10], poor substrates are the beta-1,3- or beta-1,6-derivatives of beta-D-galactosyl-1,4-N-acetyl-beta-D-glucosamine [9]) [3, 7, 9, 10]
3 CMP-N-acetylneuraminate + lactose (bovine or goat enzyme [7], poor substrate [3, 9]) [3, 4, 7, 9]
4 CMP-N-acetylneuraminate + asialo-alpha$_1$-acid glycoprotein (about half as effective as asialofetuin [10], CMP-N-acetylneuraminate can be replaced with comparable or even higher transfer rates by CMP-N-glycolyl-neuraminate, CMP-9-O-acetyl-N-acetylneuraminate [14], CMP-9-amino-N-acetylneuraminate, CMP-9-acetamido-N-acetylneuraminate, CMP-9-benzamido-N-acetylneuraminate, CMP-9-hexanoylamido-N-acetyl-neuraminate or CMP-9-azido-N-acetylneuraminate [11]) [1, 3, 4, 8–12, 14]
5 CMP-N-acetylneuraminate + beta-D-galactosyl-1,4-N-acetyl-beta-D-glucosaminyl-1,2-alpha-mannosyl-1,6-beta-mannosyl-1,4-N-acetylglucosamine [13]

Product spectrum

1 CMP + alpha-N-acetylneuraminyl-2,6-beta-D-galactosyl-1,4-N-acetyl-beta-D-glucosaminyl-R [1–4, 7–12]
2 CMP + alpha-N-acetylneuraminyl-2,6-beta-D-galactosyl-1,4-N-acetyl-beta-D-glucosamine [7, 9]

3 CMP + 6'-sialyllactose [3, 7]
4 CMP + alpha$_1$-acid glycoprotein [1]
5 CMP + alpha-N-acetylneuraminyl-2,6-beta-D-galactosyl-1,4-N-acetyl-beta-
   D-glucosaminyl-1,2-alpha-mannosyl-1,6-beta-mannosyl-1,4-N-acetylglu-
   cosamine [13]

Inhibitor(s)

Long chain fatty acids (oleic acid, stearic acid, less efficient: palmitic acid, lauric acid, capric acid, caprylic acid or caproic acid, not: linoleic acid or linolenic acid) [1]; PCMB (weak) [2]; Ba^{2+} [6]; Hg^{2+} [6]; CDP (Mn^{2+} restores activity [3]) [3, 10]; CTP [3, 10]; CMP (less effective than CDP or CTP) [10]; UTP (0.25 mM, inhibition decreases at higher concentrations) [10]; N-Ace-tyllactosamine (with asialo-alpha$_1$ acid glycoprotein as substrate) [9]; EDTA (weak [3, 10], not [2]) [3, 10]; Aflatoxins [5]; Pb^{2+} [6]; Cu^{2+} (weak [10]) [6, 10]; Mn^{2+} (above 10 mM) [10]; Antibodies raised in rabbit [4]; More (no inhi-bition by N-ethylmaleimide) [10]

Cofactor(s)/prosthetic group(s)/activating agents

Deoxycholate (activation) [1]; Triton X-100 (activation) [10]; Triton CF-54/Tween 80 (activation) [4]; EDTA (activation, not [2, 7]) [1]

Metal compounds/salts

$MgCl_2$ (stimulation [1], not [2, 7]) [1]; More (no divalent metal cation require-ment, e.g. Ca^{2+} [7] or Mn^{2+} [3, 7]) [2, 3, 6, 7, 10]

Turnover number (min^{-1})

Specific activity (U/mg)

More [2]; 0.000833 [1]; 8.2 [4]; 26–28 [8]

K$_m$-value (mM)

More (kinetic data [3, 6], kinetic mechanism [9]) [3, 6, 9]; 0.0053 (CMP-N-acetylneuraminate) [3]; 0.0064 (asialofetuin, expressed as concentration of terminal galactosyl residues) [10]; 0.016 (CMP-N-acetylneuraminate) [3]; 0.0205 (CMP-N-acetylneuraminate) [10]; 0.03 (CMP-9-benzamido-N-ace-tylneuraminate) [11]; 0.047 (CMP-N-acetylneuraminate) [1]; 0.05 (CMP-N-acetylneuraminate (+ asialo-alpha$_1$-acid glycoprotein)) [11]; 0.078 (asia-loceruloplasmin, expressed as concentration of acceptor sites) [1]; 0.08 (CMP-N-acetylneuraminate (+ lactose)) [7]; 0.12 (CMP-9-acetamido-N-ace-tylneuraminate) [11]; 0.136 (asialo-alpha$_1$-acid glycoprotein) [3]; 0.21 (asia-lo-transferrin) [3]; 0.27 (CMP-N-acetylneuraminate) [2]; 0.3 (CMP-N-ace-tylneuraminate (+ N-acetyllactosamine)) [7]; 0.59 (asialothyroglobulin glyco-peptides) [2]; 0.72 (CMP-9-amino-N-acetylneuraminate) [11]; 1.62 (N-ace-tyllactosamine) [3]; 1.67 (lacto-N-neotetraose) [3]; 12 (N-acetyllactosamine) [9]; 129 (lactose) [3]

pH-optimum

6 (broad [10]) [2, 10]; 6–6.5 (2-(N-morpholino)ethane sulfonic acid buffer) [1]; 6.3 [6]; 6.5 (broad, goat or bovine, optimal buffer concentration: 0.05–0.15 M, decreasing activity above 0.15 M) [7]; 7.5 (Tris-HCl buffer) [1]

pH-range

5–9.2 (about half-maximal activity at pH 5 and 9.2) [2]; 5–9.7 (about 65% of maximal activity at pH 5 and about half-maximal activity at pH 9.7) [1]; 5.2–7.4 (about half-maximal activity at pH 5.2 and 7.4) [10]; 5.4–8 (about half-maximal activity at pH 5.4 and 8, goat) [7]

Temperature optimum (°C)

28 [10]; 37 (assay at) [1–4, 7–15]

Temperature range (°C)

---

3 ENZYME STRUCTURE

Molecular weight

42900 (bovine low molecular weight form, sedimentation equilibrium centrifugation, two enzyme forms differing in molecular weight when submitted to gel filtration, the smaller one is probably a degradation product of the larger one) [8]

45000 (bovine low molecular weight form, gel filtration, two enzyme forms differing in molecular weight when submitted to gel filtration, the smaller one is probably a degradation product of the larger one) [8]

57900 (bovine high molecular weight form, sedimentation equilibrium centrifugation, two enzyme forms differing in molecular weight when submitted to gel filtration, the smaller one is probably a degradation product of the larger one) [8]

80000 (bovine high molecular weight form, gel filtration, two enzyme forms differing in molecular weight when submitted to gel filtration, the smaller one is probably a degradation product of the larger one) [8]

Subunits

Monomer (1 × 40500, rat, SDS-PAGE under reducing conditions [4], 1 × 41500, bovine low molecular weight form, SDS-PAGE under non-reducing conditions [8], 1 × 44000, bovine low molecular weight form, SDS-PAGE under reducing conditions [8], 1 × 53500, bovine high molecular weight form, SDS-PAGE under non-reducing conditions [8], 1 × 56000, bovine high molecular weight form, SDS-PAGE under reducing conditions [8]) [4, 8]

Glycoprotein/Lipoprotein

–

4

4 ISOLATION/PREPARATION

Source organism
Rat [1, 3–5, 11, 14]; Bovine (calf [2]) [2, 7–9, 14, 15]; Guinea pig [6]; Goat [7]; Human [7, 10, 12, 13]

Source tissue
Liver [1, 3–5, 11, 13, 14]; Thyroid [2]; Brain (cerebral cortex) [6]; Colostrum [7–9, 14, 15]; Cervical epithelium [10]; Placenta [13]; Platelets [12]

Localization in source
Microsomes [1, 5, 13]; Membrane-bound [1–4]; Soluble [7, 8]; Golgi complex [3, 4]; Synaptosomes [6]

Purification
Rat (partial, solubilized with deoxycholate [1], CDP-hexanolamine-agarose affinity chromatography [4]) [1, 4]; Bovine (partial [2, 7], solubilized by ultrasonic treatment [2], CDP-hexanolamine-agarose affinity chromatography [8]) [2, 7, 8]; Guinea pig (partial) [6]; Goat (partial) [7]; Human (solubilized with Triton X-100) [12]

Crystallization
–

Cloned
–

Renatured
–

---

5 STABILITY

pH
5.2–5.5 (most stable) [8]; 5.3 (and below, storage stability decreases with $t_{1/2}$: about 3 months) [4]; 6 (full stability for at least 1 year) [4]

Temperature (°C)
67 (30 s, 90% loss of activity) [3]; 100 (5 min, complete inactivation) [2]

Oxidation

Organic solvent

General stability information
High protein concentrations stabilize [8]; High NaCl concentrations stabilize during storage [4]; Use of plastic containers instead of glass stabilizes [8]; Glycerol, 50%, stabilizes during storage [8]; Freeze-thawing, crude preparation, stable to [10]; Glycerol, 20%, v/v, stabilizes [12]; Fractionation on Ultrogel AcA34 decreases activity, bovine serum albumin restores [12]

Enzyme Handbook © Springer-Verlag Berlin Heidelberg 1996

Storage
−20°C, crude enzyme preparation, at least 3 months [10]; −20°C, in 50% glycerol, 12 mM cacodylate, pH 5.3, 2 months [8]; Frozen, crude, solubilized enzyme preparation, several months [2]; −20°C, in 50% glycerol, 35 mM cacodylate, pH 6, 0.45 M NaCl, 0.08% Triton CF-54, at least 1 year [4]; 0–4°C, partially purified enzyme preparation, 2–4 weeks [7]; 3°C, crude, solubilized enzyme preparation, 2 weeks [1]; 4°C, partially purified enzyme preparation in 20% glycerol, v/v, at least 1 week [12]

6 CROSSREFERENCES TO STRUCTURE DATABANKS

PIR/MIPS code
PIR2:A41734 (human); PIR2:JH0286 (human); PIR2:A33424 (human (fragment)); PIR2:A28451 (rat); PIR2:B34465 (A renal rat (fragments)); PIR2:A34465 (E renal rat (fragments))

Brookhaven code

7 LITERATURE REFERENCES

[1] Hickman, J., Ashwell, G., Morell, A.G., Van den Hamer, C.J.A., Scheinberg, I.H.: J. Biol. Chem.,245,759–766 (1970)
[2] Spiro, M.J., Spiro, R.G.: J. Biol. Chem.,243,6520–6528 (1968)
[3] Weinstein, J., de Souza-e-Silva, U., Paulson, J.C.: J. Biol. Chem.,257,13845–13853 (1982)
[4] Weinstein, J., de Souza-e-Silva, U., Paulson, J.C.: J. Biol. Chem.,257,13835–13844 (1982)
[5] Bernacki, R.J., Gurtoo, H.L.: Res. Commun. Chem. Pathol. Pharmacol.,10,681–692 (1975)
[6] Bosmann, H.B.: J. Neurochem.,20,1037–1049 (1973)
[7] Bartholomew, B.A., Jourdian, G.W., Roseman, S.: J. Biol. Chem.,248,5751–5762 (1973)
[8] Paulson, J.C., Beranek, W.E., Hill, R.L.: J. Biol. Chem.,252,2356–2362 (1977)
[9] Paulson, J.C., Rearick, J.I., Hill, R.L.: J. Biol. Chem.,252,2363–2371 (1977)
[10] Scudder, P.R., Chantler, E.N.: Biochim. Biophys. Acta,660,136–141 (1981)
[11] Gross, H.J., Rose, U., Krause, J.M., Paulson, J.C., Schmid, K., Feeney, R.E., Brossmer, R.: Biochemistry,28,7386–7392 (1989)
[12] Bauvois, B., Montreuil, J., Verbert, A.: Biochim. Biophys. Acta,788,234–240 (1984)
[13] Nemansky, M., Schiphorst, W.E.C.M., Koeleman, C.A.M., Van den Eijnden, D.H.: FEBS Lett.,312,31–36 (1992)
[14] Higa, H.H., Paulson, J.C.: J. Biol. Chem.,260,8838–8849 (1985)
[15] Beyer, T.A., Rearick, J.I., Paulson, J.C., Prieels, J.-P., Sadler, J.E., Hill, R.L.: J. Biol. Chem.,254,12531–12541 (1979)

1 NOMENCLATURE

EC number
2.4.99.2

Systematic name
CMP-N-acetylneuraminate:D-galactosyl-N-acetyl-D-galactosaminyl-(N-acetyl-neuraminyl)-D-galactosyl-D-glucosylceramide N-acetylneuraminyltransferase

Recommended name
Monosialoganglioside sialyltransferase

Synonyms
Sialyltransferase, cytidine monophosphoacetylneuraminate-monosialo-ganglioside
SAT-4 [7]
GD1a-synthase [6]
CMP-NeuAc:GM1alpha2–3-sialyltransferase [7]
More (cf. EC 2.4.99.9, pig enzyme may be identical with EC 2.4.99.4 [8])

CAS Reg. No.
60202-12-2

---

2 REACTION AND SPECIFICITY

Catalysed reaction
CMP-N-acetylneuraminate + D-galactosyl-N-acetyl-D-galactosaminyl-(N-ace-tylneuraminyl)-D-galactosyl-D-glucosylceramide →
→ CMP + N-acetylneuraminyl-D-galactosyl-N-acetyl-D-galactosaminyl-(N-acetylneuraminyl)-D-galactosyl-D-glucosylceramide

Reaction type
Glycosyl group transfer

Natural substrates
CMP-N-acetylneuraminate + ganglioside GM1 (involved in ganglioside bio-synthesis) [10]

Substrate spectrum

1 CMP-N-acetylneuraminate + D-galactosyl-N-acetyl-D-galactosaminyl-
   (N-acetylneuraminyl)-D-galactosyl-D-glucosylceramide (i.e. ganglioside
   GM1, specific for galactosyl-1,3-N-acetyl-D-galactosamine structure, also
   in alpha-1-O-linkage to Thr/Ser [8], other substrates are antifreeze glycoprotein, porcine submaxillary asialomucin, asialo- or agalactofetuin [8], no
   substrates are galactosyl-, lactosylceramide, globoside, Gb3, Forssman
   antigen, gangliosides GM3 [6, 10], GD3 [6], GM2 [6, 8], GD1a, GL-4,
   beta-galactosyl-1,3-galactosylamine, beta-galactosylamine-1,3-beta-ga-
   lactosyl-1,4-beta-galactosyl-1,4-glucosylceramide, asialotransferrin, asia-
   lo-alpha$_1$ acid glycoprotein, ovine submaxillary asialomucin [8]) [1–10]
2 CMP-N-acetylneuraminate + ganglioside GM1-amide (better substrate
   than GM1) [5]
3 CMP-N-acetylneuraminate + ganglioside GA1 [2, 10]
4 CMP-N-acetylneuraminate + ganglioside GD1b (sialylated at 120% the
   rate of GM1 [6]) [2, 6, 10]
5 CMP-N-acetylneuraminate + asialo-ganglioside GM1 (sialylated at 61%
   the rate of GM1) [6]

Product spectrum

1 CMP + N-acetylneuraminyl-D-galactosyl-N-acetyl-D-galactosaminyl-
   (N-acetylneuraminyl)-D-galactosyl-D-glucosylceramide (i.e. ganglioside
   D1a) [1–10]
2 CMP + ganglioside GD1a-amide [5]
3 CMP + ganglioside GM1b [2, 10]
4 CMP + ganglioside GT1b [2, 6, 10]
5 CMP + ?

Inhibitor(s)

CDP [9]; CTP [9]; Cu^{2+} [1]; Cd^{2+} [1]; Co^{2+} [1]; Ca^{2+} [1]; Ni^{2+} [1]; Ba^{2+} (weak)
[1]; Zn^{2+} (weak) [1]; Ganglioside GM3 (in the presence of Triton CF-54) [2];
CMP-N-acetylneuraminate (above 0.2 M) [3]; Antifreeze glycoprotein (i.e.
Galbeta(1–3)GalNAcThr) [8]; More (no inhibition by CMP [9], Mg^{2+}, EDTA [1,
3], Mn^{2+}, Fe^{2+}, IAA, 2-mercaptoethanol [1] or sialic acid [3]) [1, 3, 9]

Cofactor(s)/prosthetic group(s)/activating agents

Triton X-100 (activation [1], slight [2]) [1, 2]; Triton CF-54 (activation [2, 3,
10], can partially be replaced by cutscum or Triton X-100, not by sodium
taurocholate or deoxycholate, Tween 80 or cholic acid [3]) [2, 3, 10]

Metal compounds/salts

More (no Mg^{2+} requirement [3], no metal ion requirement [1]) [1, 3]

Turnover number (min⁻¹)

Specific activity (U/mg)
0.0000046 [4]; 0.00118 (ganglioside GM1 as substrate) [5]; 0.00191 (ganglioside GM1-amide as substrate) [5]; 0.008 [6]

K_m-value (mM)
More (kinetic study) [8]; 0.000196 (ganglioside GD1b) [7]; 0.000538 (ganglioside GM1) [7]; 0.065 (CMP-N-acetylneuraminate) [6]; 0.16 (ganglioside GM1a) [2]; 0.35–0.39 (antifreeze glycoprotein) [8]; 0.5 (CMP-N-acetylneuraminate, ganglioside GM1) [1]; 75 (ganglioside GM1) [6]

pH-optimum
6.4 [1]; 6.5 (cacodylate buffer) [3]

pH-range

Temperature optimum (°C)
37 (assay at) [1–9]

Temperature range (°C)

---

3 ENZYME STRUCTURE

Molecular weight

Subunits
? (x × 44000, rat, SDS-PAGE [6], x × 50000, pig, SDS-PAGE [8]) [6, 8]

Glycoprotein/Lipoprotein
Glycoprotein [6]

---

4 ISOLATION/PREPARATION

Source organism
Rat (female Wistar [1], Sprague-Dawley strain [3]) [1–3, 5–7, 9, 10]; Human [4]; Pig [8]

Source tissue
Liver [1, 2, 9, 10]; Brain [3, 6]; Submaxillary gland [8]; HeLa-cells (strain R) [4]

Localization in source
Membrane-bound [1, 6]; Golgi apparatus (distribution [7]) [1, 2, 5, 7, 9, 10]; Microsomes [3]

Purification
Rat (CDP-Sepharose and 'GM1-acid'-Sepharose affinity chromatography) [6]

Crystallization

–

Cloned

–

Renatured

–

---

5 STABILITY

pH

Temperature (°C)

56 ($t_{1/2}$: 60 s) [2]

Oxidation

Organic solvent

General stability information

Storage

---

6 CROSSREFERENCES TO STRUCTURE DATABANKS

PIR/MIPS code

Brookhaven code

---

7 LITERATURE REFERENCES

[1] Busam, K., Decker, K.: Eur. J. Biochem.,160,23–30 (1986)
[2] Iber, H., van Echten, G., Sandhoff, K.: Eur. J. Biochem.,195,115–120 (1991)
[3] Yip, M.C.M.: Biochim. Biophys. Acta,306,298–306 (1973)
[4] Fishman, P.H., Bradley, R.M., Henneberry, R.C.: Arch. Biochem. Biophys.,
      172,618–626 (1976)
[5] Klein, D., Pohlentz, G., Schwarzmann, G., Sandhoff, K.: Eur. J. Biochem.,167,
      417–424 (1987)
[6] Gu, T.-J., Gu, X.-B., Ariga, T., Yu, R.K.: FEBS Lett.,275,83–86 (1990)
[7] Trinchera, M., Pirovano, B., Ghidoni, R.: J. Biol. Chem.,265,18242–18247 (1990)
[8] Rearick, J.I., Sadler, J.E., Paulson, J.C., Hill, R.L.: J. Biol. Chem.,254,4444–4451
      (1979)
[9] Eppler, C.M., Morré, D.J., Keenan, T.W.: Biochim. Biophys. Acta,619,332–343
      (1980)
[10] Pohlentz, G., Klein, D., Schwarzmann, G., Schmitz, D., Sandhoff, K.: Proc. Natl.
      Acad. Sci. USA,85,7044–7048 (1988)

1 NOMENCLATURE

EC number
2.4.99.3

Systematic name
CMP-N-acetylneuraminate:glycano-1,3-(N-acetyl-alpha-D-galactosaminyl)-glycoprotein alpha-2,6-N-acetylneuraminyltransferase

Recommended name
alpha-N-Acetylgalactosaminide alpha-2,6-sialyltransferase

Synonyms
Sialyltransferase, cytidine monophosphoacetylneuraminate-alpha-acetyl-galactosaminide alpha2→6-
Mucin sialyltransferase [2]
alpha-N-Acetylgalactosaminylprotein alpha2→6 sialyltransferase [4]
More (not identical with EC 2.4.99.7)

CAS Reg. No.
71124-50-0

---

2 REACTION AND SPECIFICITY

Catalysed reaction
CMP-N-acetylneuraminate + glycano-1,3-(N-acetyl-alpha-D-galactosaminyl)-glycoprotein →
→ CMP + glycano-(2,6-alpha-N-acetylneuraminyl)-(N-acetyl-D-galactos-aminyl)-glycoprotein (mechanism [1])

Reaction type
Glycosyl group transfer

Natural substrates
CMP-N-acetylneuraminate + glycano-1,3-(N-acetyl-alpha-D-galactos-aminyl)-glycoprotein (involved in biosynthesis of O-linked oligosaccharide chains of glycoproteins [1, 7], pathway of biosynthesis of glycans present in mucins [4], one of the last steps of glycoprotein biosynthesis [5]) [1, 4, 5, 7]

Substrate spectrum

1 CMP-N-acetylneuraminate + R-1,3-(N-acetyl-alpha-D-galactosaminyl)-glycoprotein (ir [11], R: H or beta-galactoside [1], transfers sialic acid to the core region of O-linked glycans [10], specific for alpha-N-acetyl-galactosamine attached through alpha-glycosidic linkage to threonine or serine in polypeptide chain [1, 11], specificity [3], substrates are porcine [1, 4] or ovine submaxillary asialomucin (with simple O-linked GalNAc-alpha-Thr/Ser) [1–5, 11], asialofetuin [1, 2, 6, 10], asialoorosomucoid [6], antifreeze glycoprotein (best substrate [1], from antarctic fish [3]) [1, 3, 5, 10], sialyl-2,3-antifreeze glycoprotein (better than native form) [3], erythrocyte hemagglutination inhibitor, milk glycoprotein [11]. CMP-glyco-lylneuraminate can replace CMP-N-acetylneuraminate (with the same efficiency [5]) [5, 11], no donors are CMP-9-O-acetyl-N-acetylneuraminate [5] or CMP-9-amino-acetylneuraminate [10]. CMP-9-acetamidoneuraminate, CMP-9-benzamidoneuraminate (less effective), CMP-9-hexanoyla-midoneuraminate (less effective), CMP-9-azido-N-acetylneuraminate can replace CMP-N-acetylneuraminate with asialofetuin or antifreeze glyco-protein as acceptor, CMP-9-amino-acetylneuraminate only with asia-lofetuin as acceptor [10]. No substrates are transferrin [3], asialo-col-lacalia mucoid, asialo-proteose peptone, asialo-prothrombin, asialo-thyro-globulin, asialo-transferrin, asialo-chondromucoprotein fractions from car-tilage, human blood group substance A or B [11], glycoproteins with Asn-linked oligosaccharides, glycosides, mono- or disaccharides [1], var-ious sugars [11]) [1–11]

2 CMP-N-acetylneuraminate + N-acetylneuraminyl-beta-galactosyl-1,3-N-acetyl-D-galactosylaminyl-R (R: protein or p-nitrophenol) [4]

3 CMP-N-acetylneuraminate + beta-galactosyl-1,3-N-acetyl-D-galactos-aminyl-R (R: protein [4, 9], R: p-nitrophenol (not [4, 9]) [8]) [4, 8, 9]

4 CMP-N-acetylneuraminate + N-acetyl-D-galactosaminyl-R (R: protein, not p-nitrophenol) [4]

5 CMP-N-acetylneuraminate + lactose (rat enzyme [2], not [11]) [2]

Product spectrum

1 CMP + R-(2,6-alpha-N-acetylneuraminyl)-(N-acetyl-D-galactosaminyl)-glycoprotein [1–11]

2 ?

3 CMP + beta-galactosyl-1,3-(2,6-alpha-N-acetylneuraminyl)-N-acetyl-D-galactosamine [9]

4 ?

5 CMP + 6'-sialyllactose [2]

Inhibitor(s)

CTP (strong [7]) [1, 7]; CDP (less effective than CTP) [7]; CMP (less effective than CTP) [7]; Antifreeze glycoprotein (non-competitive to CMP-N-acetylneuraminate) [1]; PCMB (weak) [8]; Triton X-100 (above critical micelle concentration, reversible) [1]; Dithioerythritol [6]; Ca^{2+} [6, 11]; More (no inhibition by native ovine submaxillary mucin [11], EDTA [1, 11], Zn^{2+}, Mn^{2+} or Mg^{2+} [1]) [1, 11]

Cofactor(s)/prosthetic group(s)/activating agents

Metal compounds/salts

More (no divalent metal cation requirement [11], e.g. Mn^{2+} or Mg^{2+} [1]) [1, 11]

Turnover number (min^{-1})

Specific activity (U/mg)

28.3 (bovine) [5]; 44.6 [1]

K_m-value (mM)

0.08 (CMP-9-O-acetyl-N-acetylneuraminate, bovine) [5]; 0.1 (CMP-9-O-acetyl-N-acetylneuraminate, pig) [5]; 0.15 (CMP-N-acetylneuraminate, bovine) [5]; 0.21 (CMP-N-glycolylneuraminate, bovine) [5]; 0.27 (CMP-N-acetylneuraminate, pig) [5]; 0.52 (CMP-N-glycolylneuraminate, pig) [5]; 0.57 (CMP-N-acetylneuraminate) [11]; 1.9 (ovine submaxillary asialomucin, calculated in terms of N-acetylgalactosamine acceptor sites) [11]; 4.7 (ovine submaxillary asialomucin (+ CMP-N-acetylneuraminate), bovine) [5]

pH-optimum

6–6.1 [11]; 6–6.5 [7]; 7.6 [6]

pH-range

6.2–8.5 (about half-maximal activity at pH 6.2 and 8.5) [6]

Temperature optimum (°C)

25 (broad) [8]; 37 [6]

Temperature range (°C)

17–38 (about half-maximal activity at 17°C and 38°C) [8]; 20–40 (about half-maximal activity at 20°C and 40°C) [6]

3 ENZYME STRUCTURE

Molecular weight

More (pig enzyme exists in several active molecular weight forms from 100000–172000, gel filtration or PAGE) [1]

Subunits
    ? (x × 120000, bovine, SDS-PAGE) [5]

Glycoprotein/Lipoprotein
    –

---

4 ISOLATION/PREPARATION

Source organism
    Pig [1–5, 7, 10]; Bovine [2, 5]; Sheep [2, 4, 9, 11]; Rat [2]; Human [6];
    Mouse (strain OF1) [8]

Source tissue
    Submaxillary gland [1–5, 7, 9–11]; Platelets [6]; Liver [8]; More (tissue distri-
    bution) [11]

Localization in source
    Membrane-bound [1, 3, 7]; Microsomes [4]; Mitochondria (outer membrane)
    [8]

Purification
    Pig (solubilized with Triton X-100, CDP-agarose affinity chromatography [1],
    partial [7]) [1, 7]; Bovine (solubilized with Triton X-100, CDP-agarose affinity
    chromatography) [5]; Sheep [11]

Crystallization
    –

Cloned
    –

Renatured
    –

---

5 STABILITY

pH
    6–6.5 (maximal stability) [7]

Temperature (°C)
    4 ($t_{1/2}$: about 60 min) [6]

Oxidation

Organic solvent

General stability information
    50 mM NaCl stabilizes during purification [1]; Glycerol, 50% w/v, stabilizes
    during storage [1]; Glycerol, 25% w/v, stabilizes during purification [3]

Storage

−20°C, 10 mM sodium cacodylate, 50 mM NaCl, 50% glycerol, 0.02% NaN_3, at least 6 months [5]; Frozen, crude preparation, at least 2 months [11]; 4°C, partially purified, at least 6 months [7]

6 CROSSREFERENCES TO STRUCTURE DATABANKS

PIR/MIPS code

Brookhaven code

7 LITERATURE REFERENCES

[1] Sadler, J.E., Rearick, J.I., Hill, R.L.: J. Biol. Chem.,254,5934–5941 (1979)

[2] Sherblom, A.P., Bourassa, C.R.: Biochim. Biophys. Acta,761,94–102 (1983)

[3] Beyer, T.A., Rearick, J.I., Paulson, J.C., Prieels, J.-P., Sadler, J.E., Hill, R.L.: J. Biol. Chem.,254,12531–12541 (1979)

[4] Bergh, M.L.E., van den Eijnden, D.H.: Eur. J. Biochem.,136,113–118 (1983)

[5] Higa, H.H., Paulson, J.C.: J. Biol. Chem.,260,8838–8849 (1985)

[6] Bauvois, B., Cacan, R., Fournet, B., Caen, J., Montreuil, J., Verbert, A.: Eur. J. Biochem.,121,567–572 (1982)

[7] Sadler, J.E., Rearick, J.I., Paulson, J.C., Hill, R.L.: J. Biol. Chem.,254,4434–4443 (1979)

[8] Gasnier, F., Baubichon-Cortay, H., Louisot, P., Gateau-Roesch, O.: J. Biochem., 110,702–707 (1991)

[9] Bergh, M.L.E., Koppen, P.L., Van den Eijnden, D.H.: Biochem. J.,201,411–415 (1982)

[10] Gross, H.J., Rose, U., Krause, J.M., Paulson, J.C., Schmid, K., Feeney, R.E., Brossmer, R.: Biochemistry,28,7386–7392 (1989)

[11] Carlson, D.M., McGuire, E.J., Jourdian, G.W., Roseman, S.: J. Biol. Chem.,248, 5763–5773 (1973)

1 NOMENCLATURE

EC number
2.4.99.4

Systematic name
CMP-N-acetylneuraminate:beta-D-galactoside alpha-2,3-N-acetylneuraminyltransferase

Recommended name
beta-Galactoside alpha-2,3-sialyltransferase

Synonyms
Sialyltransferase, cytidine monophosphoacetylneuraminate-beta-galactoside alpha2→3-
alpha2→3 Sialyltransferase
NeuAc alpha-2,3-sialyltransferase
More (may be identical with EC 2.4.99.2)

CAS Reg. No.
71124-51-1

2 REACTION AND SPECIFICITY

Catalysed reaction
CMP-N-acetylneuraminate + beta-D-galactosyl-1,3-N-acetyl-alpha-D-galactosaminyl-R →
→ CMP + alpha-N-acetylneuraminyl-2,3-beta-D-galactosyl-1,3-N-acetyl-alpha-D-galactosaminyl-R

Reaction type
Glycosyl group transfer

Natural substrates
CMP-N-acetylneuraminate + beta-D-galactosyl-1,3-N-acetyl-alpha-D-galactosaminyl-R (pathway in mucin biosynthesis) [7, 13]

Substrate spectrum
1 CMP-N-acetylneuraminate + beta-D-galactosyl-1,3-N-acetyl-alpha-D-galactosaminyl-R (R: H, a threonine or serine residue in a glycoprotein [1], or a glycolipid, transfers sialic acid to the core region of O-linked glycans [7], no transfer to N-acetylgalactosamine residues [13], specificity [16], acceptor substrates are asialoglycoproteins, e.g. asialoorosomucoid (poor substrate [4]) [2, 4], asialofetuin [1, 2, 4, 8], agalactofetuin [1], ovine [3, 8] or bovine [15] submaxillary asialomucin (not [1, 13]), porcine

submaxillary asialo/afucomucin (rat: poor substrate, dog: no substrate [13]), antifreeze glycoprotein (best substrate [1]) [1, 8, 13, 16], native or asialo-kappa-casein [14] or ganglioside GM1 [1]. CMP-N-glycolyl-N-acetylneuraminate, CMP-9-O-acetyl-N-acetylneuraminate [5], CMP-9-acetamido-N-acetylneuraminate, CMP-9-benzamido-N-acetylneuraminate, CMP-9-hexanoylamido-N-acetylneuraminate or CMP-9-azido-N-acetylneuraminate can replace CMP-N-acetylneuraminate with asialofetuin [7] or antifreeze glycoprotein [5, 7] as acceptor, not CMP-9-amino-acetylneuraminate [7]. No substrates are the native glycoproteins orosomucoid, fetuin or transferrin [4], asialotransferrin, asialo-alpha$_1$ acid glycoprotein [1]) [1–16]

  2 CMP-N-acetylneuraminate + beta-D-galactosyl-1,3-N-acetyl-alpha-D-glucosamine [10, 11]

  3 CMP-N-acetylneuraminate + p-nitrophenyl-beta-D-galactoside [2, 4, 9]

  4 CMP-N-acetylneuraminate + lactose [1, 3, 5]

Product spectrum

  1 CMP + alpha-N-acetylneuraminyl-2,3-beta-D-galactosyl-1,3-N-acetyl-alpha-D-galactosaminyl-R [1–13]

  2 CMP + alpha-N-acetylneuraminyl-2,3-beta-D-galactosyl-1,3-N-acetyl-alpha-D-galactosamine [10, 11]

  3 CMP + alpha-N-acetylneuraminyl-2,3-p-nitrophenyl-beta-D-galactoside [2]

  4 CMP + 3'-sialyllactose [1, 3, 5]

Inhibitor(s)

beta-D-Galactosyl-1,3-N-acetylgalactosylamide (with porcine submaxillary asialomucin as substrate) [1]; CTP (competitive to CMP-N-acetylneuraminate [1], strong [3]) [1, 3, 10]; Zn^{2+} (weak) [14]; Dithioerythritol (above 1 mM, stimulates up to 0.5 mM) [2]; Ca^{2+} (above 10 mM, stimulates up to 2.5 mM) [2]; EDTA [14]; CDP (less effective than CTP) [3]; CMP (less effective than CTP) [3]; PCMB [4]; Lysophosphatidylserine (enzyme form A) [10]; Lysophosphatidylglycerol (enzyme forms A and B) [10]; 1-Palmitoyl-sn-glycero-3-phosphorylcholine [15]; Octylglucoside (irreversible, at higher concentrations) [10]

Cofactor(s)/prosthetic group(s)/activating agents

Triton X-100 (activation, not when Golgi-membranes are frozen and thawed prior to assay [14]) [3, 14]; Triton detergents (activation, enzyme form A, only above critical micelle concentration, not form B) [10]; Lysophosphatidylcholine (activation, enzyme form A, not B) [10]; Incorporation into phospholipid/octylglucoside liposomes (activation, enzyme form A) [11]; Dithiothreitol (activation) [14]; Dithioerythritol (stimulation, up to 0.5 mM, inhibitory above 1 mM) [2]

Metal compounds/salts

Mn^{2+} (activation, 10 mM [14], solubilized enzyme: requirement, K_m-value: 1.1 mM [15]) [4, 14, 15]; Ca^{2+} (activation [14], stimulation, up to 2.5 mM, inhibitory above 10 mM [2]) [2, 14]; Mg^{2+} (activation) [14]; Sr^{2+} (activation) [14]; Ba^{2+} (activation) [14]; Cu^{2+} (activation) [14]

Turnover number (min^{-1})

Specific activity (U/mg)
More [10, 13]; 10.6 [3]

K_m-value (mM)
More (kinetic study [1, 11], of porcine enzyme: purified, membrane-bound or included in liposomes [11]) [1, 11]; 0.0008 (bovine submaxillary asialomucin) [15]; 0.0011 (CMP-N-acetylneuraminate) [15]; 0.00434 (CMP-N-acetylneuraminate, enzyme form A, with 1% detergent) [10]; 0.00482 (CMP-N-acetylneuraminate, enzyme form B) [10]; 0.01 (asialo-kappa-casein) [14]; 0.0123 (CMP-N-acetylneuraminate, enzyme form A, no detergent) [10]; 0.32 (beta-D-galactosyl-1,3-N-acetylgalactosylamide, enzyme form A, with 1% Triton X-100 or enzyme form B) [10]; 0.37 (beta-D-galactosyl-1,3-N-acetylgalactosylamide, enzyme form A, with 1% lysophosphatidylcholine or without detergent) [10]

pH-optimum
5.5 (2-(N-morpholino)ethane sulfonic acid buffer) [14]; 6 [15]; 6–6.5 [3]; 6.1 [4]; 6.7 [2]

pH-range
5.8–8.0 (about half-maximal activity at pH 5.8 and 8.0) [2]

Temperature optimum (°C)
20–25 [4]; 30 [2, 15]; 37 [14]

Temperature range (°C)
15–34 (about 85% of maximal activity at 15°C and about half-maximal activity at 34°C) [4]; 23–45 (about half-maximal activity at 23°C and 45°C) [14]

3 ENZYME STRUCTURE

Molecular weight
44000 (pig enzyme form B, sucrose density gradient centrifugation, gel filtration yields two different porcine enzyme forms, form A and B, form A presumably binds a large amount of low densitiy lipid or detergent) [3]
50000 (pig enzyme form B, gel filtration, gel filtration yields two different porcine enzyme forms, form A and B, form A presumably binds a large amount of low densitiy lipid or detergent) [3]

220000 (pig enzyme form A, gel filtration, gel filtration yields two different porcine enzyme forms, form A and B, form A presumably binds a large amount of low densitiy lipid or detergent) [3]

Subunits
? (x × 50000, pig [1, 3], both enzyme forms [3], SDS-PAGE [1, 3], under reducing and non-reducing conditions [3]) [1, 3]

Glycoprotein/Lipoprotein
–

---

4 ISOLATION/PREPARATION

Source organism
Pig [1, 3, 5, 7, 10, 11, 13, 16]; Human (healthy individuals and Wiscott-Aldrich syndrom patients [9]) [2, 9, 12, 13]; Mouse (strain OF1) [4, 15]; Sheep [6]; Bovine (fetal calf [8], lactating Holstein cow [14]) [8, 12, 14]; Rat [12, 13]; Dog [13]; Rabbit [12]

Source tissue
Submaxillary glands [1, 3, 5–7, 10, 11, 16]; Platelets [2, 9]; Liver [4, 8, 12, 13, 15]; Lymphocytes (T-lymphocytes, Epstein-Barr virus immortilized B-lymphocytes) [9]; Placenta (human) [12]; Mammary gland [14]

Localization in source
Golgi apparatus (about 85% of total activity, presumably enzyme form A [10]) [10, 11, 14]; Membrane-bound (about 85% of total activity, presumably enzyme form A [10]) [1–4, 9–11, 13–15]; Mitochondria (outer membrane) [4]; Soluble (about 15% of total activity, presumably enzyme form B) [10]; More (subcellular distribution) [10, 14]

Purification
Pig (solubilized with Triton X-100, CDP-hexanolamine-affinity chromatography [3, 10], the purification yields two enzyme forms, A and B, separable by gel filtration which are distinct with respect to detergent requirement but identical when submitted to SDS-PAGE [10]) [3, 10]

Crystallization
–

Cloned
–

Renatured
–

5 STABILITY

pH
6–6.5 (maximal stability) [3]

Temperature (°C)

Oxidation

Organic solvent

General stability information

Storage
–20°C, concentrated enzyme preparation in 50% v/v glycerol, more than 6
months [3]; 4°C, at least 24 h [2]

6 CROSSREFERENCES TO STRUCTURE DATABANKS

PIR/MIPS code
PIR3:S36824 (mouse); PIR2:A54420 (ST3GalA.2 mouse); PIR2:B54420
(ST3GalA.2 rat)

Brookhaven code

7 LITERATURE REFERENCES

[1] Rearick, J.I., Sadler, J.E., Paulson, J.C., Hill, R.L.: J. Biol. Chem.,254,4444–4451
(1979)
[2] Bauvois, B., Cacan, R., Fournet, B., Caen, J., Montreuil, J., Verbert, A.: Eur. J.
Biochem.,121,567–572 (1982)
[3] Sadler, J.E., Rearick, J.I., Paulson, J.C., Hill, R.L.: J. Biol. Chem.,254,4434–4443
(1979)
[4] Gasnier, F., Baubichon-Cortay, H., Louisot, P., Gateau-Roesch, O.: J. Biochem.,
110,702–707 (1991)
[5] Higa, H.H., Paulson, J.C.: J. Biol. Chem.,260,8838–8849 (1985)
[6] Bergh, M.L.E., Koppen, P.L., Van den Eijnden, D.H.: Biochem. J.,201,411–415
(1982)
[7] Gross, H.J., Rose, U., Krause, J.M., Paulson, J.C., Schmid, K., Feeney, R.E.,
Brossmer, R.: Biochemistry,28,7386–7392 (1989)
[8] Bergh, M.L.E., Hooghwinkel, G.J.M., Van den Eijnden, D.H.: J. Biol. Chem.,258,
7430–7436 (1983)
[9] Higgins, E.A., Siminovitch, K.A., Zhuang, D., Brockhausen, I., Dennis, J.W.: J. Biol.
Chem.,266,6280–6290 (1991)
[10] Westcott, K.R., Wolf, C.C., Hill, R.L.: J. Biol. Chem.,260,13109–13115 (1985)
[11] Westcott, K.R., Hill, R.L.: J. Biol. Chem.,260,13116–13121 (1985)
[12] De Heij, H.T., Koppen, P.L., Van den Eijnden, D.H.: Carbohydr. Res.,149,85–100
(1986)

[13] Van den Eijnden, D.H., Bergh, M.L.E., Dieleman, B., Schiphorst, W.E.C.M.: Hoppe-Seyler's Z. Physiol. Chem.,362,113–124 (1981)
[14] Keller, S.J., Keenan, T.W., Eigel, W.N.: Biochim. Biophys. Acta,566,266–273 (1979)
[15] Bador, H., Morelis, R., Louisot, P.: Biochim. Biophys. Acta,706,36–41 (1982)
[16] Beyer, T.A., Rearick, J.I., Paulson, J.C., Prieels, J.-P., Sadler, J.E., Hill, R.L.: J. Biol. Chem.,254,12531–12541 (1979)

1 NOMENCLATURE

EC number
2.4.99.5

Systematic name
CMP-N-acetylneuraminate:1,2-diacyl-3-beta-D-galactosyl-sn-glycerol N-acetylneuraminyltransferase

Recommended name
Galactosyldiacylglycerol alpha-2,3-sialyltransferase

Synonyms
Sialyltransferase, cytidine monophosphoacetylneuraminate-galactosyl-diacylglycerol

CAS Reg. No.
80237-98-5

2 REACTION AND SPECIFICITY

Catalysed reaction
CMP-N-acetylneuraminate + 1,2-diacyl-3-beta-D-galactosyl-sn-glycerol →
→ CMP + 1,2-diacyl-3-[3-(alpha-D-N-acetylneuraminyl)-beta-D-galactosyl]-sn-glycerol

Reaction type
Glycosyl group transfer

Natural substrates

Substrate spectrum
1 CMP-N-acetylneuraminate + 1,2-diacyl-3-beta-D-galactosyl-sn-glycerol
(i.e. galactosyldiacylglycerol, transfers N-acetylneuraminic acid to position C-3 of the galactosyl residue) [1]

Product spectrum
1 CMP + 1,2-diacyl-3-[3-(alpha-D-N-acetylneuraminyl)-beta-D-galactosyl]-sn-glycerol (i.e. sialosylgalactosyldiacylglycerol) [1]

Inhibitor(s)

Cofactor(s)/prosthetic group(s)/activating agents

Metal compounds/salts

Turnover number (min^{-1})

Specific activity (U/mg)

K$_m$-value (mM)
    0.13 (galactosyldiacylglycerol) [1]; 0.78 (CMP-acetylneuraminate) [1]

pH-optimum
    6.2 [1]

pH-range

Temperature optimum (°C)
    37 (assay at) [1]

Temperature range (°C)

---

3 ENZYME STRUCTURE

Molecular weight

Subunits

Glycoprotein/Lipoprotein
    –

---

4 ISOLATION/PREPARATION

Source organism
    Mouse (Swiss-Webster albino) [1]

Source tissue
    Brain [1]

Localization in source
    Membrane-bound [1]; Microsomes [1]

Purification

Crystallization
    –

Cloned
    –

Renatured
    –

5 STABILITY

pH

Temperature (°C)

Oxidation

Organic solvent

General stability information

Storage

6 CROSSREFERENCES TO STRUCTURE DATABANKS

PIR/MIPS code

Brookhaven code

7 LITERATURE REFERENCES

[1] Pieringer, J., Keech, S., Pieringer, R.A.: J. Biol. Chem.,256,12306–12309 (1981)

1 NOMENCLATURE

EC number
2.4.99.6

Systematic name
CMP-N-acetylneuraminate:beta-D-galactosyl-1,4-N-acetyl-D-glucosaminyl-glycoprotein alpha-2,3-N-acetylneuraminyltransferase

Recommended name
N-Acetyllactosaminide alpha-2,3-sialyltransferase

Synonyms
Sialyltransferase
Sialyltransferase, cytidine monophosphoacetylneuraminate-beta-galactosyl(1→4)acetylglucosaminide alpha2→3-
alpha2→3 Sialyltransferase

CAS Reg. No.
77537-85-0

2 REACTION AND SPECIFICITY

Catalysed reaction
CMP-N-acetylneuraminate + beta-D-galactosyl-1,4-N-acetyl-D-glucos-aminyl-glycoprotein →
→ CMP + alpha-N-acetylneuraminyl-2,3-beta-D-galactosyl-1,4-N-acetyl-D-glucosaminyl-glycoprotein

Reaction type
Glycosyl group transfer

Natural substrates

Substrate spectrum
1 CMP-N-acetylneuraminate + beta-D-galactosyl-1,4-N-acetyl-D-glucosaminyl-glycopeptide (i.e. type 2 chain, preferred substrate, type 1 chain acceptor is a less effective substrate, oligosaccharides or glycopeptides are better substrates than glycoproteins, preferred structure of branched oligosaccharide substrates are triantennary forms, poor substrates are those with bisected N-acetyl-D-glucosaminyl structures [3]) [1–4]
2 CMP-N-acetylneuraminate + asialo-alpha$_1$ acid glycoprotein (transfers sialic acid to hydroxyl group of galactosyl residue of beta-galactosyl-1,4-acetylglucosaminyl structure [4]) [1, 4]

3 CMP-N-acetylneuraminate + beta-D-galactosyl-1,4-N-acetyl-D-glucos-
aminyl-1,2-alpha-mannosyl-1,6(3)-beta-mannosyl-1,4-N-acetyl-D-glucos-
amine [2]

Product spectrum
1 CMP + alpha-N-acetylneuraminyl-2,3-beta-D-galactosyl-1,4-N-acetyl-D-
glucosaminyl-glycopeptide [3]
2 ?
3 CMP + alpha-N-acetylneuraminyl-2,3-beta-D-galactosyl-1,4-N-acetyl-
D-glucosaminyl-1,2-alpha-mannosyl-1,6(3)-beta-mannosyl-1,4-N-acetyl-
D-glucosamine [2]

Inhibitor(s)

Cofactor(s)/prosthetic group(s)/activating agents
Triton X-100 (activation, 1% w/v) [4]

Metal compounds/salts

Turnover number (min^{-1})

Specific activity (U/mg)

K$_m$-value (mM)

pH-optimum

pH-range

Temperature optimum (°C)
37 (assay at) [4]

Temperature range (°C)

3 ENZYME STRUCTURE

Molecular weight

Subunits

Glycoprotein/Lipoprotein
–

4 ISOLATION/PREPARATION

Source organism
Bovine (calf) [1]; Chicken (embryo) [1]; Human [1–4]

Source tissue
Brain (chicken) [1]; Liver (calf [1]) [1, 2]; Placenta (human [1]) [1–3]; Platelets [4]

Localization in source
Membrane-bound [1]

Purification
Human (partial, CDP-ethanolamine-Sepharose affinity chromatography) [4]

Crystallization
–

Cloned
–

Renatured
–

5 STABILITY

pH

Temperature (°C)

Oxidation

Organic solvent

General stability information
Fractionation on Ultrogel AcA34 decreases activity, bovine serum albumin restores [4]

Storage
4°C, partially purified, in 20% v/v glycerol, at least 1 week, after Ultrogel AcA34 fractionation, $t_{1/2}$: 24 h [4]

6 CROSSREFERENCES TO STRUCTURE DATABANKS

PIR/MIPS code

Brookhaven code

7 LITERATURE REFERENCES

[1] Van den Eijnden, D.H., Schiphorst, W.E.C.M.: J. Biol. Chem.,256,3159–3162 (1981)
[2] Nemansky, M., Schiphorst, W.E.C.M., Koeleman, C.A.M., Van den Eijnden, D.H.: FEBS Lett.,312,31–36 (1992)
[3] Nemansky, M., Van den Eijnden, D.H.: Glycoconjugate J.,10,99–108 (1993)
[4] Bauvois, B., Montreuil, J., Verbert, A.: Biochim. Biophys. Acta,788,234–240 (1984)

1 NOMENCLATURE

EC number
2.4.99.7

Systematic name
CMP-N-acetylneuraminate:(alpha-N-acetylneuraminyl-2,3-beta-D-galactosyl-1,3)-N-acetyl-D-galactosaminide alpha-2,6-N-acetylneuraminyltransferase

Recommended name
(alpha-N-Acetylneuraminyl-2,3-beta-galactosyl-1,3)-N-acetylgalactosaminide alpha-2,6-sialyltransferase

Synonyms
Sialyltransferase
Sialyltransferase, cytidine monophosphoacetylneuraminate-(alpha-N-acetyl-neuraminyl-2,3-beta-galactosyl-1,3)-N-acetylgalactosaminide-alpha-2,6-
More (not identical with EC 2.4.99.3)

CAS Reg. No.
129924-24-9

---

2 REACTION AND SPECIFICITY

Catalysed reaction
CMP-N-acetylneuraminate + alpha-N-acetylneuraminyl-2,3-beta-D-galac-tosyl-1,3-N-acetyl-D-galactosaminyl-R →
→ CMP + alpha-N-acetylneuraminyl-2,3-beta-D-galactosyl-1,3-(N-acetyl-neuraminyl-2,6)-N-acetyl-D-galactosaminyl-R

Reaction type
Glycosyl group transfer

Natural substrates
CMP-N-acetylneuraminate + alpha-N-acetylneuraminyl-2,3-beta-D-galac-tosyl-1,3-N-acetyl-D-galactosaminyl-R (pathway in glycoprotein biosynthesis) [1–3]

Substrate spectrum
1 CMP-N-acetylneuraminate + alpha-N-acetylneuraminyl-2,3-beta-D-galac-
   tosyl-1,3-N-acetyl-D-galactosaminyl-R (R can be a protein or p-nitrophenol
   [3], attaches sialic acid in alpha-2,6-linkage to N-acetylgalactosamine only when present in the structure of alpha-N-acetylneuraminyl-2,3-
   beta-galactosyl-1,3-N-acetylgalactosaminyl-R, substrates are sialylated
   antifreeze glycoprotein, N-acetyl-D-galactosamine (not bovine enzyme),
   beta-galactosyl-1,3-N-acetyl-D-galactosamine (not bovine enzyme) [1], no
   substrates are beta-D-galactosyl-1,3-N-acetyl-D-galactosaminyl- or N-ace-
   tylgalactosylaminyl-R [3]) [1–3]

Product spectrum
1 CMP + alpha-N-acetylneuraminyl-2,3-beta-D-galactosyl-1,3-(N-acetyl-
   neuraminyl-2,6)-N-acetyl-D-galactosaminyl-R [1–3]

Inhibitor(s)

Cofactor(s)/prosthetic group(s)/activating agents

Metal compounds/salts

Turnover number (min^{-1})

Specific activity (U/mg)

K$_m$-value (mM)

pH-optimum

pH-range

Temperature optimum (°C)
   37 (assay at) [3]

Temperature range (°C)

3 ENZYME STRUCTURE

Molecular weight

Subunits

Glycoprotein/Lipoprotein
   –

4 ISOLATION/PREPARATION

Source organism
   Bovine (calf) [1, 3]; Human [2]; Sheep [1]

Source tissue
Liver (bovine, fetal [3]) [1, 3]; Submaxillary gland (sheep) [1]; T-Lymphocytes [2]; Platelets [2]; Epstein-Barr virus transformed B-cell line [2]

Localization in source
Microsomes [1–3]; Membrane-bound [1, 2]

Purification

Crystallization
–

Cloned
–

Renatured
–

---

5 STABILITY

pH

Temperature (°C)

Oxidation

Organic solvent

General stability information

Storage

---

6 CROSSREFERENCES TO STRUCTURE DATABANKS

PIR/MIPS code

Brookhaven code

---

7 LITERATURE REFERENCES

[1] Bergh, M.L.E., Hooghwinkel, G.J.M., Van den Eijnden, D.H.: J. Biol. Chem.,258,
    7430–7436 (1983)
[2] Higgins, E.A., Siminovitch, K.A., Zhuang, D., Brockhausen, I., Dennis, J.W.: J. Biol.
    Chem.,266,6280–6290 (1991)
[3] Bergh, M.L.E., van den Eijnden, D.H.: Eur. J. Biochem.,136,113–118 (1983)

1 NOMENCLATURE

EC number
   2.4.99.8

Systematic name
   CMP-N-acetylneuraminate:alpha-N-acetylneuraminyl-2,3-beta-D-galactoside
   alpha-2,8-N-acetylneuraminyltransferase

Recommended name
   alpha-N-Acetylneuraminate alpha-2,8-sialyltransferase

Synonyms
   Sialyltransferase, cytidine monophosphoacetylneuraminate-
   ganglioside GM3
   alpha-2,8-Sialyltransferase
   Ganglioside GD3 synthase
   Ganglioside GD3 synthetase
   CMP-NeuAc:LM1(alpha2–8) sialyltranferase [1]
   GD3 synthase [3]
   SAT-2 [1]

CAS Reg. No.
   67339-00-8

___

2 REACTION AND SPECIFICITY

Catalysed reaction
   CMP-N-acetylneuraminate + alpha-N-acetylneuraminyl-2,3-beta-D-galac-
   tosyl-R →
   → CMP + alpha-N-acetylneuraminyl-2,8-alpha-N-acetylneuraminyl-2,3-
   beta-D-galactosyl-R

Reaction type
   Glycosyl group transfer

Natural substrates
   CMP-alpha-N-acetylneuraminate + alpha-N-acetylneuraminyl-2,3-beta-D-
   galactosyl-1,4-beta-D-glucosylceramide (branch-point enzyme in ganglioside biosynthetic sequence) [3, 5]

Substrate spectrum

1 CMP-N-acetylneuraminate + alpha-N-acetylneuraminyl-2,3-beta-D-galactosyl-1,4-beta-D-glucosylceramide (i.e. sialosyllactosylceramide or ganglioside GM3, specificity is determined by the substrate's negative charge and the acyl-residue in amide bond to the amino group of neuraminic acid [5], poor substrates are GM3 methyl ester, GM3 amide or GM3 methyl amide, no substrates are neuraminyllactosylceramide or N-biotinylneuraminyllactosylceramide [5]) [1–11]

2 CMP-N-acetylneuraminate + N-acyl-lyso-GM3 derivatives (N-acetyl derivative is a better substrate than GM3 (detergent-like effect)) [5]

3 CMP-N-acetylneuraminate + alpha-N-acetylneuraminyl-2,3-beta-D-galactosyl-1,4-N-acetyl-beta-D-glucosaminyl-1,3-beta-D-galactosyl-1,4-D-glucosylceramide (i.e. sialosylneolactotetraosylceramide or ganglioside LM1) [1]

4 CMP-N-acetylneuraminate + disialoganglioside GD1a [7]

5 CMP-N-acetylneuraminate + trisialoganglioside GT1b [10]

6 CMP-N-acetylneuraminate + alpha-N-glycosylneuraminyl lactosylceramide [5]

7 CMP-N-acetylneuraminate + alpha-N-butyrylneuraminyllactosylceramide (sialylated at about 60% the rate of GM3) [5]

Product spectrum

1 CMP + alpha-N-acetylneuraminyl-2,8-alpha-N-acetylneuraminyl-2,3-beta-D-galactosyl-1,4-beta-D-glucosylceramide (i.e. disialosyllactosylceramide or ganglioside GD3) [1–10]

2 CMP + N-acyl-lyso-GD3 derivatives [5]

3 CMP + alpha-N-acetylneuraminyl-2,8-alpha-N-acetylneuraminyl-2,3-beta-D-galactosyl-1,4-N-acetyl-beta-D-glucosaminyl-1,3-beta-D-galactosyl-1,4-D-glucosylceramide (i.e. disialosylneolactotetraosylceramide or ganglioside LD1c) [1]

4 CMP + trisialoganglioside GT1a [7]

5 CMP + ganglioside GQ1b [10]

6 CMP + ?

7 CMP + ?

Inhibitor(s)

CMP (strong) [3]; CDP (partially relieved by excess Mg^{2+}) [3]; CTP [3]; GMP (as strong as CMP) [3]; GDP [3]; GTP [3]; AMP (less effective than CMP or GMP) [3]; ADP [3]; ATP [3]; TMP (less effective than AMP) [3]; TDP [3]; TTP [3]; UMP (weak) [3]; UDP (weak) [3]; UTP (weak) [3]; Ca^{2+} [7, 8]; Cd^{2+} (strong) [8]; Co^{2+} [8]; Cu^{2+} (strong) [8, 10]; Ni^{2+} [8]; Fe^{2+} (not [8]) [10]; Mn^{2+} (not [7, 8]) [10]; Ba^{2+} (weak) [8]; Zn^{2+} (weak) [8]; N-Ethylmaleimide [1]; EDTA (Mg^{2+} protects [4], not [7, 8, 10]) [4]; Ganglioside LM1 (at higher concentrations, substrate inhibition) [1]; Ganglioside Q1b (strong) [6]; Ganglioside T1b [6]; Ganglioside D1a [6]; Lysophospholipids [4]; More (no inhibition by Mg^{2+}, CMP-N-acetylneuraminate [7], IAA, 2-mercaptoethanol [8] or ganglioside GM2 [6]) [6–8]

Cofactor(s)/prosthetic group(s)/activating agents
Myrj 59 (activation, most potent activator, can be replaced by the following detergents (descending efficiency): sodium deoxycholate, Triton CF-54, Tween 20, Tween 80/Triton CF-54 (ratio 1:2), Triton X-100 or Tween 80) [2]; Triton CF-54 (activation, further enhanced by supplementation with diacyl phospholipids [4], can be replaced by Triton X-100, Tween 80 or Tween 20 with 25%, 11% or 9% efficiency, respectively [7]) [1, 2, 4, 7, 10, 11]; Triton X-100 (activation) [1, 2, 4, 6–8]; Nonidet P-40 (activation) [1]; Histone (slight activation) [7]; More (the presence of detergents is essential for activity, no activation by phosphatidylglycerol [6], digitonin [4], zwittergent 3–10 or 3–14 [1]) [1, 4, 6]

Metal compounds/salts
Mg^{2+} (slight activation [10], stimulation [4]) [4, 10]; More (no metal ion requirement) [8, 10]

Turnover number (min^{-1})

Specific activity (U/mg)
0.00028–0.00055 [5]; 0.00105 [4]; 0.0189 [2]

K_m-value (mM)
0.063 (ganglioside LM1, soluble enzyme preparation) [1]; 0.07 (CMP-N-acetylneuraminate) [1]; 0.078 (ganglioside GM3) [1]; 0.1 (ganglioside GM3) [8]; 0.2 (ganglioside GM3 [4], CMP-N-acetylneuraminate [8]) [4, 8]; 0.8 (CMP-N-acetylneuraminate) [4]; 1.0 (ganglioside GD1a) [7]

pH-optimum
5.8 [8]; 6–7.2 (trisialoganglioside formation) [7]; 6.2 [4]; 6.5 (GM3 or GT1b as substrate) [10]

pH-range

Temperature optimum (°C)
37 (assay at) [1–8, 11]

Temperature range (°C)

3 ENZYME STRUCTURE

Molecular weight

Subunits
? (x × 55000, rat, SDS-PAGE) [2]

Glycoprotein/Lipoprotein
Glycoprotein [4]

4 ISOLATION/PREPARATION

Source organism
Chicken (embryo) [1, 7]; Rat (female Wistar [8], 12–14 days old [2]) [2–6, 8, 10, 11]; Human [9]

Source tissue
Brain [1, 2, 7]; Liver [2–6, 8, 10, 11]; Heart [2]; Lung [2]; Kidney [2]; Pancreas [2]; HeLa-cells [9]

Localization in source
Membrane-bound (tightly bound) [1–8]; Golgi apparatus [3–6, 8, 11]

Purification
Rat (solubilized with Triton X-100, CDP-Sepharose affinity chromatography) [2]

Crystallization
–

Cloned
–

Renatured
–

---

5 STABILITY

pH

Temperature (°C)
56 (20% loss of activity after 120 s) [11]

Oxidation

Organic solvent

General stability information

Storage

---

6 CROSSREFERENCES TO STRUCTURE DATABANKS

PIR/MIPS code

Brookhaven code

7 LITERATURE REFERENCES

[1] Higashi, H., Basu, M., Basu, S.: J. Biol. Chem.,260,824–828 (1985)

[2] Gu, X.-B., Gu, T.-J., Yu, R.K.: Biochem. Biophys. Res. Commun.,166,387–393 (1990)

[3] Eppler, C.M., Morré, D.J., Keenan, T.W.: Biochim. Biophys. Acta,619,332–343 (1980)

[4] Eppler, C.M., Morré, D.J., Keenan, T.W.: Biochim. Biophys. Acta,619,318–331 (1980)

[5] Klein, D., Pohlentz, G., Schwarzmann, G., Sandhoff, K.: Eur. J. Biochem., 167,417–424 (1987)

[6] Yusuf, H.K.M., Schwarzmann, G., Pohlentz, G., Sandhoff, K.: Biol. Chem. Hoppe-Seyler,368,455–462 (1987)

[7] Yohe, H.C., Yu, R.K.: J. Biol. Chem.,255,608–613 (1980)

[8] Busam, K., Decker, K.: Eur. J. Biochem.,160,23–30 (1986)

[9] Fishman, P.H., Bradley, R.M., Henneberry, R.C.: Arch. Biochem. Biophys., 172,618–626 (1976)

[10] Trinchera, M., Pirovano, B., Ghidoni, R.: J. Biol. Chem.,265,18242–18247 (1990)

[11] Iber, H., van Echten, G., Sandhoff, K.: Eur. J. Biochem.,195,115–120 (1991)

1 NOMENCLATURE

EC number
2.4.99.9

Systematic name
CMP-N-acetylneuraminate:lactosylceramide alpha-2,3-N-acetylneuraminyl-transferase

Recommended name
Lactosylceramide alpha-2,3-sialyltransferase

Synonyms
Sialyltransferase, cytidine monophosphoacetylneuraminate-lactosylceramide alpha2,3-
CMP-acetylneuraminate-lactosylceramide-sialyltransferase
CMP-acetylneuraminic acid:lactosylceramide sialyltransferase
CMP-sialic acid:lactosylceramide-sialyltransferase
Cytidine monophosphoacetylneuraminate-lactosylceramide sialyltransferase
Ganglioside GM3 synthetase
GM3 synthase
GM3 synthetase
SAT 1
More (cf. EC 2.4.99.2)

CAS Reg. No.
125752-90-1

---

2 REACTION AND SPECIFICITY

Catalysed reaction
CMP-N-acetylneuraminate + beta-D-galactosyl-1,4-beta-D-glucosylceramide →
→ CMP + alpha-N-acetylneuraminyl-2,3-beta-D-galactosyl-1,4-beta-D-glucosylceramide

Reaction type
Glycosyl group transfer

Natural substrates
CMP-N-acetylneuraminate + lactosylceramide (synthesizes the first ganglioside of the gangliotetraose series [2], involved in sequential addition of monosaccharides from sugar nucleotides to non-reducing end of oligosaccharide chains of glycosphingolipids [11]) [2, 11]

Substrate spectrum

1   CMP-N-acetylneuraminate + beta-D-galactosyl-1,4-beta-D-glucosylcer-
    amide (i.e. lactosylceramide, high specificity [9], preferred acceptors
    have the general structure beta1-O-ceramide with disaccharide preferred
    to monosaccharide [6], no acceptors are fetuin, mucin, alpha$_1$-acid glyco-
    protein, glycophorin or their respective asialo-derivatives, gangliosides
    GD1b, GM3, GM2, GM1 or GD1a [6]) [1–9, 11–15]

2   CMP-N-acetylneuraminate + glucosylceramide (sialylated at about 65%
    the rate of lactosylceramide) [6]

3   CMP-N-acetylneuraminate + galactosylceramide (sialylated at about 40%
    the rate of lactosylceramide) [6]

4   CMP-N-acetylneuraminate + asialo-ganglioside GM1 (sialylated at about
    85% the rate of lactosylceramide) [6]

Product spectrum

1   CMP + N-acetylneuraminyl-2,3-beta-D-galactosyl-1,4-beta-D-glucosylcer-
    amide (i.e. ganglioside GM3) [1–9, 11–15]

2   CMP + ?

3   CMP + ?

4   CMP + ?

Inhibitor(s)

CDP [6, 14]; CTP [6, 14]; CMP [2, 6, 14]; UDP [2]; UDPdialdehyde (kinetics)
[2]; CMPdialdehyde (kinetics) [2]; UDPgalactose (weak) [2]; UDP-N-acetylgalactosamine (weak) [2]; Monoclonal antibody M12GC7 [6]; alpha$_2$-Mac-
roglobulin (not in vivo) [7]; (4-Amidinophenyl)-methanesulfonyl fluoride [7];
Aprotinin [7]; Leupeptin (not in vivo) [7]; Pepstatin A (not in vivo) [7];
L-1-Chloro-3-[4-tosylamido]-7-amino-2-heptanone-HCl [7]; L-1-Chloro-3-
[4-tosylamido]-4-phenyl-2-butanone [7]; Thiol protease inhibitors (e.g. E-64
(not in vivo), Ep-459 or Ep-475) [7]; EDTA (not [1, 12]) [7]; Tris buffer (weak)
[1]; DTT [7]; N_3^- [7]; Ionic detergents [11]; Triton CF-54 (at higher concentrations) [13]; Ganglioside GM3 [11]; Ganglioside GM1a [13]; Globotetraosylceramide (weak) [11]; Asialofetuin (weak) [11]; Cu^{2+} (strong) [12];
Cd^{2+} (strong) [12]; Ni^{2+} (strong) [12]; Co^{2+} [12]; Ca^{2+} (weak) [12]; Mn^{2+}
(weak) [12]; Zn^{2+} (weak) [12]; Ba^{2+} (weak) [12]; More (inhibition in vivo (stable microsomes) not as dramatic as in vitro [7], no inhibition by Fe^{2+}, Mg^{2+},
IAA, 2-mercaptoethanol [12], glucosylceramide or globotriaosylceramide
[11]) [7, 11, 12]

Cofactor(s)/prosthetic group(s)/activating agents

Lauryldimethylamine oxide (i.e. Ammonyx LO, nonionic/cationic detergent,
activation, better than Triton CF-54, Triton X-100 or beta-octylglucoside) [8];
Triton CF-54 (activation, most potent activator [11], inhibits at higher concentrations [13]) [1, 8–11, 13]; Myrj 59 (activation, most potent activator, can
be replaced by sodium deoxycholate, Triton CF-54, Tween 20, Triton

CF-54/Tween 80 (ratio 2:1), Triton X-100 or Tween 80, with descending efficiency) [10]; Cutscum (activation, most potent activator [1], can be replaced by Triton CF-54, Triton X-100 or Triton CF-54/Tween 80 (ratio 2:1) with
descending efficiency [1]) [1, 11]; 2-Mercaptoethanol (activation, stronger in
vivo than in vitro) [7]; Triton X-100 (activation) [1, 8, 10–12]; Triton
CF-54/Tween 80 (ratio 2:1, activation, about 50% as efficient as Cutscum
[1], less than 50% as efficient as Myrj 59 [10]) [1, 10, 11]; Cardiolipin (activation) [1]; Nonidet P-40 (slight activation) [1]; beta-Octylglucoside [8]; More
(no activation by Tween 20 or 80) [11]

Metal compounds/salts
Mn^{2+} (activation) [6, 7, 9, 11]; Mg^{2+} (activation, slightly less efficient than
Mn^{2+} [11]) [11, 15]; Ca^{2+} (activation, slightly less efficient than Mn^{2+}) [11];
More (no metal ion requirement) [1, 12]

Turnover number (min^{-1})

Specific activity (U/mg)
0.0000045 [1]; 0.0055 [6]

K_m-value (mM)
0.00011 (lactosylceramide) [6]; 0.00026 (CMP-N-acetylneuraminate) [6];
0.035 (lactosylceramide) [1]; 0.068 (lactosylceramide) [13]; 0.075 (lactosylceramide) [15]; 0.08 (lactosylceramide) [9, 12]; 0.0815 (lactosylceramide) [2]; 0.11 (lactosylceramide) [11]; 0.16 (CMP-N-acetylneuraminate)
[11]; 0.19 (CMP-N-acetylneuraminate) [15]; 0.21 (CMP-N-acetylneuraminate)
[9]; 1.5 (CMP-N-acetylneuraminate) [1, 12]

pH-optimum
More (pI: 5.7–6.2 [6], no effects of buffer on activity [11]) [6, 11]; 5.7 [12]; 6
[1]; 6.2 [15]; 6.5 (broad [11]) [6, 9, 11]

pH-range
5–8 [11]; 6.1–7.6 (detectable activity) [6]

Temperature optimum (°C)
37 [15]

Temperature range (°C)

3 ENZYME STRUCTURE

Molecular weight

Subunits
? (x × 60000, rat, SDS-PAGE [6, 8], x × 76000, rat, SDS-PAGE [9]) [6, 8, 9]

Glycoprotein/Lipoprotein
  Glycoprotein (N- and O-linked carbohydrate side chains, containing sialic
  acids in alpha-2,3 and 2,6-linkage, galactose and galactosamine, branched
  N-glycans containing mannose) [6]

4 ISOLATION/PREPARATION

Source organism
  Human [1, 15]; Chicken (15 days old embryo [2], 7–11 days old embryo [4])
  [2, 4]; Hamster [11]; Rat (female Wistar [12]) [3, 5–10, 12–14]

Source tissue
  HeLa-cells (strain R, elevated enzyme level if cultured in butyrate containing
  media) [1]; Brain [2, 4, 9, 10]; Liver [3, 5–8, 12–14]; Cell suspension culture
  [11]; Fibroblasts (contact-inhibited cell-line NIL-8) [11]; Lymphocytes (from
  peripheral blood mononuclear cells, enzyme level stimulated by growth in
  the presence of phytohemagglutinin) [15]

Localization in source
  Membrane-bound [10, 12]; Golgi apparatus (distribution [3], luminal side
  [5]) [3, 5–7, 12–14]; Microsomes [2, 7]

Purification
  Rat (partial [10], solubilized with lauryldimethylamine oxide [6, 8],
  CMP-hexanolamine Sepharose and lactosylceramide aldehyde Sepharose
  4B affinity chromatography [6], immunoaffinity chromatography on
  M12GC7-gel 10 [6, 7]) [6–10]

Crystallization
  –

Cloned
  –

Renatured
  –

5 STABILITY

pH

Temperature (°C)
  56 ($t_{1/2}$: 60 s) [13]; 100 (5 min, inactivation) [1]

Oxidation

Organic solvent

General stability information
Lauryldimethylamine oxide solubilizes and stabilizes solubilized enzyme during purification and storage [8]; PMSF, leupeptin or pepstatin A stabilizes during purification [7]

Storage
-80°C, in 25 mM sodium cacodylate, pH 6.5, 15% w/v lauryldimethylamine oxide, 6–12 months [8]

---

6 CROSSREFERENCES TO STRUCTURE DATABANKS

PIR/MIPS code

Brookhaven code

---

7 LITERATURE REFERENCES

[1] Fishman, P.H., Bradley, R.M., Henneberry, R.C.: Arch. Biochem. Biophys.,172, 618–626 (1976)
[2] Cambron, L.D., Leskawa, K.C.: Biochem. Biophys. Res. Commun.,193,585–590 (1993)
[3] Trinchera, M., Pirovano, B., Ghidoni, R.: J. Biol. Chem.,265,18242–18247 (1990)
[4] Higashi, H., Basu, M., Basu, S.: J. Biol. Chem.,260,824–828 (1985)
[5] Trinchera, M., Fabbri, M., Ghidoni, R.: J. Biol. Chem.,266,20907–20912 (1991)
[6] Melkerson-Watson, L.J., Sweeley, C.C.: J. Biol. Chem.,266,4448–4457 (1991)
[7] Melkerson-Watson, L.J., Sweeley, C.C.: Biochem. Biophys. Res. Commun.,175, 325–332 (1991)
[8] Melkerson-Watson, L.J., Sweeley, C.C.: Biochem. Biophys. Res. Commun.,172, 165–171 (1990)
[9] Preuss, U., Gu, X., Gu, T., Yu, R.K.: J. Biol. Chem.,268,26273–26278 (1993)
[10] Gu, X.-B., Gu, T.-J., Yu, R.K.: Biochem. Biophys. Res. Commun.,166,387–393 (1990)
[11] Burczak, J.D., Fairley, J.L., Sweeley, C.C.: Biochim. Biophys. Acta,804,442–449 (1984)
[12] Busam, K., Decker, K.: Eur. J. Biochem.,160,23–30 (1986)
[13] Iber, H., van Echten, G., Sandhoff, K.: Eur. J. Biochem.,195,115–120 (1991)
[14] Eppler, C.M., Morré, D.J., Keenan, T.W.: Biochim. Biophys. Acta,619,332–343 (1980)
[15] Basu, S.K., Whisler, R.L., Yates, A.J.: Biochemistry,25,2577–2581 (1986)

1 NOMENCLATURE

EC number
2.4.99.10

Systematic name
CMP-N-acetylneuraminate:neolactotetraosylceramide alpha-2,3-sialyltransferase

Recommended name
Neolactotetraosylceramide alpha-2,3-sialyltransferase

Synonyms
Sialyltransferase
Sialyltransferase, cytidine monophosphoacetylneuraminate-neolactotetraosylceramide
Sialyltransferase 3 [1]
SAT-3 [4]

CAS Reg. No.
83745-06-6

---

2 REACTION AND SPECIFICITY

Catalysed reaction
CMP-N-acetylneuraminate + beta-D-galactosyl-1,4-N-acetyl-beta-D-glucosaminyl-1,3-beta-D-galactosyl-1,4-D-glucosylceramide →
→ CMP + alpha-N-acetylneuraminyl-2,3-beta-D-galactosyl-1,4-N-acetyl-beta-D-glucosaminyl-1,3-beta-D-galactosyl-1,4-D-glucosylceramide

Reaction type
Glycosyl group transfer

Natural substrates
CMP-N-acetylneuraminate + beta-D-galactosyl-1,4-N-acetyl-beta-D-glucosaminyl-1,3-beta-D-galactosyl-1,4-D-glucosylceramide (involved in glycosphingolipid biosynthesis) [1]

Substrate spectrum
1 CMP-N-acetylneuraminate + beta-D-galactosyl-1,4-N-acetyl-beta-D-glucosaminyl-1,3-beta-D-galactosyl-1,4-D-glucosylceramide (i.e. neolactotetraosylceramide or ganglioside LC4, transfers sialic acid to the O-3-position of the terminal galactosyl residue [1, 3], acceptor and donor specificity [3]) [1–4]

---

2 CMP-N-acetylneuraminate + ganglioside GM1 [3]
3 CMP-N-acetylneuraminate + gangliotetraosylceramide [1]
4 CMP-N-acetylneuraminate + lacto-N-neohexaosylceramide [3]

Product spectrum
1 CMP + alpha-N-acetylneuraminyl-2,3-beta-D-galactosyl-1,4-N-acetyl-beta-D-glucosaminyl-1,3-beta-D-galactosyl-1,4-D-glucosylceramide (i.e. ganglioside LM1) [1–4]
2 CMP + ganglioside GM3 [3]
3 CMP + N-acetylneuraminyl-alpha-2,3-gangliotetraosylceramide [1]
4 CMP + N-acetylneuraminyl-lacto-N-neohexaosylceramide [3]

Inhibitor(s)
5'-CMP [1, 3]; EDTA [3]; Sialic acid (i.e. acetylneuraminate) [3]; AMP (not [1]) [3]; CMP-N-acetylneuraminate (i.e. CMP-sialic acid, at higher concentrations, substrate inhibition) [3]; Neolactotetraosylceramide (at higher concentrations, substrate inhibition) [3]; N-Ethylmaleimide (weak) [4]; Co^{2+} (strong) [3]; Fe^{2+} (strong) [3]; Hg^{2+} (strong) [3]; Zn^{2+} (strong) [3]; Ba^{2+} [3]; Cu^{2+} [3]; Fe^{3+} [3]; K^+ [3]; Ni^{2+} [3]; Pb^{2+} [3]

Cofactor(s)/prosthetic group(s)/activating agents
Triton CF-54 (activation) [1, 3]; Triton X-100 (activation, 50% as efficient as CF-54 [3]) [1, 3]; Triton DF-16 (activation, can replace Triton CF-54) [3]; Triton N-57 (activation, can replace Triton CF-54) [3]; Triton CF-54/X-100 (activation, ratio 2:1, can replace Triton CF-54 [3], only 80% as efficient as CF-54 or X-100 alone [1]) [1, 3]; Tween 20 (activation, 40% as efficient as Triton CF-54) [3]; Tween 80 (activation, 40% as efficient as Triton CF-54) [3]; Tween 60 (activation, 27% as efficient as Triton CF-54) [3]; More (no activation by Triton GR-7M, TW-30 or sodium taurocholate) [3]

Metal compounds/salts
Mg^{2+} (slight stimulation [1], activation [2, 3]) [1–3]; Mn^{2+} (activation) [3]

Turnover number (min^{-1})

Specific activity (U/mg)

K_m-value (mM)
0.027 (neolactotetraosylceramide) [1]; 0.5 (gangliotetraosylceramide) [1]; 0.67 (CMP-N-acetylneuraminate (+ lacto-N-hexaosylceramide)) [3]; 0.9 (lacto-N-hexaosylceramide) [3]

pH-optimum
6.0 [2]; 6.8 (HEPES-buffer preferred to cacodylate, phosphate, Tris-maleate or morpholinoethane sulfonic acid buffer) [3]

pH-range
    5.8–7.2 (about half-maximal activity at pH 5.8 and about 90% of maximal ac-
    tivity at pH 7.2) [3]

Temperature optimum (°C)
    37 [4]

Temperature range (°C)

---

3 ENZYME STRUCTURE

Molecular weight

Subunits

Glycoprotein/Lipoprotein
    –

---

4 ISOLATION/PREPARATION

Source organism
    Chicken (embryo, 7–12 days old [1], 10–12 days old [3]) [1, 3, 4]; Human
    [2]

Source tissue
    Brain [1, 4]; Lymphocytes (from peripheral blood mononuclear cells, en-
    zyme level stimulated by growth in the presence of phytohemagglutinin) [2]

Localization in source
    Membrane-bound [1–4]

Purification

Crystallization
    –

Cloned
    –

Renatured
    –

5 STABILITY

pH

Temperature (°C)

Oxidation

Organic solvent

General stability information

Storage

6 CROSSREFERENCES TO STRUCTURE DATABANKS

PIR/MIPS code

Brookhaven code

7 LITERATURE REFERENCES

[1] Basu, M., Basu, S., Stoffyn, A., Stoffyn, P.: J. Biol. Chem.,257,12765–12769 (1982)
[2] Basu, S.K., Whisler, R.L., Yates, A.J.: Biochemistry,25,2577–2581 (1986)
[3] Dasgupta, S., Chien, J.-L., Hogan, E.L.: Biochim. Biophys. Acta,876,363–370 (1986)
[4] Higashi, H., Basu, M., Basu, S.: J. Biol. Chem.,260,824–828 (1985)

1 NOMENCLATURE

EC number
2.4.99.11

Systematic name
CMP-N-acetylneuraminate:lactosylceramide alpha-2,6-N-acetylneuraminyl-transferase

Recommended name
Lactosylceramide alpha-2,6-N-sialyltransferase

Synonyms
Sialyltransferase, cytidine monophosphoacetylneuraminate-lactosylceramide
CMP-acetylneuraminate-lactosylceramide-sialyltransferase
CMP-N-acetylneuraminic acid:lactosylceramide sialyltransferase
CMP-sialic acid:lactosylceramide sialyltransferase
Cytidine monophosphoacetylneuraminate-lactosylceramide sialyltransferase

CAS Reg. No.
55071-95-9

---

2 REACTION AND SPECIFICITY

Catalysed reaction
CMP-N-acetylneuraminate + beta-D-galactosyl-1,4-beta-D-glucosyl-ceramide →
→ CMP + alpha-N-acetylneuraminyl-2,6-beta-D-galactosyl-1,4-beta-D-glucosylceramide

Reaction type
Glycosyl group transfer

Natural substrates

Substrate spectrum
1 CMP-N-acetylneuraminate + beta-D-galactosyl-1,4-beta-D-glucosylcer-amide (i.e. lactosylceramide) [1]

Product spectrum
1 CMP + alpha-N-acetylneuraminyl-2,6-beta-D-galactosyl-1,4-beta-D-glucosylceramide (i.e. ganglioside GM3) [1]

Inhibitor(s)
  Endogenous acidic peptide (heat-stable, amino acid composition) [1]

Cofactor(s)/prosthetic group(s)/activating agents

Metal compounds/salts

Turnover number (min^{-1})

Specific activity (U/mg)

K_m-value (mM)

pH-optimum

pH-range

Temperature optimum (°C)

Temperature range (°C)

---

3 ENZYME STRUCTURE

Molecular weight

Subunits

Glycoprotein/Lipoprotein
  –

---

4 ISOLATION/PREPARATION

Source organism
  Rat (15 days old albino) [1]; Chicken (20 days old) [1]

Source tissue
  Brain (cerebellum excluded) [1]

Localization in source
  Membrane-bound [1]

Purification

Crystallization
  –

Cloned
  –

Renatured
  –

5 STABILITY

pH

Temperature (°C)

Oxidation

Organic solvent

General stability information

Storage

6 CROSSREFERENCES TO STRUCTURE DATABANKS

PIR/MIPS code

Brookhaven code

7 LITERATURE REFERENCES

[1] Albarracin, I., Lassaga, F.E., Caputto, R.: Biochem. J.,254,559–565 (1988)